D1408581

CONSTRUCTION CONTRACT DOCUMENTS AND SPECIFICATIONS

THOMAS C. JELLINGER

Iowa State University, Ames, Iowa

ADDISON-WESLEY PUBLISHING COMPANY

Reading, Massachusetts • Menlo Park, California

London • Amsterdam • Don Mills, Ontario • Sydney

CONSTRUCTION CONTRACT DOCUMENTS AND SPECIFICATIONS

Library of Congress Cataloging in Publication Data

Jellinger, Thomas C
Construction contract documents and specifications.

Includes index.
1. Buildings—Contracts and specifications—
United States. I. Title.
KF902.J44 343.73'07869 80-27250
ISBN 0-201-03785-8

Copyright © 1981 by Addison-Wesley Publishing Company, Inc. Philippines copyright 1981 by Addison-Wesley Publishing Company, Inc.

All rights reserved. No part of this publication may be reproduced, stored in a retrieval system, or transmitted, in any form or by any means, electronic, mechanical, photocopying, recording, or otherwise, without the prior written permission of the publisher. Printed in the United States of America. Published simultaneously in Canada.

ISBN 0-201-03785-8
BCDEFGHIJ-A-93210/898765

This book is dedicated to BILL KLINGER *and* HARRY ROCKEL

in recognition of past obligations

ACKNOWLEDGMENTS

It would be presumptuous to claim that this book was the result of only one person's efforts and thoughts. While only my name is listed as author, many people have actually contributed. I am fortunate to have been associated for the last thirty-plus years with the construction industry. During this time information has been gleaned from many sources—conversations, meetings, seminars, and literally hundreds of books, plus fellow architects, engineers and contractors with whom I have been in constant contact. After such exposure it becomes virtually impossible to identify a particular idea with a specific source. Two deceased contractor friends, however, have had measurable influence on my life. For that reason, this book is dedicated to Bill Klinger and Harry Rockel.

Thanks also to Iowa State University and the faculty of Construction Engineering for understanding and assistance. Special recognition goes to Teresa Olson and Kay Peterson for doing the typing and exercising much patience. Lastly, I thank my wife, Ro, for the encouragement that was so important throughout the entire effort.

CONTENTS

DIVISION VI
CONSTRUCTION SPECIFICATIONS

DIVISION VII
CONSTRUCTION DRAWINGS

DOCUMENTS

INTRODUCTION

The construction industry provides employment for a great number of people and makes the largest of all dollar contributions to the gross national product. It may be said that construction activity began when primitive men first decided to improve the caves in which they lived by partially closing entrances with rocks of various sizes and shapes. When it was discovered that some of these early builders possessed more ability in this activity than their fellows, they were persuaded to perform this task for others in return for food, weapons, or some other form of compensation—and the embryo of the construction industry thus had its conception.

To date, most things that mankind has done to change and modify living conditions on this planet have involved construction. Construction builds the houses in which we live, the airports from which we travel, the factories and offices where we work, the churches where we worship, the theatres and other places of recreation where we are entertained, the schools where we study, the highways we drive on, the lakes we boat on—an almost unlimited and constantly increasing list of facilities. The construction industry requires money, machines, and personnel to accomplish its assigned responsibilities. It is also true that these responsibilities could not be fulfilled without a great amount of paper work to facilitate and regulate construction activities. This paperwork is, at least in part, made up of a large variety of documents. Each has its own purpose and function without which the construction process would become more difficult.

Good management is essential to the success of any construction enterprise. The construction manager who does not do a thorough job of project planning and project control is taking the first step towards financial disaster. All of the activities essential to the successful completion of a construction venture are closely connected with one or more construction contract documents. The manager who wants to perform at the highest possible level with the greatest resulting benefit to the firm will possess an accurate and thorough knowledge of each of these documents.

Intelligent application and interpretation of these construction contract documents is as important as an accurate and impartial preparation of them. Like most other facets of human activity, documents can be slanted or shaded to provide an unfair advantage to one of the affected parties. In keeping with fair and ethical business practices, this should be avoided. Most successful managers would agree that honesty is not just a desirable personal trait—it is also good business practice.

The purpose of this text will be to present information regarding the content, preparation, application, and interpretation of the various construction contract documents. The interrelationship of the documents will also be discussed. There are pitfalls that exist for the unwise or unknowing user of construction contract documents and many of these will be brought to the reader's attention. Above all, it is hoped that this text will make some contribution to the improvement of the construction industry by helping both students and practitioners become better informed regarding construction contract documents.

DIVISION

PREVIEW OF DOCUMENTS

1 INTRODUCTION TO CONSTRUCTION CONTRACT DOCUMENTS

1.1 PURPOSE OF DOCUMENTS

Just as it is true that there are several construction contract documents, it is also true that these documents serve several purposes. Each document is prepared to aid in achieving a specific goal or objective; each document contains certain specific types of information. From necessity and practice, some information will be included in more than one of the contract documents.

Some documents have as their major function the detailed description of the work that is to be undertaken. This description (presented in both written and graphic form) represents the design of the project. It also serves as a documentation of the agreement, which has been reached between the owner and the design professional, regarding the work to be done. Also included within this group are those documents that are connected with and supportive to the design phase.

Another group of documents is connected with the procedure of securing for the project a projected cost that will be used as the basis for a contract. These documents will often be used by persons and firms that will not remain a part of the project team. They are the bidders who submit proposals, only a few of which will be accepted. The successful bidders will become part of the project team and utilize other contract documents, while the unsuccessful bidders will repeat this process until their proposal has been accepted for another contract.

Another group of documents is primarily used during the actual construction of the project and has as its major purpose the aiding of the field forces in the performance of their duties. It is these documents that make it possible for the ideas and concepts of the design professional to be translated and transformed into physical reality. Until this has been achieved, the design is of no great value to the owner. No persons or activities can be housed by pieces of paper, and no pieces of paper can be used to carry traffic or control flood waters. If designs are not given physical substance, they remain only an academic exercise. This group of documents is used to aid in this transformation.

The last group of construction contract documents is used after the project has been completed. The documents serve to guarantee or ensure the quality of work that has been performed. In this manner the owner is assured of receiving that for which payment has been made.

1.2 DEFINITION AND PREPARATION

A document may be defined as an official paper that can be relied upon as the basis, proof, or support of something. It may also be defined as a piece of writing that conveys information. In the case of construction contract documents, both definitions are applicable.

Most documents used in the construction process present information. This information may pertain to the construction of the project or to the duties and responsibilities of individual members of the project team. These documents also serve as the basis for a contract or establish a base for the enforcement of a contract.

In most instances these documents are prepared by, or under the supervision of, a licensed individual. This is most often the architect in the case of building projects and the engineer in the case of civil or engineering projects. There may also be times when a specific document is prepared or reviewed by another professional or specialist, such as an attorney, a bonding agent, or an insurance consultant. Outside professionals or specialists are more often utilized by smaller design firms that do not have sufficient "in-house" expertise. Since many of the larger firms have experts within their own firm, the use of outside personnel is the exception rather than the rule.

It should also be noted that if a given project is neither complex nor of great size, these documents are sometimes prepared by individuals who do not have the formal training and subsequent licensing that is usually desired. This is especially true of agricultural and residential projects. Such projects are often specifically excluded under state licensing laws and documents are prepared by persons not possessing professional registration.

1.3 PREPARED FOR WHOM

Although construction contract documents are used by a large number and variety of people during the life of a project, the major beneficiary is the owner. It is the owner's need that initiates the project. The design is prepared and the construction is performed to answer or fulfill the determined need. In achieving this goal, the documents are provided to inform the owner of the scope and quality of work that can reasonably be expected and to provide some assurance that such performance will be delivered. This procedure helps to provide a constantly improving project at the lowest possible cost. This is one of the primary goals of the construction contracting industry. While it is easy to lose sight of this goal during periods of economic inflation, it remains a major force in the construction industry.

In addition to the owner, the list of persons who might use one or more of the construction contract documents in connection with a project includes:

- Code officials
- Lending officers from financial institutions
- Prospective investors and/or tenants
- Estimators for both general and specialty contractors
- Material suppliers
- Field workers, ranging from superintendent to laborer
- Inspectors
- Governmental compliance personnel
- Bonding officials
- Attorneys

1.4 MEANS OF COMMUNICATION

One of the prime purposes of the construction contract documents is to serve as a means of communication. At its best, accurate communication is a difficult process. Any device or procedure that may improve the process should be given serious consideration. In some instances, the documents present information that may assist various persons using them. In this respect, it is important for the preparer to keep in mind the educational level of the user and to structure the documents accordingly. In other instances, the documents convey orders and directives that must be fulfilled. There should be no doubt left about the imperative nature of this part of the communication.

The documents also serve as the basis for a legal agreement between the owner and the contractor and they should be prepared in such a way that legal interpretation and enforcement will be enhanced. Throughout the preparation of the documents, every effort should be made to make the communication as straightforward as possible to avoid confusion or doubt in the interpretation of the material.

1.5 LEGAL STATUS

The legal status of the construction contract documents should always be kept in mind during their preparation and also during their use. Of primary importance is the fact that they are used to form a legal agreement between the owner and the contractor. In this regard, they should be constructed so as to avoid, or at least minimize, the possibility of future litigation. Many attorneys regard litigation of civil matters an indication of failure.

Although trials are necessary in criminal cases, disputes in civil matters can more often be equitably resolved without resorting to a trial. Such concern would recommend the use of standard phrases and terminology whose meaning and interpretation have already been tested in the courts. Originality of language in the preparation of construction contract documents is not a desirable trait. What was intended by the preparer may be changed by court interpretation at the expense of one of the parties involved. This would constitute a disservice to the affected party on the part of the preparer.

Unfortunately, the possibility of legal action exists for all construction projects. As their complexity and size increases, the potential for litigation also increases. Once court action begins, the information contained within the construction contract documents becomes extremely important. These documents are used as a basis for the proof or disproof of claims and reasonable expectations. Many a case has been won or lost because of the quality of the construction contract documents. As a safeguard, many design professionals seek legal advice in the preparation of documents.

1.6 OWNERSHIP OF DOCUMENTS

Most court decisions have held that ownership of the construction contract documents is retained by the design professional who has prepared them. While it is usual for the owner to be furnished with a permanent copy of the documents, there have been instances where even this practice has not been followed. The basis for these court decisions is that the documents represent instruments of service and that the documents themselves are not the end product. As a corollary, it may be stated that when you have an operation you are not buying the surgeon's scalpel, and when you hire a carpenter to hang a door you have no claim on the saw and screwdriver.

SURVEY OF DOCUMENTS 2

FIGURES

2.1 AN OVERVIEW

It may be helpful at this point to provide an overview of the more widely used construction contract documents. Figure 2.1 illustrates this in graphic form. The documents are divided into four major use areas:

1. Design phase
2. Bidding phase
3. Award phase
4. Construction phase

Note that some documents are present in more than one phase and that not all of the documents illustrated here will be used in every construction project. A fairly typical example has been selected where a contract is awarded as a result of competitive bidding for a project requiring bonds. The information illustrated in Fig. 2.1 should be regarded as general, not absolute, in nature. Furthermore, only the more widely used documents have been illustrated, and the text will discuss others not shown. There is a general sequence in the use of documents that should be understood.

2.2 BASIC CONTRACT DOCUMENTS

Among the many construction contract documents there are four that can be referred to as "basic" or "essential" documents because they must always be present, at least in function, for all construction projects. The documents are:

Agreement
General conditions of the contract
Drawings
Specifications

In some instances the drawings and the specifications may be combined by incorporating a varying number of written notes on the drawings. In this case the specifications still exist and perform their function even though they are not present as a separate and distinct document. Similarly, the general conditions may be presented as part of the agreement rather than as a separate entity. In other situations, it may be that the agreement does not exist as a document since no signed contract has been executed. If a verbal agreement has been reached, however, a legal contract may have been created and the function of the agreement is therefore in existence. In an extreme situation, none of the four documents may exist as actual documents. In such a case, however, the function of each of the documents is understood and agreed to. If this were not true, there could be no understanding about the nature of the project or even that the project was to be constructed. It is for this reason that we refer to these as the four basic contract documents.

2.2.1 AGREEMENT

The agreement is the document that represents and reflects the legal contract between the parties involved with the construction project. Traditionally, this refers to the contract between the owner and the contractor. Contracts for construction differ from those used for most other business ventures in that they are usually longer and more complex, and they are seldom independent instruments. Other documents, such as drawings and specifications, are usually

FIG. 2.1 Time sequence of documents

Key: P = Preparation; I = Issuance; U = Use

DOCUMENT	PRE-DESIGN	DESIGN	BIDDING	AWARD	CONSTRUCTION	POST-CONSTRUCTION
Owner-design prof. agreement	II					
Agency permissions		PPP III				
Drwgs, specs and conditions		PPPPPPPPPP	IUUUUUUUUUU		UUUUUUUUUUUUUUUUU	
Permits (zoning, etc.)		PP II				
Advertisement/invitation to bid		PP	I			
Prequalification of bidders			PIU			
Instructions to bidders		PP	I UUUUUUUUU			
Quotation requests			PI UUUUUUUU			
Addenda			PIU PIU			
Proposal w/bid security			PI			
Letter of intent				PIU		
Agreement		PP		I	UUUUUUUUUUUUUUUUU	
Contract bonds				I	UUUUUUUUUUUUUUUUU	UU
Affirmative action plan				PPP I	P I UUUUUUUUUUUUUUU	
Building permit					I	
Subcontracts-purchase orders					PIUUUUUUUUUUUUUUU	
Schedules (progress and paym't)				P	PIUUUUUUUUUUUUUUU	
Notice to proceed					PI	
Change orders					PIU PIU	
Application/certificate for paym't					PI PI PI PI PI	
Application for final paym't					PI	
Warranty bonds					I	UU

made a part of the agreement by reference. This means that they actually become a part of the agreement and thereby help to define and describe the contract conditions.

A great variety of forms of contracts are used within the construction industry. These can vary from a simple offer and acceptance to more complicated forms, such as "cost plus fixed fee with upset price and profit sharing." Each of the many forms of construction contracts provides a range of advantages and disadvantages under certain conditions. Owners, contractors, and design professionals need to be well acquainted with each of these forms and when they should be used.

2.2.2 GENERAL CONDITIONS

The construction process is a complex one involving many different materials, expensive equipment, and a large number of people with a wide range of talents. It is also true that each construction project is unique and is, at least in part, a "once in a lifetime" or custom situation. However, most construction contracts have many things in common. A document called the "General Conditions of the Contract" is used to define the responsibilities of the parties affected by the contract and to describe the guidelines that will be used in the administration of the contract. The "parties affected by the contract" include not only the owner and contractor, who are signatory to the contract, but also the architect/engineer, subcontractors, and others.

It is the purpose of the general conditions to define the duties and responsibilities of each of the parties as well as the relationships that will exist between them. This document also describes and delineates the procedures that must be used during construction of the project as dictated by contract conditions. Several standard forms of general conditions exist. These are published and promulgated by various industry and professional groups as well as by different governmental agencies. Some of these groups and agencies are mentioned in Chapter 11.

2.2.3 DRAWINGS

The drawings consist of a varying number of sheets showing in graphic form the work that is to be undertaken. They include plans covering general layout, size, shape, and dimensions, as well as a number of sheets giving more detailed information on specific parts of the project. In an architectural project the sheets will be grouped into subsets of architectural, structural, mechanical, and electrical drawings. In civil works the subsets will represent layout, structural, profiles, and possibly mechanical, depending upon the nature and scope of the project.

Drawings are constructed to scale, with the scale applicable to the particular plan or detail being noted. Building plans usually use an architectural scale that calls for a certain fraction of an inch to be equal to a foot. Civil projects utilize an engineering scale that sets a foot equal to a decimal part of an inch. It is a general rule that anything that can be better shown graphically is included on the drawings. This rule is followed because a graphic form of communication is less likely to be misunderstood or misinterpreted.

2.2.4 SPECIFICATIONS

The specifications are written instructions describing in detail the work that is to be undertaken. They are used as a supplement to the drawings and therefore pertain to items of information that cannot easily be shown in graphic terms. This would include information regarding quality of material and workmanship.

The drawings and the specifications should perform as a pair of complementary documents that together will completely describe and define the work to be accomplished. Given this condition, there may be occasions when the two documents are not in agreement. The general rule is that in case of conflict, the specifications will govern.

2.3 SUPPLEMENTARY DOCUMENTS

There are many other construction contract documents that may be used in connection with a construction project. However, since these documents, are not requried on all projects, they can be classified as supplementary contract documents. While the requirements of an individual project may dictate the use of one or more of these documents, there are instances when none of them will be used.

2.3.1 ADVERTISEMENT AND INVITATION TO BID

One of the first supplementary contract documents to be used on a project is the advertisement. This document is required on projects using public funds and serves to inform potential bidders regarding the upcoming work. The advertisement presents a description of the proposed project and information regarding the submission of a proposal. Distribution of the advertisement is influenced by the size of the project.

The invitation to bid serves the same purpose as the advertisement. The advertisement is used on projects involving public funds; the invitation to bid is used for projects involving private financing. Although the invitation to bid need not include information regarding the financing of the project, this is a requirement for the advertisement. All other elements of content and the use of the two documents are similar. Although titles of the two documents are sometimes interchanged, use of the advertisement title is recommended on public funded projects.

2.3.2 INSTRUCTIONS TO BIDDERS

The instructions to bidders is a document given to each firm that has taken out drawings and specifications. It presents detailed information regarding the submission of a proposal. In addition, most of the information that was presented in the advertisement or invitation to bid is included. Its purpose is to enhance and ensure uniformity in the form of proposal so that a valid comparison of the bids is possible.

2.3.3 ADDENDA

Most sets of bidding documents are found to contain some errors or discrepancies. A design firm issues an addendum to correct or clarify this situation. Addenda are used to correct errors, add equivalent products to the approved list, answer questions, and clarify portions of the bidding documents. An acknowledgment procedure should be used for all addenda. There are also time restrictions to their issuance that are recommended.

2.3.4 PROPOSAL

The proposal is an offer on the part of the bidder to furnish the labor, equipment, material, and expertise necessary to construct the proposed project. The offer may be submitted as a fixed amount or a unit price, or in a variety of other forms. It is tendered to the owner or the owner's representative. The proposal is usually valid for a fixed period of time and the owner has the power of acceptance within the stated time. It forms the basis upon which the contract is drawn.

2.3.5 BID SECURITY

Bid security is often required as a guarantee of the sincerity of the proposal. It may be compared with "earnest money" in a real estate transaction. Bid security is furnished to protect the owner's right to accept a bidder's offer. If the bidder refuses to enter into a contract consistent with the offer, the bids security is forfeited to the owner. Bid security is most often furnished as a bond or a certified check.

2.3.6 CONTRACT BONDS

Just as a bid bond is used to furnish the owner security in connection with an offer, contract bonds are used to ensure the owner's reasonable expectations from the contract. Performance and payment bonds are furnished to guarantee the owner that the project will be constructed in accordance with the contract documents and that it will be delivered to the owner free from liens and encumbrances. Bonds are issued to the owner on behalf of the contractor. They are furnished by a surety who receives a fee or premium for the service. In case of default by the contractor, the surety is required to fulfill the contract in accordance with the original terms.

2.3.7 SUBCONTRACTS

Because of the size and complexity of construction projects today, very few firms construct the entire job with their own forces. This is especially true in the case of building construction projects. Accordingly, most firms award subcontracts for various branches of the work. This practice permits the general contractor to delegate the responsibility for parts of the project to firms that are more expert in a particular branch. However, the general contractor retains overall responsibility to the owner for performance by the subcontractors.

2.3.8 CHANGE ORDERS

A change order performs functions similar to those of an addendum. The addendum effects changes in the documents during the bidding stage, the change order effects changes to the contract documents during the construction stage. A change order may be executed to effect corrections, to add or delete work from the original contract, or to authorize a change in completion time. Not all change orders concern a change in contract price. Some may be concerned with changes in contract time or conditions. Since they are a modification to the contract, change orders should be signed by the parties who were signatory to the contract.

2.3.9 APPLICATION AND CERTIFICATE FOR PAYMENT

According to normal practice within the construction industry, payments are made monthly as the work progresses. The contractor collects all charges for the preceding month. Typically, application for payment would be submitted by the tenth of the month following the month when work was performed. After the design professional has reviewed the request, a certificate for payment is issued to the owner if the application is approved. The owner would normally be expected to pay the request by the end of the month. Late payments by the owner are subject to interest charges.

2.4 OTHER DOCUMENTS

The foregoing discussion has covered only the four "essential" contract documents and the supplemental documents that are in more common use. Those documents shown in Fig. 2.1 that are not discussed in this chapter are examined in detail in later sections of the text.

DIVISION 

DESIGN ACTIVITIES

3 DESIGN PHASE DOCUMENTS

FIGURES

3.1 SELECTING THE DESIGN PROFESSIONAL

An owner retains a design professional in connection with a construction project for a number of reasons, the most obvious and best known being the preparation of the drawings and specifications. Less well known, but probably of equal importance, is for the design professional to serve as an adviser and confidant. In many projects the design professional is involved with varied activities, such as market studies, economic feasibility studies, financial arrangements, and promotional activities. In addition, the design professional is frequently retained for continuing services with respect to inspection or "observation" during the construction phase.

3.1.1 FOR PRIVATE PROJECTS

There are various methods an owner may use to select a design professional. Most private owners will interview a number of design firms and then negotiate an agreement with them. Among the items of interest to such an owner would be the following:

Experience of the firm in the type of work similar to the project being proposed. If the firm has designed this type of project before, the size of that project should be considered.

Reputation of the firm. This would include such matters as ethical conduct, promptness in execution of work, and how well previous clients have been satisfied.

Qualifications and size of the work force available. This item may take into consideration the education and experience of individual members of the firm, as well as how much work the firm already has on hand.

Familiarity with and access to financial lending institutions. This is becoming increasingly critical to private owners, especially on building construction projects.

Amount of fee that is being proposed. While this is usually relatively standard, sufficient variation may exist to cause it to be a major factor.

3.1.2 FOR PUBLIC PROJECTS

For public owners, the method of selecting the design professional may be controlled by statute. Depending upon the law and individual preference, selection may be by negotiation, by competition administered by a jury, or by competitive bids after a prequalification process. Under negotiation with individual firms, the process is almost identical to that used by the private owner. In some instances, efforts are made to be sure that the work is evenly distributed among all qualified firms. This principle is sometimes violated to permit political favoritism. If it can be shown that such favoritism is the result of bribery, both parties are susceptible to fines and/or imprisonment.

3.1.3 SELECTION BY COMPETITIVE BIDS

Court decisions of recent years have tended to encourage public bodies to utilize the competitive bid process in the selection of design professionals. This has been strongly opposed by the professional associations, who feel it will lead to a diminution in the quality of design. Nevertheless, some federal agencies have mandated the process for various periods of time. There have been conflicting reports as to its effectiveness. While there are other factors in addition to fee that are to be used in the selection process, most awards would appear to be strictly based on a low bidder criterion. Efforts will no doubt continue on the part of public bodies to improve and modify the design professional selection process so as to ensure fair dealing.

3.1.4 DESIGN COMPETITIONS

On specifically designated major works, a design competition may be held to select the design professional. It is usually necessary for the competition to be approved by the respective professional association, which then establishes guidelines for conducting the competition and may even furnish competition administration. It is common practice for a jury to be the judge of the designs submitted, with the winner being offered the design commission. The majority of jury members should be qualified, disinterested parties who would not be influenced by vested interests.

3.2 NEED FOR A CONTRACT

It is unfortunate that many design professionals begin work on a project without a signed agreement with the owner. Although in most cases involving a dispute regarding payment the courts uphold the right of the design professional to be paid for services performed, much difficulty could be avoided if design professionals would insist upon a signed contract before beginning work. All too often it is the design professional's own ego that refuses to "jeopardize" the relationship by insisting upon a signed agreement.

3.3 PROFESSIONAL ASSOCIATIONS

A large percentage of design professionals are members of professional associations. Associations such as the American Institute of Architects and the American Consulting Engineers Council are active in establishing registration requirements in cooperation with the various state licensing boards, setting and policing codes of ethics, conducting continuing education programs, providing guidelines and information related to practices that may be of aid to the members, lobbying for legislation beneficial to the profession, and many other similar activities. These associations and others may also produce standard forms of some of the construction contract documents for possible use by their members as well as by other persons involved in the construction process.

3.4 OWNER-ARCHITECT AGREEMENT

The owner-architect agreement is one of the forms used by owners to contract for design services. There are various types of forms available, the scope and nature of the proposed project often dictating the type of form that should be used. Such an agreement may require the architect to provide any one or all of the following services:

Preparation of preliminary designs.

Preparation of a preliminary cost estimate (usually based upon a square foot or cubic foot price).

Preparation of the drawings and specifications.

Preparation of all other documents used during the bidding and construction phases.

Receiving and evaluating proposals along with recommendations to the owner relative to contract award.

Administration of the contract during the construction phase. This may include approval of shop drawings, observation of construction, handling of change orders.

Acting in the capacity of construction manager.

Performing any of the many duties permitted in the current agreement form.

The American Institute of Architects' Document B141, shown in Fig. 3.1, is one of the more widely used forms for this type of agreement. It is interesting to note that the major part of this document is devoted to covering the terms and conditions of the agreement, with only a small part covering the specifics for a particular project.

3.4.1 METHODS OF PAYMENT

The most common form of payment for architectural services continues to be the percentage of construction costs system. This system has some distinct disadvantages for both the architect and the owner. There is no incentive for the architect to control the costs of the project since the higher the project costs, the greater the amount of fee received. This in turn works to the disadvantage of the owner. On the other hand, if the architect does a better job and is able to design a better building at a lower cost, the "reward" is a reduction in the amount of the fee. In spite of these disadvantages, the percentage of cost contract continues to be widely used. The percentage used in an individual contract will vary according to the type and size of the structure. It will also be influenced by whether the project is relatively standard in nature or new and innovative. In the past the American Institute of Architects published recommended minimum fee schedules. Such schedules are now suspect in light of court decisions during recent years. The courts have held fee schedules to be a "constraint of trade." Most associations have therefore ceased publishing such fee recommendations.

Another system in use is the multiple of direct costs. Under this method the anticipated direct costs of design are defined, such as a certain price per hour for the principals, with other hourly costs for production personnel. A multiplier of direct costs is then agreed upon. This multiplier may be two, or two and one-half, or more. This method reduces the amount of risk that the design firm must take and transfers it to the owner. On some projects the multiple of direct costs will be used for the schematic phase, and the percentage of construction costs used during the phase when working drawings and specifications are developed.

Some architectural firms provide services under a system where hourly rates are quoted. Here again different rates will be quoted for different classifications of employees. Basically, this is a variation on the multiple of direct costs system except that the owner is not aware of the multiplier being used. Sometimes a guaranteed maximum is also furnished to the owner.

Another method used for architectural services is the lump sum or fixed fee. Some private owners as well as some governmental agencies prefer to know in advance the amount of money required for design services. Under this system the architectural firm submits a single price (total) for the performance of all services. The determination of this figure is based upon past experience on the same type of project and perhaps with the same owner. Such a fee sys-

(text continues on page 32)

FIG. 3.1 Owner-architect agreement

THE AMERICAN INSTITUTE OF ARCHITECTS

AIA Document B141

Standard Form of Agreement Between Owner and Architect

1977 EDITION

THIS DOCUMENT HAS IMPORTANT LEGAL CONSEQUENCES; CONSULTATION WITH AN ATTORNEY IS ENCOURAGED WITH RESPECT TO ITS COMPLETION OR MODIFICATION

AGREEMENT

made as of the day of in the year of Nineteen Hundred and

BETWEEN the Owner:

and the Architect:

For the following Project:
(Include detailed description of Project location and scope.)

The Owner and the Architect agree as set forth below.

Copyright 1917, 1926, 1948, 1951, 1953, 1958, 1961, 1963, 1966, 1967, 1970, 1974, © 1977 by The American Institute of Architects, 1735 New York Avenue, N.W., Washington, D.C. 20006. Reproduction of the material herein or substantial quotation of its provisions without permission of the AIA violates the copyright laws of the United States and will be subject to legal prosecution.

This document has been reproduced with the permission of the American Institute of Architects under application number 79082. Further reproduction, in part or in whole, is not authorized. Because AIA documents are revised from time to time, users should ascertain from AIA the current edition of the document reproduced above.

FIG. 3.1 *(continued)*

TERMS AND CONDITIONS OF AGREEMENT BETWEEN OWNER AND ARCHITECT

ARTICLE 1

ARCHITECT'S SERVICES AND RESPONSIBILITIES

BASIC SERVICES

The Architect's Basic Services consist of the five phases described in Paragraphs 1.1 through 1.5 and include normal structural, mechanical and electrical engineering services and any other services included in Article 15 as part of Basic Services.

1.1 SCHEMATIC DESIGN PHASE

1.1.1 The Architect shall review the program furnished by the Owner to ascertain the requirements of the Project and shall review the understanding of such requirements with the Owner.

1.1.2 The Architect shall provide a preliminary evaluation of the program and the Project budget requirements, each in terms of the other, subject to the limitations set forth in Subparagraph 3.2.1.

1.1.3 The Architect shall review with the Owner alternative approaches to design and construction of the Project.

1.1.4 Based on the mutually agreed upon program and Project budget requirements, the Architect shall prepare, for approval by the Owner, Schematic Design Documents consisting of drawings and other documents illustrating the scale and relationship of Project components.

1.1.5 The Architect shall submit to the Owner a Statement of Probable Construction Cost based on current area, volume or other unit costs.

1.2 DESIGN DEVELOPMENT PHASE

1.2.1 Based on the approved Schematic Design Documents and any adjustments authorized by the Owner in the program or Project budget, the Architect shall prepare, for approval by the Owner, Design Development Documents consisting of drawings and other documents to fix and describe the size and character of the entire Project as to architectural, structural, mechanical and electrical systems, materials and such other elements as may be appropriate.

1.2.2 The Architect shall submit to the Owner a further Statement of Probable Construction Cost.

1.3 CONSTRUCTION DOCUMENTS PHASE

1.3.1 Based on the approved Design Development Documents and any further adjustments in the scope or quality of the Project or in the Project budget authorized by the Owner, the Architect shall prepare, for approval by the Owner, Construction Documents consisting of Drawings and Specifications setting forth in detail the requirements for the construction of the Project.

1.3.2 The Architect shall assist the Owner in the preparation of the necessary bidding information, bidding forms, the Conditions of the Contract, and the form of Agreement between the Owner and the Contractor.

1.3.3 The Architect shall advise the Owner of any adjustments to previous Statements of Probable Construction Cost indicated by changes in requirements or general market conditions.

1.3.4 The Architect shall assist the Owner in connection with the Owner's responsibility for filing documents required for the approval of governmental authorities having jurisdiction over the Project.

1.4 BIDDING OR NEGOTIATION PHASE

1.4.1 The Architect, following the Owner's approval of the Construction Documents and of the latest Statement of Probable Construction Cost, shall assist the Owner in obtaining bids or negotiated proposals, and assist in awarding and preparing contracts for construction.

1.5 CONSTRUCTION PHASE—ADMINISTRATION OF THE CONSTRUCTION CONTRACT

1.5.1 The Construction Phase will commence with the award of the Contract for Construction and, together with the Architect's obligation to provide Basic Services under this Agreement, will terminate when final payment to the Contractor is due, or in the absence of a final Certificate for Payment or of such due date, sixty days after the Date of Substantial Completion of the Work, whichever occurs first.

1.5.2 Unless otherwise provided in this Agreement and incorporated in the Contract Documents, the Architect shall provide administration of the Contract for Construction as set forth below and in the edition of AIA Document A201, General Conditions of the Contract for Construction, current as of the date of this Agreement.

1.5.3 The Architect shall be a representative of the Owner during the Construction Phase, and shall advise and consult with the Owner. Instructions to the Contractor shall be forwarded through the Architect. The Architect shall have authority to act on behalf of the Owner only to the extent provided in the Contract Documents unless otherwise modified by written instrument in accordance with Subparagraph 1.5.16.

1.5.4 The Architect shall visit the site at intervals appropriate to the stage of construction or as otherwise agreed by the Architect in writing to become generally familiar with the progress and quality of the Work and to determine in general if the Work is proceeding in accordance with the Contract Documents. However, the Architect shall not be required to make exhaustive or continuous on-site inspections to check the quality or quantity of the Work. On the basis of such on-site observations as an architect, the Architect shall keep the Owner informed of the progress and quality of the Work, and shall endeavor to guard the Owner against defects and deficiencies in the Work of the Contractor.

1.5.5 The Architect shall not have control or charge of and shall not be responsible for construction means, methods, techniques, sequences or procedures, or for safety precautions and programs in connection with the Work, for the acts or omissions of the Contractor, Sub-

FIG. 3.1 *(continued)*

contractors or any other persons performing any of the Work, or for the failure of any of them to carry out the Work in accordance with the Contract Documents.

1.5.6 The Architect shall at all times have access to the Work wherever it is in preparation or progress.

1.5.7 The Architect shall determine the amounts owing to the Contractor based on observations at the site and on evaluations of the Contractor's Applications for Payment, and shall issue Certificates for Payment in such amounts, as provided in the Contract Documents.

1.5.8 The issuance of a Certificate for Payment shall constitute a representation by the Architect to the Owner, based on the Architect's observations at the site as provided in Subparagraph 1.5.4 and on the data comprising the Contractor's Application for Payment, that the Work has progressed to the point indicated; that, to the best of the Architect's knowledge, information and belief, the quality of the Work is in accordance with the Contract Documents (subject to an evaluation of the Work for conformance with the Contract Documents upon Substantial Completion, to the results of any subsequent tests required by or performed under the Contract Documents, to minor deviations from the Contract Documents correctable prior to completion, and to any specific qualifications stated in the Certificate for Payment); and that the Contractor is entitled to payment in the amount certified. However, the issuance of a Certificate for Payment shall not be a representation that the Architect has made any examination to ascertain how and for what purpose the Contractor has used the moneys paid on account of the Contract Sum.

1.5.9 The Architect shall be the interpreter of the requirements of the Contract Documents and the judge of the performance thereunder by both the Owner and Contractor. The Architect shall render interpretations necessary for the proper execution or progress of the Work with reasonable promptness on written request of either the Owner or the Contractor, and shall render written decisions, within a reasonable time, on all claims, disputes and other matters in question between the Owner and the Contractor relating to the execution or progress of the Work or the interpretation of the Contract Documents.

1.5.10 Interpretations and decisions of the Architect shall be consistent with the intent of and reasonably inferable from the Contract Documents and shall be in written or graphic form. In the capacity of interpreter and judge, the Architect shall endeavor to secure faithful performance by both the Owner and the Contractor, shall not show partiality to either, and shall not be liable for the result of any interpretation or decision rendered in good faith in such capacity.

1.5.11 The Architect's decisions in matters relating to artistic effect shall be final if consistent with the intent of the Contract Documents. The Architect's decisions on any other claims, disputes or other matters, including those in question between the Owner and the Contractor, shall be subject to arbitration as provided in this Agreement and in the Contract Documents.

1.5.12 The Architect shall have authority to reject Work which does not conform to the Contract Documents. Whenever, in the Architect's reasonable opinion, it is necessary or advisable for the implementation of the intent of the Contract Documents, the Architect will have authority to require special inspection or testing of the Work in accordance with the provisions of the Contract Documents, whether or not such Work be then fabricated, installed or completed.

1.5.13 The Architect shall review and approve or take other appropriate action upon the Contractor's submittals such as Shop Drawings, Product Data and Samples, but only for conformance with the design concept of the Work and with the information given in the Contract Documents. Such action shall be taken with reasonable promptness so as to cause no delay. The Architect's approval of a specific item shall not indicate approval of an assembly of which the item is a component.

1.5.14 The Architect shall prepare Change Orders for the Owner's approval and execution in accordance with the Contract Documents, and shall have authority to order minor changes in the Work not involving an adjustment in the Contract Sum or an extension of the Contract Time which are not inconsistent with the intent of the Contract Documents.

1.5.15 The Architect shall conduct inspections to determine the Dates of Substantial Completion and final completion, shall receive and forward to the Owner for the Owner's review written warranties and related documents required by the Contract Documents and assembled by the Contractor, and shall issue a final Certificate for Payment.

1.5.16 The extent of the duties, responsibilities and limitations of authority of the Architect as the Owner's representative during construction shall not be modified or extended without written consent of the Owner, the Contractor and the Architect.

1.6 PROJECT REPRESENTATION BEYOND BASIC SERVICES

1.6.1 If the Owner and Architect agree that more extensive representation at the site than is described in Paragraph 1.5 shall be provided, the Architect shall provide one or more Project Representatives to assist the Architect in carrying out such responsibilities at the site.

1.6.2 Such Project Representatives shall be selected, employed and directed by the Architect, and the Architect shall be compensated therefor as mutually agreed between the Owner and the Architect as set forth in an exhibit appended to this Agreement, which shall describe the duties, responsibilities and limitations of authority of such Project Representatives.

1.6.3 Through the observations by such Project Representatives, the Architect shall endeavor to provide further protection for the Owner against defects and deficiencies in the Work, but the furnishing of such project representation shall not modify the rights, responsibilities or obligations of the Architect as described in Paragraph 1.5.

1.7 ADDITIONAL SERVICES

The following Services are not included in Basic Services unless so identified in Article 15. They shall be provided if authorized or confirmed in writing by the Owner, and they shall be paid for by the Owner as provided in this Agreement, in addition to the compensation for Basic Services.

AIA DOCUMENT B141 • OWNER-ARCHITECT AGREEMENT • THIRTEENTH EDITION • JULY 1977 • AIA® • © 1977
THE AMERICAN INSTITUTE OF ARCHITECTS, 1735 NEW YORK AVENUE, N.W., WASHINGTON, D.C. 20006

FIG. 3.1 *(continued)*

1.7.1 Providing analyses of the Owner's needs, and programming the requirements of the Project.

1.7.2 Providing financial feasibility or other special studies.

1.7.3 Providing planning surveys, site evaluations, environmental studies or comparative studies of prospective sites, and preparing special surveys, studies and submissions required for approvals of governmental authorities or others having jurisdiction over the Project.

1.7.4 Providing services relative to future facilities, systems and equipment which are not intended to be constructed during the Construction Phase.

1.7.5 Providing services to investigate existing conditions or facilities or to make measured drawings thereof, or to verify the accuracy of drawings or other information furnished by the Owner.

1.7.6 Preparing documents of alternate, separate or sequential bids or providing extra services in connection with bidding, negotiation or construction prior to the completion of the Construction Documents Phase, when requested by the Owner.

1.7.7 Providing coordination of Work performed by separate contractors or by the Owner's own forces.

1.7.8 Providing services in connection with the work of a construction manager or separate consultants retained by the Owner.

1.7.9 Providing Detailed Estimates of Construction Cost, analyses of owning and operating costs, or detailed quantity surveys or inventories of material, equipment and labor.

1.7.10 Providing interior design and other similar services required for or in connection with the selection, procurement or installation of furniture, furnishings and related equipment.

1.7.11 Providing services for planning tenant or rental spaces.

1.7.12 Making revisions in Drawings, Specifications or other documents when such revisions are inconsistent with written approvals or instructions previously given, are required by the enactment or revision of codes, laws or regulations subsequent to the preparation of such documents or are due to other causes not solely within the control of the Architect.

1.7.13 Preparing Drawings, Specifications and supporting data and providing other services in connection with Change Orders to the extent that the adjustment in the Basic Compensation resulting from the adjusted Construction Cost is not commensurate with the services required of the Architect, provided such Change Orders are required by causes not solely within the control of the Architect.

1.7.14 Making investigations, surveys, valuations, inventories or detailed appraisals of existing facilities, and services required in connection with construction performed by the Owner.

1.7.15 Providing consultation concerning replacement of any Work damaged by fire or other cause during construction, and furnishing services as may be required in connection with the replacement of such Work.

1.7.16 Providing services made necessary by the default of the Contractor, or by major defects or deficiencies in the Work of the Contractor, or by failure of performance of either the Owner or Contractor under the Contract for Construction.

1.7.17 Preparing a set of reproducible record drawings showing significant changes in the Work made during construction based on marked-up prints, drawings and other data furnished by the Contractor to the Architect.

1.7.18 Providing extensive assistance in the utilization of any equipment or system such as initial start-up or testing, adjusting and balancing, preparation of operation and maintenance manuals, training personnel for operation and maintenance, and consultation during operation.

1.7.19 Providing services after issuance to the Owner of the final Certificate for Payment, or in the absence of a final Certificate for Payment, more than sixty days after the Date of Substantial Completion of the Work.

1.7.20 Preparing to serve or serving as an expert witness in connection with any public hearing, arbitration proceeding or legal proceeding.

1.7.21 Providing services of consultants for other than the normal architectural, structural, mechanical and electrical engineering services for the Project.

1.7.22 Providing any other services not otherwise included in this Agreement or not customarily furnished in accordance with generally accepted architectural practice.

1.8 TIME

1.8.1 The Architect shall perform Basic and Additional Services as expeditiously as is consistent with professional skill and care and the orderly progress of the Work. Upon request of the Owner, the Architect shall submit for the Owner's approval a schedule for the performance of the Architect's services which shall be adjusted as required as the Project proceeds, and shall include allowances for periods of time required for the Owner's review and approval of submissions and for approvals of authorities having jurisdiction over the Project. This schedule, when approved by the Owner, shall not, except for reasonable cause, be exceeded by the Architect.

ARTICLE 2

THE OWNER'S RESPONSIBILITIES

2.1 The Owner shall provide full information regarding requirements for the Project including a program, which shall set forth the Owner's design objectives, constraints and criteria, including space requirements and relationships, flexibility and expandability, special equipment and systems and site requirements.

2.2 If the Owner provides a budget for the Project it shall include contingencies for bidding, changes in the Work during construction, and other costs which are the responsibility of the Owner, including those described in this Article 2 and in Subparagraph 3.1.2. The Owner shall, at the request of the Architect, provide a statement of funds available for the Project, and their source.

FIG. 3.1 *(continued)*

2.3 The Owner shall designate, when necessary, a representative authorized to act in the Owner's behalf with respect to the Project. The Owner or such authorized representative shall examine the documents submitted by the Architect and shall render decisions pertaining thereto promptly, to avoid unreasonable delay in the progress of the Architect's services.

2.4 The Owner shall furnish a legal description and a certified land survey of the site, giving, as applicable, grades and lines of streets, alleys, pavements and adjoining property; rights-of-way, restrictions, easements, encroachments, zoning, deed restrictions, boundaries and contours of the site; locations, dimensions and complete data pertaining to existing buildings, other improvements and trees; and full information concerning available service and utility lines both public and private, above and below grade, including inverts and depths.

2.5 The Owner shall furnish the services of soil engineers or other consultants when such services are deemed necessary by the Architect. Such services shall include test borings, test pits, soil bearing values, percolation tests, air and water pollution tests, ground corrosion and resistivity tests, including necessary operations for determining subsoil, air and water conditions, with reports and appropriate professional recommendations.

2.6 The Owner shall furnish structural, mechanical, chemical and other laboratory tests, inspections and reports as required by law or the Contract Documents.

2.7 The Owner shall furnish all legal, accounting and insurance counseling services as may be necessary at any time for the Project, including such auditing services as the Owner may require to verify the Contractor's Applications for Payment or to ascertain how or for what purposes the Contractor uses the moneys paid by or on behalf of the Owner.

2.8 The services, information, surveys and reports required by Paragraphs 2.4 through 2.7 inclusive shall be furnished at the Owner's expense, and the Architect shall be entitled to rely upon the accuracy and completeness thereof.

2.9 If the Owner observes or otherwise becomes aware of any fault or defect in the Project or nonconformance with the Contract Documents, prompt written notice thereof shall be given by the Owner to the Architect.

2.10 The Owner shall furnish required information and services and shall render approvals and decisions as expeditiously as necessary for the orderly progress of the Architect's services and of the Work.

ARTICLE 3

CONSTRUCTION COST

3.1 DEFINITION

3.1.1 The Construction Cost shall be the total cost or estimated cost to the Owner of all elements of the Project designed or specified by the Architect.

3.1.2 The Construction Cost shall include at current market rates, including a reasonable allowance for overhead and profit, the cost of labor and materials furnished by the Owner and any equipment which has been designed, specified, selected or specially provided for by the Architect.

3.1.3 Construction Cost does not include the compensation of the Architect and the Architect's consultants, the cost of the land, rights-of-way, or other costs which are the responsibility of the Owner as provided in Article 2.

3.2 RESPONSIBILITY FOR CONSTRUCTION COST

3.2.1 Evaluations of the Owner's Project budget, Statements of Probable Construction Cost and Detailed Estimates of Construction Cost, if any, prepared by the Architect, represent the Architect's best judgment as a design professional familiar with the construction industry. It is recognized, however, that neither the Architect nor the Owner has control over the cost of labor, materials or equipment, over the Contractor's methods of determining bid prices, or over competitive bidding, market or negotiating conditions. Accordingly, the Architect cannot and does not warrant or represent that bids or negotiated prices will not vary from the Project budget proposed, established or approved by the Owner, if any, or from any Statement of Probable Construction Cost or other cost estimate or evaluation prepared by the Architect.

3.2.2 No fixed limit of Construction Cost shall be established as a condition of this Agreement by the furnishing, proposal or establishment of a Project budget under Subparagraph 1.1.2 or Paragraph 2.2 or otherwise, unless such fixed limit has been agreed upon in writing and signed by the parties hereto. If such a fixed limit has been established, the Architect shall be permitted to include contingencies for design, bidding and price escalation, to determine what materials, equipment, component systems and types of construction are to be included in the Contract Documents, to make reasonable adjustments in the scope of the Project and to include in the Contract Documents alternate bids to adjust the Construction Cost to the fixed limit. Any such fixed limit shall be increased in the amount of any increase in the Contract Sum occurring after execution of the Contract for Construction.

3.2.3 If the Bidding or Negotiation Phase has not commenced within three months after the Architect submits the Construction Documents to the Owner, any Project budget or fixed limit of Construction Cost shall be adjusted to reflect any change in the general level of prices in the construction industry between the date of submission of the Construction Documents to the Owner and the date on which proposals are sought.

3.2.4 If a Project budget or fixed limit of Construction Cost (adjusted as provided in Subparagraph 3.2.3) is exceeded by the lowest bona fide bid or negotiated proposal, the Owner shall (1) give written approval of an increase in such fixed limit, (2) authorize rebidding or renegotiating of the Project within a reasonable time, (3) if the Project is abandoned, terminate in accordance with Paragraph 10.2, or (4) cooperate in revising the Project scope and quality as required to reduce the Construction Cost. In the case of (4), provided a fixed limit of Construction Cost has been established as a condition of this Agreement, the Architect, without additional charge, shall modify the Drawings and Specifications as necessary to comply

AIA DOCUMENT B141 • OWNER-ARCHITECT AGREEMENT • THIRTEENTH EDITION • JULY 1977 • AIA® • © 1977
THE AMERICAN INSTITUTE OF ARCHITECTS, 1735 NEW YORK AVENUE, N.W., WASHINGTON, D.C. 20006

FIG. 3.1 *(continued)*

with the fixed limit. The providing of such service shall be the limit of the Architect's responsibility arising from the establishment of such fixed limit, and having done so, the Architect shall be entitled to compensation for all services performed, in accordance with this Agreement, whether or not the Construction Phase is commenced.

ARTICLE 4

DIRECT PERSONNEL EXPENSE

4.1 Direct Personnel Expense is defined as the direct salaries of all the Architect's personnel engaged on the Project, and the portion of the cost of their mandatory and customary contributions and benefits related thereto, such as employment taxes and other statutory employee benefits, insurance, sick leave, holidays, vacations, pensions and similar contributions and benefits.

ARTICLE 5

REIMBURSABLE EXPENSES

5.1 Reimbursable Expenses are in addition to the Compensation for Basic and Additional Services and include actual expenditures made by the Architect and the Architect's employees and consultants in the interest of the Project for the expenses listed in the following Subparagraphs:

5.1.1 Expense of transportation in connection with the Project; living expenses in connection with out-of-town travel; long distance communications; and fees paid for securing approval of authorities having jurisdiction over the Project.

5.1.2 Expense of reproductions, postage and handling of Drawings, Specifications and other documents, excluding reproductions for the office use of the Architect and the Architect's consultants.

5.1.3 Expense of data processing and photographic production techniques when used in connection with Additional Services.

5.1.4 If authorized in advance by the Owner, expense of overtime work requiring higher than regular rates.

5.1.5 Expense of renderings, models and mock-ups requested by the Owner.

5.1.6 Expense of any additional insurance coverage or limits, including professional liability insurance, requested by the Owner in excess of that normally carried by the Architect and the Architect's consultants.

ARTICLE 6

PAYMENTS TO THE ARCHITECT

6.1 PAYMENTS ON ACCOUNT OF BASIC SERVICES

6.1.1 An initial payment as set forth in Paragraph 14.1 is the minimum payment under this Agreement.

6.1.2 Subsequent payments for Basic Services shall be made monthly and shall be in proportion to services performed within each Phase of services, on the basis set forth in Article 14.

6.1.3 If and to the extent that the Contract Time initially established in the Contract for Construction is exceeded or extended through no fault of the Architect, compensation for any Basic Services required for such extended period of Administration of the Construction Contract shall be computed as set forth in Paragraph 14.4 for Additional Services.

6.1.4 When compensation is based on a percentage of Construction Cost, and any portions of the Project are deleted or otherwise not constructed, compensation for such portions of the Project shall be payable to the extent services are performed on such portions, in accordance with the schedule set forth in Subparagraph 14.2.2, based on (1) the lowest bona fide bid or negotiated proposal or, (2) if no such bid or proposal is received, the most recent Statement of Probable Construction Cost or Detailed Estimate of Construction Cost for such portions of the Project.

6.2 PAYMENTS ON ACCOUNT OF ADDITIONAL SERVICES

6.2.1 Payments on account of the Architect's Additional Services as defined in Paragraph 1.7 and for Reimbursable Expenses as defined in Article 5 shall be made monthly upon presentation of the Architect's statement of services rendered or expenses incurred.

6.3 PAYMENTS WITHHELD

6.3.1 No deductions shall be made from the Architect's compensation on account of penalty, liquidated damages or other sums withheld from payments to contractors, or on account of the cost of changes in the Work other than those for which the Architect is held legally liable.

6.4 PROJECT SUSPENSION OR TERMINATION

6.4.1 If the Project is suspended or abandoned in whole or in part for more than three months, the Architect shall be compensated for all services performed prior to receipt of written notice from the Owner of such suspension or abandonment, together with Reimbursable Expenses then due and all Termination Expenses as defined in Paragraph 10.4. If the Project is resumed after being suspended for more than three months, the Architect's compensation shall be equitably adjusted.

ARTICLE 7

ARCHITECT'S ACCOUNTING RECORDS

7.1 Records of Reimbursable Expenses and expenses pertaining to Additional Services and services performed on the basis of a Multiple of Direct Personnel Expense shall be kept on the basis of generally accepted accounting principles and shall be available to the Owner or the Owner's authorized representative at mutually convenient times.

ARTICLE 8

OWNERSHIP AND USE OF DOCUMENTS

8.1 Drawings and Specifications as instruments of service are and shall remain the property of the Architect whether the Project for which they are made is executed or not. The Owner shall be permitted to retain copies, including reproducible copies, of Drawings and Specifications for information and reference in connection with the Owner's use and occupancy of the Project. The Drawings and Specifications shall not be used by the Owner on

FIG. 3.1 *(continued)*

other projects, for additions to this Project, or for completion of this Project by others provided the Architect is not in default under this Agreement, except by agreement in writing and with appropriate compensation to the Architect.

8.2 Submission or distribution to meet official regulatory requirements or for other purposes in connection with the Project is not to be construed as publication in derogation of the Architect's rights.

ARTICLE 9

ARBITRATION

9.1 All claims, disputes and other matters in question between the parties to this Agreement, arising out of or relating to this Agreement or the breach thereof, shall be decided by arbitration in accordance with the Construction Industry Arbitration Rules of the American Arbitration Association then obtaining unless the parties mutually agree otherwise. No arbitration, arising out of or relating to this Agreement, shall include, by consolidation, joinder or in any other manner, any additional person not a party to this Agreement except by written consent containing a specific reference to this Agreement and signed by the Architect, the Owner, and any other person sought to be joined. Any consent to arbitration involving an additional person or persons shall not constitute consent to arbitration of any dispute not described therein or with any person not named or described therein. This Agreement to arbitrate and any agreement to arbitrate with an additional person or persons duly consented to by the parties to this Agreement shall be specifically enforceable under the prevailing arbitration law.

9.2 Notice of the demand for arbitration shall be filed in writing with the other party to this Agreement and with the American Arbitration Association. The demand shall be made within a reasonable time after the claim, dispute or other matter in question has arisen. In no event shall the demand for arbitration be made after the date when institution of legal or equitable proceedings based on such claim, dispute or other matter in question would be barred by the applicable statute of limitations.

9.3 The award rendered by the arbitrators shall be final, and judgment may be entered upon it in accordance with applicable law in any court having jurisdiction thereof.

ARTICLE 10

TERMINATION OF AGREEMENT

10.1 This Agreement may be terminated by either party upon seven days' written notice should the other party fail substantially to perform in accordance with its terms through no fault of the party initiating the termination.

10.2 This Agreement may be terminated by the Owner upon at least seven days' written notice to the Architect in the event that the Project is permanently abandoned.

10.3 In the event of termination not the fault of the Architect, the Architect shall be compensated for all services performed to termination date, together with Reimbursable Expenses then due and all Termination Expenses as defined in Paragraph 10.4.

10.4 Termination Expenses include expenses directly attributable to termination for which the Architect is not otherwise compensated, plus an amount computed as a percentage of the total Basic and Additional Compensation earned to the time of termination, as follows:

.1 20 percent if termination occurs during the Schematic Design Phase; or

.2 10 percent if termination occurs during the Design Development Phase; or

.3 5 percent if termination occurs during any subsequent phase.

ARTICLE 11

MISCELLANEOUS PROVISIONS

11.1 Unless otherwise specified, this Agreement shall be governed by the law of the principal place of business of the Architect.

11.2 Terms in this Agreement shall have the same meaning as those in AIA Document A201, General Conditions of the Contract for Construction, current as of the date of this Agreement.

11.3 As between the parties to this Agreement: as to all acts or failures to act by either party to this Agreement, any applicable statute of limitations shall commence to run and any alleged cause of action shall be deemed to have accrued in any and all events not later than the relevant Date of Substantial Completion of the Work, and as to any acts or failures to act occurring after the relevant Date of Substantial Completion, not later than the date of issuance of the final Certificate for Payment.

11.4 The Owner and the Architect waive all rights against each other and against the contractors, consultants, agents and employees of the other for damages covered by any property insurance during construction as set forth in the edition of AIA Document A201, General Conditions, current as of the date of this Agreement. The Owner and the Architect each shall require appropriate similar waivers from their contractors, consultants and agents.

ARTICLE 12

SUCCESSORS AND ASSIGNS

12.1 The Owner and the Architect, respectively, bind themselves, their partners, successors, assigns and legal representatives to the other party to this Agreement and to the partners, successors, assigns and legal representatives of such other party with respect to all covenants of this Agreement. Neither the Owner nor the Architect shall assign, sublet or transfer any interest in this Agreement without the written consent of the other.

ARTICLE 13

EXTENT OF AGREEMENT

13.1 This Agreement represents the entire and integrated agreement between the Owner and the Architect and supersedes all prior negotiations, representations or agreements, either written or oral. This Agreement may be amended only by written instrument signed by both Owner and Architect.

AIA DOCUMENT B141 • OWNER-ARCHITECT AGREEMENT • THIRTEENTH EDITION • JULY 1977 • AIA® • © 1977
THE AMERICAN INSTITUTE OF ARCHITECTS, 1735 NEW YORK AVENUE, N.W., WASHINGTON, D.C. 20006

FIG. 3.1 *(continued)*

ARTICLE 14

BASIS OF COMPENSATION

The Owner shall compensate the Architect for the Scope of Services provided, in accordance with Article 6, Payments to the Architect, and the other Terms and Conditions of this Agreement, as follows:

14.1 AN INITIAL PAYMENT of dollars ($)

shall be made upon execution of this Agreement and credited to the Owner's account as follows:

14.2 BASIC COMPENSATION

14.2.1 FOR BASIC SERVICES, as described in Paragraphs 1.1 through 1.5, and any other services included in Article 15 as part of Basic Services, Basic Compensation shall be computed as follows:

(Here insert basis of compensation, including fixed amounts, multiples or percentages, and identify Phases to which particular methods of compensation apply, if necessary.)

14.2.2 Where compensation is based on a Stipulated Sum or Percentage of Construction Cost, payments for Basic Services shall be made as provided in Subparagraph 6.1.2, so that Basic Compensation for each Phase shall equal the following percentages of the total Basic Compensation payable:

(Include any additional Phases as appropriate.)

Schematic Design Phase: percent (%)
Design Development Phase: percent (%)
Construction Documents Phase: percent (%)
Bidding or Negotiation Phase: percent (%)
Construction Phase: percent (%)

14.3 FOR PROJECT REPRESENTATION BEYOND BASIC SERVICES, as described in Paragraph 1.6, Compensation shall be computed separately in accordance with Subparagraph 1.6.2.

FIG. 3.1 *(continued)*

14.4 COMPENSATION FOR ADDITIONAL SERVICES

14.4.1 FOR ADDITIONAL SERVICES OF THE ARCHITECT, as described in Paragraph 1.7, and any other services included in Article 15 as part of Additional Services, but excluding Additional Services of consultants, Compensation shall be computed as follows:

(Here insert basis of compensation, including rates and/or multiples of Direct Personnel Expense for Principals and employees, and identify Principals and classify employees, if required. Identify specific services to which particular methods of compensation apply, if necessary.)

14.4.2 FOR ADDITIONAL SERVICES OF CONSULTANTS, including additional structural, mechanical and electrical engineering services and those provided under Subparagraph 1.7.21 or identified in Article 15 as part of Additional Services, a multiple of ( ) times the amounts billed to the Architect for such services.

(Identify specific types of consultants in Article 15, if required.)

14.5 FOR REIMBURSABLE EXPENSES, as described in Article 5, and any other items included in Article 15 as Reimbursable Expenses, a multiple of ( ) times the amounts expended by the Architect, the Architect's employees and consultants in the interest of the Project.

14.6 Payments due the Architect and unpaid under this Agreement shall bear interest from the date payment is due at the rate entered below, or in the absence thereof, at the legal rate prevailing at the principal place of business of the Architect.

(Here insert any rate of interest agreed upon.)

(Usury laws and requirements under the Federal Truth in Lending Act, similar state and local consumer credit laws and other regulations at the Owner's and Architect's principal places of business, the location of the Project and elsewhere may affect the validity of this provision. Specific legal advice should be obtained with respect to deletion, modification, or other requirements such as written disclosures or waivers.)

14.7 The Owner and the Architect agree in accordance with the Terms and Conditions of this Agreement that:

14.7.1 IF THE SCOPE of the Project or of the Architect's Services is changed materially, the amounts of compensation shall be equitably adjusted.

14.7.2 IF THE SERVICES covered by this Agreement have not been completed within

( ) months of the date hereof, through no fault of the Architect, the amounts of compensation, rates and multiples set forth herein shall be equitably adjusted.

AIA DOCUMENT B141 • OWNER-ARCHITECT AGREEMENT • THIRTEENTH EDITION • JULY 1977 • AIA® • © 1977
THE AMERICAN INSTITUTE OF ARCHITECTS, 1735 NEW YORK AVENUE, N.W., WASHINGTON, D.C. 20006

FIG. 3.1 (*continued*)

ARTICLE 15

OTHER CONDITIONS OR SERVICES

FIG. 3.1 *(continued)*

This Agreement entered into as of the day and year first written above.

OWNER	ARCHITECT
______________________	______________________
______________________	______________________
______________________	______________________
BY______________________	BY______________________

AIA DOCUMENT B141 • OWNER-ARCHITECT AGREEMENT • THIRTEENTH EDITION • JULY 1977 • AIA® • © 1977
THE AMERICAN INSTITUTE OF ARCHITECTS, 1735 NEW YORK AVENUE, N.W., WASHINGTON, D.C. 20006

tem places more than the normal amount of risk on the design firm. The results have been mixed, with some firms showing a profit and others experiencing a considerable loss.

3.5 ENGINEERING SERVICES

The services performed by a consulting engineer may be grouped into two broad categories: (1) consultation, investigations, and reports and (2) services for a design type project. The first category deals with a wide range of activities, such as feasibility studies, soil tests, economic considerations, inspection or testing. These activities do not normally involve the preparation of drawings and specifications.

When the construction of a project is involved (bridge, dam, highway, etc.), the services furnished by the consulting engineer closely follow those furnished by an architect, as detailed in the preceding section. The selection process used in hiring engineers for engineering construction projects is similar to that used in selecting an architect.

3.5.1 METHODS OF PAYMENT

For all projects except the design type, the engineer is usually paid on an hourly or per diem basis plus expenses. These expenses normally include such items as travel, lodging and meals while out of town, long distance calls, and testing services. If the project involves preparation of drawings and specifications for construction, the methods of payment are similar to those discussed earlier. If a percentage of construction costs is used, the percentage can vary widely, depending upon size and complexity of the project. Few firms will perform work for less than 4 percent on large, standard projects. The percentage fee on complex and innovative projects may be in the range of 10 percent or more.

3.5.2 FORMS OF AGREEMENT

Many forms of agreement may be used when contracting for engineering services. These range from a single sheet confirming acceptance of a previously submitted proposal to the 14-page form used by the General Services Administration. Regardless of the length or form used, the agreement should cover the following points:

Name of parties to the agreement.

Date of the agreement.

The project for which services are contracted. This may include a project title and a name or identification number, as well as a brief description of the project and its location.

The services that are to be performed under the agreement. This item may also include specific services that are *not* included.

Method of payment and the basis for calculating this.

Titles and signatures of the persons authorized by each party. Under some circumstances corporate seals and powers-of-attorney may be required.

Terms and conditions of the agreement. This portion of the agreement may vary widely. Many governmental agencies reflect numerous legislative and agency requirements under this section of the agreement. Topics may range from rights of termination to cost accounting standards.

An example of an owner-engineer agreement is shown in Fig. 3.2. Its content may be compared with the items listed above.

(text continues on page 38)

FIG. 3.2 Owner-engineer agreement

OWNER-ENGINEER AGREEMENT
(Example)

Proj. No.________________

County________________

I. This contract made and entered into this_____day of_____by and between the Highway Division, Iowa Department of Transportation of Ames, Iowa, hereinafter referred to as the "State" and ________________ doing business as a corporation incorporated under the laws of_____, (Home Address), ________________, hereinafter referred to as the "Consultant", ________________, a member of the firm, is a registered professional engineer in Iowa, with registration in a field appropriate to the work involved in this contract.

II. The State proposes to improve a highway known as ________________________ and the State desires to employ the Consultant in connection with the engineering work to be performed in accomplishing the objectives of the Primary Road Laws (Chapter 313, Code of Iowa) and other applicable laws and regulations of the State of Iowa and the United States, consisting of _____ miles of preparation of final design plans for construction of the proposed highway facility within the above stated limits, including all single barrel box culverts.

III. **AGREEMENT**

1. DESCRIPTION OF WORK TO BE DONE.

The following engineering services are to be performed by the Consultant and constitute the basis of this contract. All services shall be in compliance with the Detail Description of Services contained in the addendum.

A. Preparation of grading, drainage and paving plans (including any retaining walls and 20 scale scrolls of urban areas).

B. Erosion control.

C. Preparation of single barrel culvert plans.

IV. **OBLIGATION OF STATE TO CONSULTANT**

All traffic information, right-of-way data, etc., applicable to the work and in possession of the State will be made available to the Consultant without cost.

V. **CONFERENCES, VISITS TO THE SITE, INSPECTION OF WORK**

The Consultant will not begin detail design prior to a conference with the State, nor prior to being given written notice to proceed.

From time to time as the work progresses, conferences will be held at Ames at the request of the Consultant or the State to discuss details of the design. Such conferences will be arranged with the Design Engineer of the Highway Division and will be attended by the State personnel whom he designates.

VI. **CONSULTANT'S SERVICES DURING CONSTRUCTION**

If it is found necessary during the course of construction to consult with the Consultant regarding the work, the Consultant shall be available for such consultation. Such services will be considered as a part of the design contract and no separate payment will be made therefor.

In the event that plans are found to be in error during construction of the project and revision or reworking of the plans is necessary, the Consultant agrees that he shall do such revision without expense to the State, even though final payment may have been received by him. He must give immediate attention to these changes so there will be a minimum of delay to the Contractor. If the revision is made necessary by a design error, the Consultant assumes full financial responsibility for the increased costs incurred by the State beyond those costs normal to achieving the plan intent.

VII. **PLAN SUBMITTALS**

During the progress of the work, various copies of the plans and other documents concerned with the design will be required by the State. These plans will be necessary for submittal for approval of the Federal Highway Administration, for reconnaissance and field examination, for review of access compliance, for review of right-of-way requirements, and for conducting soils investigation, as well as other similar work. These plans will be furnished to the State as a part of the design work upon request. Printing costs will be reimbursed in accordance with Section IX, B, (2).

FIG. 3.2 *(continued)*

VIII. **TIME OF BEGINNING AND COMPLETION**

The Consultant shall begin work under this contract not later than ______ days after notification by the State. Plans for Right-of-Way purchase shall be furnished not later than ______. The completed plans and 20 scale scrolls shall be furnished to the State not later than ______. Prior to the execution of this contract, the Consultant and the State shall establish a schedule of work completion. Failure of the Consultant to maintain progress in accordance with this schedule will be cause for termination of the contract, and for removal of the Consultant from the list of Consultants considered qualified to do engineering work for the Iowa Department of Transportation.

IX. **FEES**

A. As full and complete compensation for engineering services performed under this Agreement, the Consultant shall be paid a fee in the amount of the Consultant's actual costs plus a fixed fee. The estimated total maximum amount payable for the design services is shown as follows:

(1) Design:
Estimated Actual Costs ______
FIXED FEE: ______

(2) Contingency Amount: ______
TOTAL Maximum Amount Payable ______

If the Consultant should, at any time during progress of the design work, determine that his actual costs will exceed the estimated actual costs and he anticipates a need to utilize the contingency fee, he shall promptly notify the State in writing and submit satisfactory justification for the request. The Consultant shall not exceed the estimated actual costs without prior written approval of the State. The contingency fee will not be used for an adjustment in the fixed fee.

The fixed fee will not be adjusted unless there is a substantial change in the scope, complexity or character of the services to be performed, or the time schedule is changed by the State.

The above Estimated Total Maximum Amount Payable can only be exceeded with written authorization from the State. If, at any time, it appears that the fees above will exceed the Estimated Total Maximum Amount Payable, the Consultant shall immediately so notify the State in writing and submit supporting documentation as to the reasons for exceeding the Estimated Total Maximum Amount Payable. The State will advise the Consultant in writing that (1) an increase in the Estimated Total Maximum Amount Payable is approved, or that (2) the support documentation does not provide adequate justification for an adjustment of the Estimated Total Maximum Amount Payable, or that (3) the Consultant should reduce the specified services.

B. The actual costs that are reimbursable are those costs attributable to the specific work covered by this Agreement and allowable under the provisions of the Federal-Aid Highway Program Manual (FAHPM), Vol, 6, Chapter 1, Section 2, Subsection 2 and Subparts 1–15.2 and 1–15.4 of the Federal Procurement Regulations. Section 1–15.201–4 and 1–15.205.6 are mandatory and not optional as indicated in subsection 1–15.1. These include the following:

(1) The Consultant shall be reimbursed for his actual costs related to salaries of his employees for the work under the terms of this Agreement and salaries of principles for time they are productively engaged in work necessary to fulfill the terms of this Agreement. Salary costs shall include direct or actual payroll for time of employees directly attributable and properly chargeable to the work plus salary related costs. The rate for salary related costs is provisionally established as______percent of direct payroll for computing partial payment of fees but, for final payment of fees, shall be adjusted to rates representative of actual salary related costs of the Consultant based on his established practices developed in accord with sound accounting principles consistent with provisions of Federal Procurement Regulations.

(2) The Consultant shall be reimbursed for his direct nonsalary costs which are directly attributable and properly allocable to the work.

(a) Direct non-salary costs include the Consultant's payments to others engaged by him on the work necessary to fulfill the terms of this Agreement.

(b) Direct non-salary costs paid by the Consultant also include in-plant and travel expenses necessary to fulfill the terms of this proposal. Travel expenses shall include meals, lodging and 17 cents per mile for regular employees and principals of the Consultant only while away from their regular place of duty and directly engaged on the work. The Consultant will not be required to submit receipts for reimbursement of in-plant and travel expenses, but will be required to submit a detailed listing of such actual expenses certified by him to be direct costs that are not included in overhead. Charges for in-plant expenses such as reproductions, printing and computer services shall be at the Consultant's standard schedule for such services.

(3) The Consultant shall be reimbursed for his actual overhead or indirect costs to the extent that they are properly allocable to the work. Such costs shall be based on the Consultant's established practice for allocating indirect costs developed in accord with sound accounting principles consistent with provisions of Federal Procurement Regulations Part 1–15. Such costs shall include the items normally and usually allowed by FHWA under provisions of FAHPM Vol. 6, Chapter 1, Section 2, Subsection 2. The rate for overhead or indirect costs is provisionally established as______percent of direct payroll for computing partial payment of fees but, for final payment of fees, shall be adjusted to rates representative of actual overhead costs of the Consultant for the fiscal year or years during which the work was accomplished based on his established practices developed in accord with sound accounting principles consistent with provisions of Federal Procurement Regulations Part 1–15.

C. Premium overtime pay shall not exceed one percent of the total direct payroll cost without written authorization.

FIG. 3.2 *(continued)*

X. **PAYMENT**

Payments shall be made to the Consultant as follows:

The Consultant shall, each month, submit to the State a progress report in triplicate disclosing the total estimated work accomplished by the Consultant during the preceding month. The progress report shall compare the work accomplished by showing graphically and in percent the actual extent of work completed and the work as originally scheduled.

With the progress report, or by separate invoice, the Consultant will submit a tabulation of the direct and indirect costs and fixed fee associated with completed work. Upon acceptance by the State, payment will be made promptly of an amount not to exceed 90 percent of the direct and indirect costs and proportional share of the fixed fee, less amounts previously paid.

Upon acceptance of the completed plans by the State and FHWA, final payment, including fixed fee, less credits for progress payments, shall be made to the Consultant within thirty days of final audit.

XI. **MISCELLANEOUS PROVISIONS**

A. The Consultant will be responsible for any necessary updating of survey to cover new improvements adjacent to the right-of-way corridor during the duration of the design agreement.

B. OWNERSHIP OF ENGINEERING DOCUMENTS All survey notes, sketches, tracings, plans, specifications, reports on special studies and other data prepared under this contract shall become the property of the State and be delivered to the State upon completion of the plans or termination of the services of the Consultant. There shall be no restriction or limitation on their further use.

C. CHANGES IN SCOPE OF WORK When there is substantial change in the scope, complexity or character of work performed, the specified fees as listed under IX.A., as the maximum total amounts payable, as well as the fixed fees, will be reappraised. In any case where the Consultant believes extra compensation will be due him for work and services not clearly covered by this agreement or supplement thereto, he shall promptly notify the State, in writing, of his intention to make claim for such extra compensation and secure written authorization from the State prior to proceeding with the work.

D. DELAYS The Consultant will notify the State of any unusual delay, including the reason therefore, to its normal progress in the preparation of plans, either actual or prospective, and request an appropriate extension of time.

If the preparation of plans is delayed by events beyond the control of the Consultant, and such plans are not substantially complete within______months after date of initial notice to proceed issued under the Agreement, the Maximum total compensation payable, fixed fee and schedules of completion, will be subject to review upon request of the Consultant to the State, accompanied by adequate substantiating data to justify a change.

E. TERMINATION If the State should desire to suspend or terminate the service to be rendered by the Consultant under this agreement, such suspension or termination may be effected by the State, giving the Consultant 30 days' written notice.

When services are suspended or terminated, payment is to be made by the State for all actual direct and indirect costs incurred by the Consultant for the work completed in accordance with this agreement, less amounts previously paid. In addition, the Consultant will be paid the proper proportional share of the fixed fee, as based on the actual costs incurred to the Consultant's estimated costs, less amounts previously paid as set forth in Section X.

F. RESPONSIBILITY FOR CLAIMS AND LIABILITY The Consultant shall indemnify and save harmless the Iowa Department of Transportation, State of Iowa, and FHWA from any and all causes of action, suits at law or in equity, or losses, damages, claims of demands, and from all liability of whatsoever nature for and on account of or due to any error, omission or negligent act of the Consultant, its members, employees, agents, subcontractors, or assigns, and arising out of or in connection with this Contract or the performance of any part thereof.

G. GENERAL COMPLIANCE WITH LAWS The Consultant shall comply with all Federal, State and Local Laws and ordinances applicable to the work.

H. SUBLETTING, ASSIGNMENT OR TRANSFER Subletting, assignment or transfer of all or part of the interest of the Consultant is prohibited unless written consent is obtained from the State.

I. DESIGN CRITERIA Design criteria shall be the applicable AASHTO Standards for the system involved and the AASHTO Policies on Geometric Design of Rural and Urban Highways, and shall conform to local requirements if within an incorporated area. Plans, specifications and all other work prepared by the Consultant shall be in accordance with Highway Division, Iowa Department of Transportation Standards.

J. FORBIDDING USE OF OUTSIDE AGENTS The Consultant warrants that he has not employed or retained any company or person, other than a bona fide employee working solely for the consultant, to solicit or secure this contract, and that he has not paid or agreed to pay any company or person, other than bona fide employee working solely for the Consultant, any fee, commission, percentage, brokerage fee, gifts, or any other consideration, contingent upon or resulting from the award or making of this contract. For breach or violation of this warranty the State shall have the right to annul this contract without liability, or, in its discretion to deduct from the contract price or consideration, or otherwise recover, the full amount of such fee, commission, percentage, brokerage fee, gift, or contingent fee.

FIG. 3.2 *(continued)*

K. **EMPLOYMENT OF STATE, COUNTY AND CITY WORKERS** The Consultant shall not engage, on a full or part-time or other basis during the period of the contract, any professional or technical personnel who are or have been at any time during the period of the contract in the employ of the Federal Highway Administration or the Iowa Department of Transportation, County or City, except regularly retired employees, without the written consent of the public employer of such person.

L. **CONSULTANT'S CERTIFICATION OF PLANS** The Consultant shall place his certification on the Title Sheet of the completed plans.

M. **COMPLIANCE WITH TITLE 49, CODE OF FEDERAL REGULATIONS** During the performance of this contract, the Consultant, for itself, its assignees and successors in interest (hereinafter referred to as the "Consultant"), agrees as follows:

(1) *Compliance with Regulations:*
The Consultant will comply with the Regulations of the Department of Transportation relative to nondiscrimination in federally assisted programs of the Department of Transportation (Title 49, Code of Federal Regulations, Part 21, hereinafter referred to as the Regulations), which are herein incorporated by reference and made a part of this contract.

(2) *Nondiscrimination:*
The Consultant, with regard to the work performed by it after award and prior to completion of the contract work, will not discriminate on the ground of race, color or national origin in the selection and retention of subcontractors, including procurements of materials and leases of equipment. The Consultant will not participate either directly or indirectly in the discrimination prohibited by Section 21.5 of the Regulations, including employment practices when the contract covers a program set forth in Appendix "A", "B", and "C" of the Regulations.

(3) *Solicitations for Subcontracts, Including Procurement of Materials and Equipment:*
In all solicitations either by competitive bidding or negotiation made by the Consultant for work to be performed under a subcontract, including procurements of materials or equipment, each potential subcontractor or supplier shall be notified by the Consultant of the Consultant's obligations under this contract and the Regulations relative to nondiscrimination on the ground of race, color, sex or national origin.

(4) *Information and Reports:*
The Consultant will provide all information and reports required by the Regulations, or orders and instructions issued pursuant thereto, and will permit access to its books, records, accounts, other sources of information, and its facilities as may be determined by the Iowa Department of Transportation or the Federal Highway Administration to be pertinent to ascertain compliance with such Regulations, orders and instructions. Where any information required of a Consultant is in the exclusive possession of another who fails or refuses to furnish this information, the Consultant shall so certify to the Iowa Department of Transportation, or the Federal Highway Administration as appropriate; and shall set forth what efforts it has made to obtain the information.

(5) *Sanctions for Noncompliance:*
In the event of the Consultant's noncompliance with the nondiscrimination provisions of this contract, the Iowa Department of Transportation shall impose such contract sanctions as it or the Federal Highway Administration may determine to be appropriate, including, but not limited to,
(a) withholding of payments to the Consultant under the contract until the Consultant complies, and/or
(b) cancellation, termination or suspension of the contract, in whole or in part.

(6) *Incorporation of Provisions:*
The Consultant will include the provisions of paragraph (1) through (6) in every subcontract, including procurements of materials and leases of equipment, unless exempt by the Regulations, order, or instructions issued pursuant thereto. The Consultant will take such action with respect to any subcontract or procurement as the Iowa Department of Transportation or the Federal Highway Administration may direct as a means of enforcing such provisions including sanctions for noncompliance: Provided, however, that, in the event a Consultant becomes involved in, or is threatened with, litigation with a subcontractor or supplier as a result of such direction, the Consultant may request the State to enter into such litigation to protect the interests of the State, and, in addition, the Consultant may request the United States to enter into such litigation to protect the interests of the United States.

N. **ACCESS TO RECORDS** The Consultant and his subcontractors are to maintain all books, documents, papers, accounting records and other evidence pertaining to cost incurred and records supporting this cost proposal in performing work covered by this contract, and to make such materials available at their respective offices at all reasonable times during the contract period and for three years from the date of final payment under the contract, for inspection by the State, FHWA or any authorized representatives of the Federal Government and copies thereof shall be furnished if requested.

NOTE:
The Consultant certifies that he is fully acquainted with the concept of the project as presently developed by the Iowa Department of Transportation, and that it is the intention of this contract with the Consultant to do work necessary to bring the plans on this project to the letting stage. Engineering decisions on this project are the responsibility of the Consultant, and he will be required to furnish, to the State, factual data supporting his decisions.

FIG. 3.2 *(continued)*

This contract expresses the entire agreement between the parties and no representations, promises or warranties have been made by either of the parties that are not fully expressed herein.

IN WITNESS WHEREOF, the parties hereto have executed this contract as of the day and year first above written.

IOWA DEPARTMENT OF TRANSPORTATION

By______________________________

CONSULTANT

____

Approved as to Form

____
Attorney

ATTEST:

____

Winneshiek Co. F-52-5(900)—20-96

CERTIFICATION OF CONSULTANT

I hereby certify that I am the______________and duly authorized representative of the firm of______________, whose address is______________, and that neither I nor the above firm I here represent has:

(a) employed or retained for a commission, percentage, brokerage, contingent fee, or other consideration, any firm or person (other than a bona fide employee working solely for me or the above consultant) to solicit or secure this contract;
(b) agreed, as an express or implied condition for obtaining this contract, to employ or retain the services of any firm or person in connection with carrying out the contract, or
(c) paid, or agreed to pay, to any firm, organization or person (other than a bona fide employee working solely for me or the above consultant) any fee, contribution, donation or consideration of any kind for, or in connection with, procuring or carrying out the contract;

except as here expressly stated (if any):

I acknowledge that this certificate is to be furnished to the State Highway Department and the Federal Highway Administration, U.S. Department of Transportation, in connection with this contract involving participation of Federal-aid highway funds, and is subject to applicable, State and Federal laws, both criminal and civil.

______________________________ Date

______________________________ (Title)

CERTIFICATION OF STATE HIGHWAY DEPARTMENT

I hereby certify that I am the______of the Highway Division, Iowa Department of Transportation, and that the above consulting firm or his representative has not been required, directly or indirectly as an express or implied condition in connection with obtaining or carrying out this contract to:

(a) employ or retain, or agree to employ or retain, any firm or person, or

(b) pay, or agree to pay, to any firm, person, or organization, any fee, contribution, donation, or consideration of any kind:

except as here expressly stated (if any):

I acknowledge that this certificate is to be furnished the Federal Highway Administration, U.S. Department of Transportation, in connection with this contract involving participation of Federal-aid highway funds, and is subject to applicable State and Federal laws, both criminal and civil.

______________________________ Date

______________________________ Signature

3.6 CONSTRUCTION MANAGEMENT

Within the construction industry there is a difference between the terms management of construction and construction management. General contractors have been managing the construction of projects for many years. As a result, they often refer to themselves as "construction managers." Their work should not be confused with construction management, which refers to a process of design and construction that is an option currently available to an owner.

Under the construction management form of contract, the planning, design, and construction of a project are integrated within an overall system. The system utilizes a construction team consisting of the owner, the design professional, and the construction manager. Under ideal conditions, all members of the team work together, with all members drawing from their individual fields of expertise. An example of the construction management form of contract is shown in Fig. 3.3. Several variations to this form may be used.

3.6.1 TEAM MEMBER ACTIVITIES

The three members of the construction management team are responsible for the planning, design, and construction of the project. It is vital to the success of the project that each member participates properly. The owner must supply information regarding problems, needs, operations, and any other factors impacting upon the solution. The owner should also be prepared to make prompt decisions regarding proposed solutions. While the design professional and the construction manager may furnish counsel and advice regarding financing, the ultimate responsibility in this area rests with the owner.

In selecting both the construction manager and the design professional under a construction management contract, serious consideration should be given to their ability to work together as well as with the owner. Although the construction manager is usually a general contracting firm, it may sometimes be a professional management organization. Some owners prefer to use one of the general contracting firms as the construction manager because of their ability to furnish such items as bonds and financial security. Under some circumstances the construction management contract may be awarded as a result of competitive bids. Such bids are extremely difficult to compare. Correspondingly, only owners with sufficient knowledge and personnel capable of proper evaluation should undertake the bidding process for this type of contract.

Construction managers are often precluded from performing construction with their own forces when a contract has been won by a competitive bid. Under a negotiated contract this restriction will usually not apply. The construction manager interacts with the owner and the design professional during the planning and design stages. Contribution will be in the area of construction materials and methods as well as that of construction economics.

The construction manager will prepare comparative budgets on design alternatives. After the design has been finalized, the construction manager will prepare a project budget, establish a schedule, oversee the taking of competitive bids, and monitor the construction process. The purpose of the last activity is to ensure that neither the project budget nor the schedule is exceeded unless changes have been made in the project. Such changes should not be put into effect without the concurrence of all of the team members.

(text continues on page 52)

FIG. 3.3 Construction management contract

THE ASSOCIATED GENERAL CONTRACTORS

STANDARD FORM OF AGREEMENT BETWEEN OWNER AND CONSTRUCTION MANAGER

(GUARANTEED MAXIMUM PRICE OPTION)

This Document has important legal consequences; consultation with an attorney is encouraged with respect to its completion or modification.

AGREEMENT

made this day of in the year
of Nineteen Hundred and

BETWEEN

the Owner, and

the Construction Manager.

For services in connection with the following described Project: (Include complete Project location and scope)

The Architect for the Project is

The Owner and the Construction Manager agree as set forth below:

Reproduced by permission of The Associated General Contractors of America.

FIG. 3.3 *(continued)*

TABLE OF CONTENTS

FIG. 3.3 *(continued)*

ARTICLE 1

The Construction Team and Extent of Agreement

THE CONSTRUCTION MANAGER accepts the relationship of trust and confidence established between him and the Owner by this Agreement. He covenants with the Owner to furnish his best skill and judgment and to cooperate with the Architect in furthering the interests of the Owner. He agrees to furnish efficient business administration and superintendence and to use his best efforts to perform the Work in the best and soundest way and in the most expeditious and economical manner consistent with the interests of the Owner.

1.1 *The Construction Team:* The Construction Manager, the Owner, and the Architect, called the "Construction Team" shall work from the beginning of design through construction completion. The Construction Manager shall provide leadership to the Construction Team on all matters relating to construction.

1.2 *Extent of Agreement:* This Agreement represents the entire agreement between the Owner and the Construction Manager and supersedes all prior negotiations, representations or agreements. When plans and specifications are complete, they shall be identified by amendment to this Agreement. This Agreement shall not be superseded by any provisions of the documents for construction and may be amended only by written instrument by both Owner and Construction Manager.

ARTICLE 2

Construction Manager's Services

2.1 The Construction Manager's services under this Agreement shall consist of the two phases described below.

DESIGN PHASE

2.1.1 *Consultation During Project Development:* Review conceptual designs during development. Advise on site use and improvements, selection of materials, building systems and equipment. Provide recommendations on construction feasibility, availability of materials and labor, time requirements for installation and construction, and factors related to cost including costs of alternative designs or materials, preliminary budgets, and possible economies.

2.1.2 *Scheduling:* Develop a Project Time Schedule that coordinates and integrates the Architect's design efforts with construction schedules. Update the Project Time Schedule incorporating a detailed schedule for the Construction operations of the Project, including realistic activity sequences and durations, allocation of labor and materials, processing of shop drawings and samples, and delivery of products requiring long lead-time procurement. Include the Owner's occupancy requirements showing portions of the Project having occupancy priority.

2.1.3 *Project Construction Budget:* Prepare a Project budget as soon as major Project requirements have been identified, and update periodically for the Owner's approval. Prepare an estimate based on a quantity survey of drawings and specifications at the end of the schematic design phase for approval by the Owner as the Project Construction Budget. Update and refine this estimate for Owner's approval as the development of the drawings and specifications proceeds, and advise the Owner and the Architect if it appears that the Project Construction Budget will not be met and make recommendations for corrective action.

2.1.4 *Coordination of Contract Documents:* Review the drawings and specifications as they are being prepared, recommending alternative solutions whenever design details affect construction feasibility or schedules without, however, assuming any of the Architect's customary responsibilities for design.

2.1.5 *Construction Planning:* Recommend for purchase and expedite the procurement of long-lead items to ensure their delivery by the required dates.

2.1.5.1 Make recommendations to the Owner and the Architect regarding the division of Work in the plans and specifications to facilitate the bidding and awarding of Trade Contracts, allowing for phased construction taking into consideration such factors as time of performance, availability of labor, overlapping trade jurisdictions, provisions for temporary facilities, and so forth.

2.1.5.2 Review plans and specifications with the Architect to eliminate areas of conflict and overlapping in the Work to be performed by the various Trade Contractors and prepare prequalification criteria for bidders.

FIG. 3.3 *(continued)*

2.1.5.3 Develop Trade Contractor interest in the Project and as working drawings and specifications are completed, take competitive bids on the Work of the various Trade Contractors. After analyzing the bids, either award contracts or recommend to the Owner that such contracts be awarded.

2.1.6 *Equal Employment Opportunity:* Determine applicable requirements for equal employment opportunity programs for inclusion in Project bidding documents.

CONSTRUCTION PHASE

2.1.7 *Project Control:* Monitor the Work of the Trade Contractors and coordinate the Work with the activities and responsibilities of the Owner, Architect and Construction Manager to complete the Project in accordance with the Owner's objectives of cost, time and quality.

2.1.7.1 Maintain a competent full-time staff at the project site to coordinate and provide general direction of the Work and progress of the Trade Contractors on the Project.

2.1.7.2 Establish on-site organization and lines of authority in order to carry out the overall plans of the Construction Team.

2.1.7.3 Establish procedures for coordination among the Owner, Architect, Trade Contractors and Construction Manager with respect to all aspects of the Project and implement such procedures.

2.1.7.4 Schedule and conduct progress meetings at which Trade Contractors, Owner, Architect and Construction Manager can discuss jointly such matters as procedures, progress, problems and scheduling.

2.1.7.5 Provide regular monitoring of the schedule as construction progresses. Identify potential variances between scheduled and probable completion dates. Review schedule for Work not started or incomplete and recommend to the Owner and Trade Contractors adjustments in the schedule to meet the probable completion date. Provide summary reports of each monitoring and document all changes in schedule.

2.1.7.6 Determine the adequacy of the Trade Contractors' personnel and equipment and the availability of materials and supplies to meet the schedule. Recommend courses of action to the Owner when requirements of a Trade Contract are not being met.

2.1.8 *Cost Control:* Develop and monitor an effective system of Project cost control. Revise and refine the initially approved Project Construction Budget, incorporate approved changes as they occur, and develop cash flow reports and forecasts as needed. Identify variances between actual and budgeted or estimated costs and advise Owner and Architect whenever projected cost exceeds budgets or estimates.

2.1.8.1 Check all materials, equipment and labor entering the Work and maintain cost accounting records on authorized Work performed under unit costs, actual costs for labor and material, or other bases requiring accounting records. Afford the Owner access to these records and preserve them for a period of three (3) years after final payment.

2.1.9 *Change Orders:* Develop and implement a system for review and processing of Change Orders. Recommend necessary or desirable changes to the Owner and the Architect, review requests for changes, submit recommendations to the Owner and the Architect, and assist in negotiating Change Orders.

2.1.10 *Payments to Trade Contractors:* Develop and implement a procedure for the review and processing of applications by Trade Contractors for progress and final payments.

2.1.11 *Permits and Fees:* Assist the Owner and Architect in obtaining all building permits and special permits for permanent improvements, excluding permits for inspection or temporary facilities required to be obtained directly by the various Trade Contractors. Verify that the Owner has paid all applicable fees and assessments for permanent facilities. Assist in obtaining approvals from all the authorities having jurisdiction.

2.1.12 *Owner's Consultants:* If required, assist the Owner in selecting and retaining professional services of a surveyor, testing laboratories and special consultants, and coordinate these services.

2.1.13 *Inspection:* Inspect the Work of Trade Contractors to guard the Owner against defects and deficiencies in the Work. This inspection of the Construction Manager during the Construction Phase shall not relieve the Trade Contractors from their responsibility for construction means, methods, techniques, sequences and procedures, nor for their responsibility to carry out the Work in accordance with the Contract Documents.

FIG. 3.3 *(continued)*

2.1.13.1 Review the safety programs of each of the Trade Contractors and make appropriate recommendations to the Owner. In making such recommendations and carrying out such reviews, he shall not be required to make exhaustive or continuous inspections to check safety precautions and programs in connection with the Work. The performance of such services by the Construction Manager shall not relieve the Trade Contractors of their responsibilities for the safety of persons and property, and for compliance with all federal, state and local statutes, rules, regulations and orders applicable to the conduct of the Work, nor shall it make the Construction Manager responsible for the conduct of Trade Contractors' safety programs and of precautions required thereunder.

2.1.14 *Contract Interpretations:* Refer all questions relative to interpretation of design intent to the Architect.

2.1.15 *Shop Drawings and Samples:* In collaboration with the Architect, establish and implement procedures for expediting the processing and approval of shop drawings and samples.

2.1.16 *Reports and Project Site Documents:* Record the progress of the Project. Submit written progress reports to the Owner and the Architect including information on the Trade Contractors' Work, and the percentage of completion. Keep a daily log available to the Owner and the Architect.

2.1.16.1 Maintain at the Project site, on a current basis: records of all necessary Contracts, shop drawings, samples, purchases, materials, equipment, maintenance and operating manuals and instructions, and any other documents and revisions thereto which arise out of the Contract or the Work. Obtain data from Trade Contractors and maintain a current set of record Drawings, Specifications and operating manuals. At the completion of the Project, deliver all such records to the Owner.

2.1.17 *Substantial Completion:* Determine Substantial Completion of the Work or designated portions thereof and prepare for the Architect a list of incomplete or unsatisfactory items and a schedule for their completion.

2.1.18 *Start-Up:* With the Owner's maintenance personnel, direct the checkout of utilities, operations systems and equipment for readiness and assist in their initial start-up and testing by the Trade Contractors.

2.1.19 *Final Completion:* Determine final completion and provide written notice to the Owner and Architect that the Work is ready for final inspection. Secure and transmit to the Architect required guarantees, affidavits, releases, bonds and waivers. Turn over to the Owner all keys, manuals, record drawings and maintenance stocks.

ARTICLE 3

The Owner's Responsibilities

3.1 The Owner shall provide full information regarding his requirements for the Project.

3.2 The Owner shall designate a representative who shall be fully acquainted with the scope of the Work, and has authority to approve Project Construction Budget, render decisions promptly and furnish information expeditiously.

3.3 The Owner shall retain an Architect for design and to prepare construction documents for the Project. The Architect's services, duties and responsibilities are described in the Agreement between the Owner and the Architect, a copy of which will be furnished to the Construction Manager. The Agreement between the Owner and the Architect shall not be modified without written notification to the Construction Manager.

3.4 The Owner shall furnish such legal services as may be necessary for the Project, and such auditing services as he may require.

3.5 The Construction Manager will be furnished without charge all copies of drawings and specifications reasonably necessary for the execution of the Work.

3.6 If the Owner becomes aware of any fault or defect in the Project or non-conformance with the Contract Documents, he shall give prompt written notice thereof to the Construction Manager.

3.7 The services, information, surveys and reports required by Paragraphs 3.3 through 3.5, inclusive, shall be furnished at the Owner's expense, and the Construction Manager shall be entitled to rely upon the accuracy and completeness thereof.

FIG. 3.3 *(continued)*

ARTICLE 4

Trade Contracts

4.1 All portions of the Work that the Construction Manager does not perform with his own forces shall be performed under Trade Contracts. The Construction Manager shall request and receive proposals from Trade Contractors and Trade Contracts will be awarded after the proposals are reviewed by the Architect and Construction Manager and Owner.

4.2 If the Owner or Architect refuse to accept a Trade Contractor recommended by the Construction Manager, the Construction Manager shall recommend an acceptable substitute and the Guaranteed Maximum Price if applicable shall be increased or decreased by the difference in cost occasioned by such substitution and an appropriate Change Order shall be issued.

4.3 The form of the Trade Contract including the General and Supplementary Conditions to the Construction Contract shall be satisfactory to both the Owner and the Construction Manager.

ARTICLE 5

Contract Time Schedule

5.1 The services and work to be performed under this Contract shall be in general accordance with the following Contract Time Schedule:

5.2 At the time a Guaranteed Maximum Price is fixed, as provided for in Article 6, a new Contract Time Schedule shall also be established.

5.3 If the Construction Manager is delayed at any time in the progress of the Work by any act or neglect of the Owner or the Architect or by any employee of either, or by any separate contractor employed by the Owner, or by changes ordered in the Work, or by labor disputes, fire, unusual delay in transportation, unavoidable casualties or any causes beyond the Construction Manager's control, or by delay authorized by the Owner pending arbitration, the Contract Time Schedule shall be extended by Change Order for a reasonable length of time.

ARTICLE 6

Guaranteed Maximum Price

6.1 When the design, plans and specifications are sufficiently complete to make the final cost estimates, the Construction Manager will, if desired by the Owner, fix a Guaranteed Maximum Price, guaranteeing the maximum cost to the Owner for the Cost of the Work and the Construction Manager's Fee. Such Guaranteed Maximum Price will be guaranteed by the Construction Manager, subject to modification for Changes in the Work as provided in Article 8 and for additional costs arising from delays caused by the Owner or the Architect.

6.2 When the Construction Manager provides a Guaranteed Maximum Price, the Trade Contracts will either be with the Construction Manager or will contain the necessary provisions to allow the Construction Manager to control the performance of the Work.

FIG. 3.3 *(continued)*

ARTICLE 7

Construction Manager's Fee

7.1 In consideration of the performance of the Contract, the Owner agrees to pay the Construction Manager in current funds as compensation for his services a Construction Manager's Fee as set forth in Paragraphs 7.1.1 and 7.1.2.

7.1.1 For the performance of the Design Phase services, a fee of which shall be paid monthly, in equal proportions, based on the scheduled Design Phase time.

7.1.2 For work or services performed during the Construction Phase, a fee of which shall be paid proportionately to the ratio the monthly payment for the Cost of the Work bears to the estimated cost. Any balance of this fee shall be paid at the time of final payment.

7.2 Adjustments in Fee shall be made as follows:

7.2.1 For Changes in the Work as provided for in Article 8, the Construction Manager's Fee shall be adjusted as follows:

7.2.2 For delays in the Work not the responsibility of the Construction Manager, there will be an equitable adjustment in the fee to compensate the Construction Manager for his increased expenses.

7.3 Included in the Construction Manager's Fee are the following:

7.3.1 Salaries or other compensation of the Construction Manager's employees at the principal office and branch offices, except employees listed in Paragraph 9.2.2.

7.3.2 General operating expenses of the Construction Manager's principal and branch offices other than the field office.

7.3.3 Any part of the Construction Manager's capital expenses, including interest on the Construction Manager's capital employed for the Work.

7.3.4 Overhead or general expenses of any kind, except as may be expressly included in Article 9.

7.3.5 Costs in excess of the Guaranteed Maximum Price, if any, as set forth in Article 6 and adjusted pursuant to Article 8.

ARTICLE 8

Changes in the Work

8.1 The Owner, without invalidating the Contract, may order Changes in the Work within the general scope of the Contract consisting of additions, deletions or other revisions, the Guaranteed Maximum Price, if established, and the Contract Time Schedule being adjusted accordingly. All such Changes in the Work shall be authorized by Change Order.

8.1.1 A Change Order is a written order to the Construction Manager signed by the Owner or his authorized agent issued after the execution of the Contract, authorizing a Change in the Work and/or an adjustment in the Guaranteed Maximum Price or the Contract Time Schedule. Each adjustment in the Guaranteed Maximum Price resulting from a Change Order shall clearly separate the amount attributable to the Cost of the Work and the Construction Manager's Fee.

8.1.2 The cost or credit to the Owner resulting from a Change in the Work shall be determined in one or more of the following ways:

.1 by mutual acceptance of a lump sum properly itemized;
.2 by unit prices stated in the Contract Documents or subsequently agreed upon; or
.3 by cost and a mutually acceptable fixed or percentage fee.

FIG. 3.3 *(continued)*

8.1.3 If none of the methods set forth in Paragraph 8.1.2 is agreed upon, the Construction Manager, provided he receives a Change Order, shall promptly proceed with the work involved. The cost of such work shall then be determined on the basis of the Construction Manager's reasonable expenditures and savings, including, in the case of an increase in the Guaranteed Maximum Price, a reasonable allowance for overhead and profit. In such case, and also under Paragraph 8.1.2.3 above, the Construction Manager shall keep and present, in such form as the Owner may prescribe, an itemized accounting together with appropriate supporting data. The amount of credit to be allowed by the Construction Manager to the Owner for any deletion or change which results in a net decrease in cost will be the amount of the actual net decrease. When both additions and credits are involved in any one change, the allowance for overhead and profit shall be figured on the basis of net increase, if any.

8.1.4 If unit prices are stated in the Contract Documents or subsequently agreed upon, and if the quantities originally contemplated are so changed in a proposed Change Order that application of the agreed unit prices to the quantities of Work proposed will create a hardship on the Owner or the Construction Manager, the applicable unit prices shall be equitably adjusted to prevent such hardship.

8.1.5 Should concealed conditions encountered in the performance of the Work below the surface of the ground be at variance with the conditions indicated by the Contract Documents or should unknown physical conditions below the surface of the ground of an unusual nature, differing materially from those ordinarily encountered and generally recognized as inherent in work of the character provided for in this Contract, be encountered, the Guaranteed Maximum Price shall be equitably adjusted by Change Order upon claim by either party made within a reasonable time after the first observance of the conditions.

8.1.6 If the Construction Manager claims that additional cost or time is involved because of (1) any written interpretation issued by the Architect, (2) any order by the Architect to stop the Work where the Construction Manager was not at fault, (3) any written order for a minor change in the Work issued pursuant to Paragraph 8.3, or (4) default by a Trade Contractor having a direct contract with the Owner, the Construction Manager shall make such claim as provided in Paragraph 8.2.

8.2 Claims for additional cost or time.

8.2.1 If the Construction Manager wishes to make a claim for Additional Services, an increase in the Guaranteed Maximum Price, or an extension in the Contract Time, he shall give the Owner written notice thereof within a reasonable time after the occurrence of the event giving rise to such claim. This notice shall be given by the Construction Manager before proceeding to execute the Work, except in an emergency endangering life or property in which case the Construction Manager shall act, at his discretion, to prevent threatened damage, injury or loss. Claims arising from delay shall be made within a reasonable time after the delay. No such claim shall be valid unless so made. If the Owner and the Construction Manager cannot agree on the amount of the adjustment in the Guaranteed Maximum Price or Contract Time Schedule, it shall be determined pursuant to the provisions of Article 5. Any change in the Guaranteed Maximum Price or Contract Time Schedule resulting from such claim shall be authorized by Change Order.

8.3 Minor Changes in the Work.

8.3.1 The Architect shall have authority to order minor Changes in the Work not involving an adjustment in the Guaranteed Maximum Price or an extension of the Contract Time Schedule and not inconsistent with the intent of the Contract Documents. Such Changes may be effected by Field Order or by other written order. Such Changes shall be binding on the Owner and the Construction Manager.

8.4 Field Orders.

8.4.1 The Architect may issue written Field Orders which interpret the Contract Documents or which order minor Changes in the Work in accordance with Paragraph 13.3.1 without change in Guaranteed Maximum Price or Contract Time.

8.5 Emergencies.

8.5.1 In any emergency affecting the safety of persons or property, the Construction Manager shall act, at his discretion, to prevent threatened damage, injury or loss. Any additional compensation or extension of time claimed by the Construction Manager on account of emergency work shall be determined as provided in this Article.

FIG. 3.3 *(continued)*

ARTICLE 9

Cost of the Work

9.1 The term Cost of the Work shall mean costs necessarily incurred in the proper performance of the Work during either the design or Construction Phase, and paid by the Construction Manager, or by the Owner if the Owner is directly paying Trade Contractors upon the Construction Manager's approval and direction. Such costs shall be at rates not higher than the standard paid in the locality of the Work except with prior consent of the Owner, and shall include the items set forth below in this Article.

9.1.1 The Owner agrees to pay the Construction Manager for the Cost of the Work as defined in Article 9. Such payment shall be in addition to the Construction Manager's Fee stipulated in Article 7.

9.2 Cost Items

9.2.1 Wages paid for labor in the direct employ of the Construction Manager in the performance of the Work under applicable collective bargaining agreements, or under a salary or wage schedule agreed upon by the Owner and Construction Manager, and including such welfare or other benefits, if any, as may be payable with respect thereto.

9.2.2 Salaries of Construction Manager's employees when stationed at the field office, in whatever capacity employed and employees in the main or branch office performing the functions listed below:

9.2.3 Cost of pension contributions, hospitalization, bonuses, vacations, medical insurance, assessments, or taxes for such items as unemployment compensation and social security, insofar as such cost is based on wages, salaries, or other remuneration paid to employees of the Construction Manager and included in the Cost of the Work under Subparagraphs 9.2.1 and 9.2.2.

9.2.4 The proportion of reasonable transportation, traveling and hotel expenses of the Construction Manager or of his officers or employees incurred in discharge of duties connected with the Work.

9.2.5 Cost of all materials, supplies and equipment incorporated in the Work, including costs of transportation thereof.

9.2.6 Payments made by the Construction Manager or Owner to Trade Contractors for work performed pursuant to contract under this Agreement.

9.2.7 Cost, including transportation and maintenance, of all materials, supplies, equipment, temporary facilities and hand tools not owned by the workmen, which are employed or consumed in the performance of the Work, and cost less salvage value on such items used but not consumed which remain the property of the Construction Manager.

9.2.8 Rental charges of all necessary machinery and equipment, exclusive of hand tools, used at the site of the Work, whether rented from the Construction Manager or other, including installation, repairs and replacements, dismantling, removal, costs of lubrication, transportation and delivery costs thereof, at rental charges consistent with those prevailing in the area.

9.2.9 Cost of the premiums for all bonds and insurance which are required by this Agreement or deemed necessary by the Construction Manager.

9.2.10 Sales, use, gross receipts or similar taxes related to the Work, imposed by any governmental authority, and for which the Construction Manager is liable.

FIG. 3.3 *(continued)*

9.2.11 Permit fees, licenses, tests, royalties, damages for infringement of patents and costs of defending suits therefor, and deposits lost for causes other than the Construction Manager's negligence.

9.2.12 Losses, expenses or damages to the extent not compensated by insurance or otherwise (including settlement made with the written approval of the Owner), the cost of correction of defective work provided the cost does not exceed the Guaranteed Maximum Price. If, however, such loss requires reconstruction and the Construction Manager is placed in charge thereof, he shall be paid for his services a fee proportionate to that stated in Article 7.

9.2.13 Minor expenses such as telegrams, long-distance telephone calls, telephone service at the site, expressage, and similar petty cash items in connection with the Work.

9.2.14 Cost of removal of all debris.

9.2.15 Costs incurred due to an emergency affecting the safety of person and property.

9.2.16 Other costs incurred in the performance of the Work if and to the extent approved in advance in writing by the Owner.

9.2.17 All costs directly incurred in the performance of the Work and not included in the Construction Manager's Fee as set forth in Paragraph 7.3.

9.2.18 Cost of data processing services as required.

9.2.19 Legal costs growing out of prosecution of the Work for the Owner.

ARTICLE 10

Discounts

All discounts for prompt payment shall accrue to the Owner to the extent the Costs of the Work are paid directly by the Owner or from a fund made available by the Owner to the Construction Manager for such payments. To the extent the Costs of the Work are paid with funds of the Construction Manager, all cash discounts shall accrue to the Construction Manager. All trade discounts, rebates and refunds, and all returns from sale of surplus materials and equipment, shall accrue to the Owner and the Construction Manager shall make provisions so that they can be secured.

ARTICLE 11

Payments to the Construction Manager

11.1 The Construction Manager shall submit monthly to the Owner a statement, sworn to if required, showing in detail all moneys paid out, costs accumulated or costs incurred on account of the Cost of the Work during the previous month and the amount of the Construction Manager's Fee due as provided in Article 7. Payment by the Owner to the Construction Manager of the statement amount shall be made within ten (10) days after it is submitted.

11.2 Final payment constituting the unpaid balance of the Cost of the Work and the Construction Manager's Fee shall be due and payable when the building is delivered to the Owner, ready for beneficial occupancy, or when the Owner occupies the building, whichever event first occurs, provided that the Work be then substantially completed and the Contract substantially performed. If there should remain minor items to be completed, the Construction Manager and Architect shall list such items and the Construction Manager shall deliver, in writing, his unconditional promise to complete said items within a reasonable time thereafter. The Owner may retain a sum equal to 150% of the estimated cost of completing any unfinished items, provided that said unfinished items are listed separately and the estimated cost of completing any unfinished items is likewise listed separately. Thereafter, Owner shall pay to Construction Manager, monthly, the amount retained for incomplete items as each of said items is completed.

11.3 Before issuance of final payment, the Construction Manager shall submit satisfactory evidence that all payrolls, materials bills and other indebtedness connected with the Work have been paid or otherwise satisfied.

11.4 If the Owner should fail to pay the Construction Manager within seven (7) days after the time the payment of any amount becomes due, then the Construction Manager may, upon seven (7) additional days' written notice to the Owner and the Architect, stop the Work until payment of the amount owing has been received.

FIG. 3.3 *(continued)*

ARTICLE 12

Insurance

12.1 Construction Manager's Liability Insurance.

12.1.1 The Construction Manager agrees to indemnify and hold the Owner harmless from all claims for bodily injury and property damage that may arise from the Construction Manager's operations under this Agreement. The Construction Manager shall purchase and maintain such insurance as will protect him, including contractual coverage, from claims set forth below which may arise out of or result from the Construction Manager's operations under this Agreement, whether such operations be by himself or by any Trade Contractor or by anyone directly or indirectly employed by any of them, or by anyone for whose acts any of them may be liable:

.1 claims under workmen's compensation, disability benefit and other similar employee benefit acts;

.2 claims for damages because of bodily injury, occupational sickness or disease, or death of his employees;

.3 claims for damages because of bodily injury, sickness or disease or death of any person other than his employees;

.4 claims for damages insured by usual personal injury liability coverage which are sustained (1) by any person as a result of an offense directly or indirectly related to the employment of such person by the Construction Manager, or (2) by any other person; and

.5 claims for damages because of injury to or destruction of tangible property, including loss of use therefrom.

12.1.2 The Construction Manager's Comprehensive General Liability Insurance, including contractual coverage, and Automobile Liability Insurance required by Paragraph 12.1.1 shall be in an amount no less than ____________ Dollars ($ ____________) for injuries, including accidental death, sustained by one or more persons in any one occurrence. The Construction Manager's Property Damage Liability Insurance shall be in an amount not less than ____________ Dollars ($ ____________). All Trade Contractors or any other contractor engaged in the Work who may have a contract with the Owner, shall agree to indemnify and hold the Owner and Construction Manager harmless from all claims, for bodily injury and property damage that may arise from the contractor's operation in the Work. That contractor shall agree to purchase and maintain such insurance as will protect him, including contractual coverage, from claims set forth in Paragraph 12.1.1 and with the limits set forth in this Paragraph.

12.1.3 Certificates of Insurance acceptable to the Owner shall be filed with the Owner and the Construction Manager prior to commencement of the Work. These certificates shall contain a provision that coverages afforded under the policies will not be cancelled until at least ten days' prior written notice has been given to the Owner.

12.2 Insurance to be Procured by Owner.

12.2.1 The Owner shall be responsible for purchasing and maintaining his own liability insurance and, at his option, may purchase and maintain such insurance as will protect him against claims which may arise from operations under this Agreement.

12.3 Insurance to Protect Project.

12.3.1 Unless otherwise provided, the Owner shall purchase and maintain property insurance upon the entire Work at the site to the full insurable value thereof. This insurance shall include the interest of the Owner, the Construction Manager, Trade Contractors and Subcontractors in the Work and shall insure against the perils of Fire, Extended Coverage, Vandalism and Malicious Mischief. If the Work covers an addition to an existing building, the Construction Manager, Trade Contractors and Subcontractors shall be named as additional insureds under the Owner's fire and extended coverage policy covering the building (including contents) to which the addition is being made.

12.3.2 The Owner shall purchase and maintain such steam boiler and machinery insurance as may be required by the Contract Documents or by law. This insurance shall include the interests of the Owner, the Construction Manager, Trade Contractors and Subcontractors in the Work.

12.3.3 Any insured loss is to be adjusted with the Owner and made payable to the Owner as trustee for the insureds, as their interests may appear.

12.3.4 If the Construction Manager requests in writing that insurance for special hazards be included in the property

FIG. 3.3 *(continued)*

insurance policy, the Owner shall, if possible, cause such hazards to be insured.

12.3.5 The Owner shall file a copy of all policies with the Construction Manager before an exposure to loss may occur. If the Owner does not intend to purchase such insurance, he shall inform the Construction Manager in writing prior to the commencement of the Work. The Construction Manager may then effect insurance which will protect the interest of himself, the Trade Contractors and their Subcontractors in the Work, the cost of which shall be a cost of the Work pursuant to Article 9, and the Guaranteed Maximum Price shall be increased by Change Order. If the Construction Manager is damaged by failure of the Owner to purchase or maintain such insurance or to so notify the Construction Manager, the Owner shall bear all reasonable costs properly attributable thereto.

12.3.6 The Owner and Construction Manager waive all rights against each other and Trade Contractors and Subcontractors for damages caused by fire or other perils to the extent covered by insurance provided under this Paragraph 12.1, except such rights as they may have to the proceeds of such insurance held by the Owner as trustee.

12.3.7 If required in writing by any party in interest, the Owner as trustee shall, upon the occurrence of an insured loss, give bond for the proper performance of his duties. He shall deposit in a separate account any money so received and he shall distribute it in accordance with such agreement as the parties in interest may reach. If after such loss no other special agreement is made, replacement of damaged Work shall be covered by an appropriate Change Order.

12.3.8 The Owner as trustee shall have power to adjust and settle any loss with the insurers.

ARTICLE 13

Termination of the Agreement

13.1 Termination by the Construction Manager.

13.1.1 If the Work is stopped for a period of thirty days under an order of any court or other public authority having jurisdiction, or as a result of an act of government, such as a declaration of a national emergency making materials unavailable, through no act or fault of the Construction Manager, or if the Work should be stopped for a period of thirty days by the Construction Manager for the Owner's failure to make payment thereon, then the Construction Manager may, upon seven days' written notice to the Owner and the Architect, terminate the Contract and recover from the Owner payment for all work executed, the Construction Manager's Fee earned to date, and for any proven loss sustained upon any materials, equipment, tools, construction equipment and machinery, including reasonable profit and damages.

13.2 Termination by the Owner.

13.2.1 If the Construction Manager is adjudged a bankrupt, or if he makes a general assignment for the benefit of his creditors, or if a receiver is appointed on account of his insolvency, or if he persistently or repeatedly refuses or fails, except in cases for which extension of time is provided, to supply enough properly skilled workmen or proper materials, or if he fails to make prompt payment to Trade Contractors or for materials or labor, or persistently disregards laws, ordinances, rules, regulations or orders of any public authority having jurisdiction, or otherwise is guilty of a substantial violation of a provision of the Contract Documents, then the Owner may, without prejudice to any right or remedy and after giving the Construction Manager and his surety, if any, seven days' written notice, during which period Construction Manager fails to cure the violation, terminate the employment of the Construction Manager and take possession of the site and of all materials, equipment, tools, construction equipment and machinery thereon owned by the Construction Manager and may finish the Work by whatever method he may deem expedient. In such case, the Construction Manager shall not be entitled to receive any further payment until the Work is finished.

13.3 If the Owner terminates the Contract, he shall reimburse the Construction Manager for any unpaid Cost of the Work due him under Article 9, plus (1) the unpaid balance of the Fee computed upon the Cost of the Work to the date of termination at the rate of the percentage named in Article 7.2.1 if the Construction Manager's Fee be stated as a fixed sum, such an amount as will increase the payment on account of his fee to a sum which bears the same ratio to the said fixed sum as the Cost of the Work at the time of termination bears to the adjusted Guaranteed Maximum Cost, if any, otherwise to a reasonable estimated Cost of the Work when completed. The Owner shall also pay to the Construction Manager fair compensation, either by purchase or rental at the election of the Owner, for any equipment retained. In case of such termination of the Contract the Owner shall further assume and become liable for obligations, commitments and unsettled claims that the Construction Manager has previously undertaken or incurred in good faith

FIG. 3.3 *(continued)*

in connection with said Work. The Construction Manager shall, as a condition of receiving the payments referred to in this Article 13, execute and deliver all such papers and take all such steps, including the legal assignment of his contractual rights, as the Owner may require for the purpose of fully vesting in him the rights and benefits of the Construction Manager under such obligations or commitments.

13.4 After the completion of the Design Phase, if the final cost estimates make the Project no longer feasible from the standpoint of the Owner, the Owner may terminate the Contract and pay the Construction Manager in accordance with Paragraph 7.1.1.

ARTICLE 14

Assignment and Governing Law

14.1 Neither the Owner nor the Construction Manager shall assign his interest in this Agreement without the written consent of the other except as to the assignment of proceeds.

14.2 This Agreement shall be governed by the law in effect at the location of the Project.

ARTICLE 15

Miscellaneous Provisions

ARTICLE 16

Arbitration

15.1 All claims, disputes and other matters in question arising out of, or relating to, this Agreement or the breach thereof, except with respect to the Architect's decision on matters relating to artistic effect, and except for claims which have been waived by the making or acceptance of final payment shall be decided by arbitration in accordance with the Construction Industry Arbitration Rules of the American Arbitration Association then obtaining unless the parties mutually agree otherwise. This Agreement to arbitrate shall be specifically enforceable under the prevailing arbitration law.

15.2 Notice of the demand for arbitration shall be filed in writing with the other party to the Contract and with the American Arbitration Association. The demand for arbitration shall be made within a reasonable time after the claim, dispute or other matter in question has arisen, and in no event shall it be made after the date when institution of legal or equitable proceedings based on such claim, dispute or other matter in question would be barred by the applicable statute of limitations.

15.3 The award rendered by the arbitrators shall be final and judgment may be entered upon it in accordance with applicable law in any court having jurisdiction thereof.

15.4 The Construction Manager shall carry on the Work and maintain the Progress Schedule during any arbitration proceedings, unless otherwise agreed by him and the Owner in writing.

This Agreement executed the day and year first written above.

ATTEST: OWNER:

ATTEST: CONSTRUCTION MANAGER:

3.6.2 ADVANTAGES AND DISADVANTAGES

One of the major advantages to the owner under the construction management form of contract is having an expert in construction contribute to the planning and design decisions. This should result in an improved and more economical project. Time should also be saved since construction can begin before the drawings and specifications are completed. The owner may still retain the advantages of competitive bidding on the actual construction.

A major disadvantage to the owner is the fact that the final cost is not guaranteed. If documents are developed sufficiently for an upset price to be developed, the major share of time savings will be lost. In addition, the owner must select the other members of the team so as to ensure that a good working relationship will result.

Under the construction manager form of contract, the contractor is relieved of many of the "traditional" management responsibilities. Some contractors may be uncomfortable in this situation and feel that there is less freedom of operation for them. It should be recognized, however, that this possible reduction in freedom is balanced by a reduction in responsibility with a corresponding lowering of supervisory and management expenses.

There are few disadvantages for the qualified construction manager firm. However, the unqualified firm may find the added managerial and financial responsibilities more than they can handle. While this situation can cause difficulty for such a firm, it may lead to more serious difficulties for the owner. Sizable cost overruns, late completion, or other unsatisfactory conditions are the common result of such a situation.

3.6.3 THE FEE

The construction manager is paid a professional fee, the amount being subject to competitive bidding but more often set as a result of negotiation with the owner. The amount of the fee will be influenced by both the size and complexity of the project. If the construction manager is a general contracting firm performing some of the construction, the firm will receive their professional management fee on the total project plus their overhead and profit included within their bid on the part of the project they are constructing.

3.7 DESIGN-BUILD

Under design-build the owner awards only one contract. This means there is only one prime contractor and this party has the undivided responsibility for both the design and construction of the project. Design-build contracts are entered into by general contracting firms and design firms, and by joint ventures made up of construction and design firms. The form of contract may be lump sum, cost plus, or one of several other types discussed in Chapter 8. Figure 3.4 illustrates an example of this form of contract.

3.7.1 METHOD OF AWARD

Although most design-build contracts are negotiated, they can be awarded as a result of competitive bidding. However, it is very difficult to develop a reasonable basis for comparison of the proposals, just as it is under the construction management method. The owner must first develop a comprehensive presentation of the problem statement, which may necessitate the preparation of preliminary plans and specifications. In addition, the owner must possess the capability of evaluating the experience and expertise of the bidders. This is often made more difficult when all bidders are not from the same type of organization.

(text continues on page 68)

FIG. 3.4 Design-build agreement

THE ASSOCIATED GENERAL CONTRACTORS

STANDARD FORM OF DESIGN-BUILD AGREEMENT BETWEEN OWNER AND CONTRACTOR

(See AGC Document 6 for Preliminary Design Agreement and AGC Document 6b for recommended General Conditions.)

This Document has important legal and insurance consequences; consultation with an attorney and insurance consultants and carriers is encouraged with respect to its completion or modification.

AGREEMENT

Made this day of in the year of Nineteen Hundred and

BETWEEN

the Owner, and

the Contractor.

For services in connection with the following described Project: (Include complete Project location and scope)

The Owner and the Contractor agree as set forth below:

Certain provisions of this document have been derived, with modifications, from the following documents published by The American Institute of Architects: AIA Document A111, Owner-Contractor Agreement, © 1976; AIA Document A201, General Conditions, © 1976 by The American Institute of Architects. Usage made of AIA language, with the permission of AIA, does not imply AIA endorsement or approval of this document. Further, reproduction of copyrighted AIA materials without separate written permission from AIA is prohibited.

© 1977 Associated General Contractors of America

Reproduced by permission of the Associated General Contractors of America.

FIG. 3.4 *(continued)*

INDEX

FIG. 3.4 *(continued)*

ARTICLE 1

The Construction Team and Extent of Agreement

THE CONTRACTOR accepts the relationship of trust and confidence established between him and the Owner by this Agreement. He agrees to furnish the architectural, engineering and construction services set forth herein and agrees to furnish efficient business administration and superintendence, and to use his best efforts to complete the Project in the best and soundest way and in the most expeditious and economical manner consistent with the interests of the Owner.

1.1 *The Construction Team:* The Contractor, the Owner and the Architect/Engineer called the "Construction Team" shall work from the beginning of design through construction completion. The services of
, as the Architect/Engineer, will be furnished by the Contractor pursuant to an agreement between the Contractor and the Architect/Engineer.

1.2 *Extent of Agreement:* This Agreement represents the entire agreement between the Owner and the Contractor and supersedes all prior negotiations, representations or agreements. When the Drawings and Specifications are complete, they shall be identified by amendment to this Agreement. This Agreement shall not be superseded by any provisions of the documents for construction and may be amended only by written instrument signed by both Owner and Contractor.

1.3 *Definitions:* The Project is the total construction to be designed and constructed of which the Work is a part. The Work comprises the completed construction required by the Drawings and Specifications. The term day shall mean calendar day unless otherwise specifically designated.

ARTICLE 2

Contractor's Responsibilities

2.1 Contractor's Services

2.1.1 The Contractor shall be responsible for furnishing the Design and for the construction of the Project. The Owner and Contractor shall develop a design and construction phase schedule and the Owner shall be responsible for prompt decisions and approvals so as to maintain the approved schedule.

2.1.2 If the working Drawings and Specifications have not been completed and a Guaranteed Maximum Price has been established, the Contractor, the Architect/Engineer and Owner will work closely together to monitor the design in accordance with prior approvals so as to ensure that the Project can be constructed within the Guaranteed Maximum Price. As these working Drawings and Specifications are being completed, the Contractor will keep the Owner advised of the effects of any Owner requested changes on the Contract Time Schedule and/or the Guaranteed Maximum Price.

2.1.3 The Contractor will assist the Owner in securing permits necessary for the construction of the Project.

2.2 Responsibilities With Respect to Construction

2.2.1 The Contractor will provide all construction supervision, inspection, labor, materials, tools, construction equipment and subcontracted items necessary for the execution and completion of the Project.

2.2.2 The Contractor will pay all sales, use, gross receipts and similar taxes related to the Work provided by the Contractor which have been legally enacted at the time of execution of this Agreement.

2.2.3 The Contractor will prepare and submit for the Owner's approval an estimated progress schedule for the Project. This schedule shall indicate the dates for the starting and completion of the various stages of the design and construction. It shall be revised as required by the conditions of the Work and those conditions and events which are beyond the Contractor's control.

2.2.4 The Contractor shall at all times keep the premises free from the accumulation of waste materials or rubbish caused by his operations. At the completion of the Work, he shall remove all of his waste material and rubbish from and around the Project as well as all his tools, construction equipment, machinery and surplus materials.

FIG. 3.4 *(continued)*

2.2.5 The Contractor will give all notices and comply with all laws and ordinances legally enacted at the date of the execution of the Agreement, which govern the proper execution of the Work.

2.2.6 The Contractor shall take necessary precautions for the safety of his employees on the Work, and shall comply with all applicable provisions of federal, state and municipal safety laws to prevent accidents or injury to persons on, about or adjacent to the Project site. He shall erect and properly maintain, at all times, as required by the conditions and progress of Work, necessary safeguards for the protection of workmen and the public. It is understood and agreed, however, that the Contractor shall have no responsibility for the elimination or abatement of safety hazards created or otherwise resulting from Work at the job site carried on by other persons or firms directly employed by the Owner as separate contractors or by the Owner's tenants, and the Owner agrees to cause any such separate contractors and tenants to abide and adhere fully to all applicable provisions of federal, state and municipal safety laws and regulations and to comply with all reasonable requests and directions of the Contractor for the elimination or abatement of any such safety hazards at the job site.

2.2.7 The Contractor shall keep such full and detailed accounts as may be necessary for proper financial management under this Agreement. The system shall be satisfactory to the Owner, who shall be afforded access to all the Contractor's records, books, correspondence, instructions, drawings, receipts, vouchers, memoranda and similar data relating to this Agreement. The Contractor shall preserve all such records for a period of three years after the final payment or longer where required by law.

2.3 Royalties and Patents

2.3.1 The Contractor shall pay all royalties and license fees. He shall defend all suits or claims for infringement of any patent rights and shall save the Owner harmless from loss on account thereof except when a particular design, process or product is specified by the Owner. In such case the Contractor shall be responsible for such loss only if he has reason to believe that the design, process or product so specified is an infringement of a patent, and fails to give such information promptly to the Owner.

2.4 Warranties and Completion

2.4.1 The Contractor warrants to the Owner that all materials and equipment furnished under this Agreement will be new, unless otherwise specified, and that all Work will be of good quality, free from improper workmanship and defective materials and in conformance with the Drawings and Specifications. The Contractor agrees to correct all Work performed by him under this Agreement which proves to be defective in material and workmanship within a period of one year from the Date of Substantial Completion as defined in Paragraph 5.2, or for such longer periods of time as may be set forth with respect to specific warranties contained in the Specifications.

2.4.2 The Contractor will secure required certificates of inspection, testing or approval and deliver them to the Owner.

2.4.3 The Contractor will collect all written warranties and equipment manuals and deliver them to the Owner.

2.4.4 The Contractor with the assistance of the Owner's maintenance personnel, will direct the checkout of utilities and operation of systems and equipment for readiness, and will assist in their initial start-up and testing.

2.5 Additional Services

2.5.1 The Contractor will provide the following additional services upon the request of the Owner. A written agreement between the Owner and Contractor shall define the extent of such additional services and the amount and manner in which the Contractor will be compensated for such additional services.

2.5.2 Services related to investigation, appraisals or evaluations of existing conditions, facilities or equipment, or verification of the accuracy of existing drawings or other Owner-furnished information.

2.5.3 Services related to Owner-furnished equipment, furniture and furnishings which are not a part of this Agreement.

2.5.4 Services for tenant or rental spaces not a part of this Agreement.

2.5.5 Obtaining and training maintenance personnel or negotiating maintenance service contracts.

FIG. 3.4 *(continued)*

ARTICLE 3

Owner's Responsibilities

3.1 The Owner shall provide full information regarding his requirements for the Project.

3.2 The Owner shall designate a representative who shall be fully acquainted with the Project, and has authority to approve changes in the scope of the Project, render decisions promptly, and furnish information expeditiously and in time to meet the dates set forth in Subparagraph 2.2.3.

3.3 The Owner shall furnish for the site of the Project all necessary surveys describing the physical characteristics, soils reports and subsurface investigations, legal limitations, utility locations, and a legal description.

3.4 The Owner shall secure and pay for necessary approvals, easements, assessments and charges required for the construction, use, or occupancy of permanent structures or for permanent changes in existing facilities.

3.5 The Owner shall furnish such legal services as may be necessary for providing the items set forth in Paragraph 3.4, and such auditing services as he may require.

3.6 If the Owner becomes aware of any fault or defect in the Project or non-conformance with the Drawings or Specifications, he shall give prompt written notice thereof to the Contractor.

3.7 The Owner shall provide the insurance for the Project as provided in Paragraph 12.4, and shall bear the cost of any bonds that may be required.

3.8 The services and information required by the above paragraphs shall be furnished with reasonable promptness at Owner's expense and the Contractor shall be entitled to rely upon the accuracy and the completeness thereof.

3.9 The Owner shall furnish reasonable evidence satisfactory to the Contractor, prior to signing the Agreement, that sufficient funds are available and committed for the entire Cost of the Project. If the Contractor elects to execute this Agreement without having received such evidence, the Owner shall provide it within a reasonable time. The Contractor may stop work upon fifteen days notice if such evidence has not been furnished within a reasonable time.

3.10 The Owner shall have no contractual obligation to the Contractor's subcontractors and shall communicate with such subcontractors only through the Contractor.

ARTICLE 4

Subcontracts

4.1 All portions of the Work that the Contractor does not perform with his own forces shall be performed under subcontracts.

4.2 A Subcontractor is a person or entity who has a direct contract with the Contractor to perform any Work in connection with the Project. The term Subcontractor does not include any separate contractor employed by the Owner or the separate contractors' subcontractors.

4.3 No contractual relationship shall exist between the Owner and any Subcontractor and the Contractor shall be responsible for the management of the Subcontractors in the performance of their Work.

FIG. 3.4 *(continued)*

ARTICLE 5

Contract Time Schedule

5.1 The Work to be performed under this Agreement shall be commenced on or about and shall be substantially completed on or about

5.2 The Date of Substantial Completion of the Project or a designated portion thereof is the date when construction is sufficiently complete in accordance with the Drawings and Specifications so the Owner can occupy or utilize the Project or designated portion thereof for the use for which it is intended. Warranties called for by this Agreement or by the Drawings and Specifications shall commence on the Date of Substantial Completion of the Project or designated portion thereof. This date shall be established by a Certificate of Substantial Completion signed by the Owner and Contractor and shall state their respective responsibilities for security, maintenance, heat, utilities, damage to the Work and insurance. This Certificate shall also list the items to be completed or corrected and fix the time for their completion and correction.

5.3 If the Contractor is delayed at any time in the progress of the Project by any act or neglect of the Owner or by any separate contractor employed by the Owner, or by changes ordered in the Project, or by labor disputes, fire, unusual delay in transportation, adverse weather conditions not reasonably anticipatable, unavoidable casualties, or any causes beyond the Contractor's control, or a delay authorized by the Owner pending arbitration, then the Date for Substantial Completion shall be extended by Change Order for the period of such delay.

ARTICLE 6

Guaranteed Maximum Price

6.1 The Contractor guarantees that the maximum price to the Owner for the Cost of the Project as set forth in Article 8, and the Contractor's Fee as set forth in Article 7, will not exceed
Dollars ($), which sum shall be called the Guaranteed Maximum Price.

6.2 The Guaranteed Maximum Price is based upon laws, codes, and regulations in existence at the date of its establishment and upon criteria, Drawings, and Specifications as set forth below:

6.3 The Guaranteed Maximum Price will be modified for delays caused by the Owner and for Changes in the Project, all pursuant to Article 9.

6.4 Allowances included in the Guaranteed Maximum Price are as set forth below:

6.5 Whenever the cost is more than or less than the Allowance, the Guaranteed Maximum Price shall be adjusted by Change Order.

ARTICLE 7

Contractor's Fee

7.1 In consideration of the performance of the Agreement, the Owner agrees to pay to the Contractor in current funds as compensation for his services a Fee as follows:

7.2 Adjustments in Fee shall be made as follows:

7.2.1 For Changes in the Project as provided in Article 9, the Contractor's Fee shall be adjusted as follows:

FIG. 3.4 *(continued)*

7.2.2 For delays in the Project not the responsibility of the Contractor, there will be an equitable adjustment in the fee to compensate the Contractor for his increased expenses.

7.2.3 In the event the Cost of the Project plus the Contractor's Fee shall be less than the Guaranteed Maximum Price as adjusted by Change Orders, the resulting savings will be shared by the Owner and the Contractor as follows:

7.2.4 The Contractor shall be paid an additional fee in the same proportion as set forth in 7.2.1 if the Contractor is placed in charge of managing the replacement of insured or uninsured loss.

7.3 The Contractor shall be paid monthly that part of his Fee proportionate to the percentage of Work completed, the balance, if any, to be paid at the time of final payment.

7.4 Included in the Contractor's Fee are the following:

7.4.1 Salaries or other compensation of the Contractor's employees at the principal office and branch offices, except employees listed in Subparagraph 8.2.3.

7.4.2 General operating expenses of the Contractor's principal and branch offices other than the field office.

7.4.3 Any part of the Contractor's capital expenses, including interest on the Contractor's capital employed for the Project.

7.4.4 Overhead or general expenses of any kind, except as may be expressly included in Article 8.

7.4.5 Costs in excess of the Guaranteed Maximum Price.

ARTICLE 8

Cost of the Project

8.1 The term Cost of the Project shall mean costs necessarily incurred in the design and construction of the Project and shall include the items set forth below in this Article. The Owner agrees to pay the Contractor for the Cost of the Project as defined in this Article. Such payment shall be in addition to the Contractor's Fee stipulated in Article 7.

8.2 Cost Items

8.2.1 All architectural, engineering and consulting fees and expenses incurred in designing and constructing the Project.

8.2.2 Wages paid for labor in the direct employ of the Contractor in the performance of the Work under applicable collective bargaining agreements, or under a salary or wage schedule agreed upon by the Owner and the Contractor, and including such welfare or other benefits, if any, as may be payable with respect thereto.

8.2.3 Salaries of Contractor's employees when stationed at the field office, in whatever capacity employed, employees engaged on the road expediting the production or transportation of material and equipment and employees from the main or branch office performing the functions listed below:

8.2.4 Cost of all employee benefits and taxes for such items as unemployment compensation and social security, insofar as such cost is based on wages, salaries, or other remuneration paid to employees of the Contractor and included in the Cost of the Project under Subparagraphs 8.2.1, 8.2.2 and 8.2.3.

8.2.5 The proportion of reasonable transportation, traveling and hotel and moving expenses of the Contractor or of his officers or employees incurred in discharge of duties connected with the Project.

8.2.6 Cost of all materials, supplies and equipment incorporated in the Project, including costs of transportation and storage thereof.

FIG. 3.4 *(continued)*

8.2.7 Payments made by the Contractor to Subcontractors for Work performed pursuant to contract under this Agreement.

8.2.8 Cost, including transportation and maintenance, of all materials, supplies, equipment, temporary facilities and hand tools not owned by the workmen, which are employed or consumed in the performance of the Work, and cost less salvage value on such items used, but not consumed, which remain the property of the Contractor.

8.2.9 Rental charges of all necessary machinery and equipment, exclusive of hand tools, used at the site of the Work, whether rented from the Contractor or others, including installations, repairs and replacements, dismantling, removal, costs of lubrication, transportation and delivery costs thereof, at rental charges consistent with those prevailing in the area.

8.2.10 Cost of the premiums for all insurance which the Contractor is required to procure by this Agreement or is deemed necessary by the Contractor.

8.2.11 Sales, use, gross receipts or similar taxes related to the Project, imposed by any governmental authority, and for which the Contractor is liable.

8.2.12 Permit fees, licenses, tests, royalties, damages for infringement of patents and costs of defending suits therefor for which the Contractor is responsible under Subparagraph 2.3.1 and deposits lost for causes other than the Contractor's negligence.

8.2.13 Losses, expenses or damages to the extent not compensated by insurance or otherwise (including settlement made with the written approval of the Owner), and the cost of corrective work.

8.2.14 Minor expenses such as telegrams, long-distance telephone calls, telephone service at the site, expressage, and similar petty cash items in connection with the Project.

8.2.15 Cost of removal of all debris.

8.2.16 Costs incurred due to an emergency affecting the safety of persons and property.

8.2.17 Cost of data processing services required in the performance of the services outlined in Article 2.

8.2.18 Legal costs reasonably and properly resulting from prosecution of the Project for the Owner.

8.2.19 All costs directly incurred in the performance of the Project and not included in the Contractor's Fee as set forth in Paragraph 7.3.

ARTICLE 9

Changes in the Project

9.1 The Owner, without invalidating this Agreement, may order Changes in the Project within the general scope of this Agreement consisting of additions, deletions or other revisions, the Guaranteed Maximum Price, if established, the Contractor's Fee, and the Contract Time Schedule being adjusted accordingly. All such Changes in the Project shall be authorized by Change Order.

9.1.1 A Change Order is a written order to the Contractor signed by the Owner or his authorized agent and issued after the execution of this Agreement, authorizing a Change in the Project and/or an adjustment in the Guaranteed Maximum Price, the Contractor's Fee or the Contract Time Schedule. Each adjustment in the Guaranteed Maximum Price resulting from a Change Order shall clearly separate the amount attributable to the Cost of the Project and the Contractor's Fee.

9.1.2 The increase or decrease in the Guaranteed Maximum Price resulting from a Change in the Project shall be determined in one or more of the following ways:

FIG. 3.4 *(continued)*

9.1.2.1 by mutual acceptance of a lump sum properly itemized and supported by sufficient substantiating data to permit evaluation; or

9.1.2.2 by unit prices stated in this Agreement or subsequently agreed upon; or

9.1.2.3 by cost to be determined as defined in Article 8 and a mutually acceptable fixed or percentage fee; or

9.1.2.4 by the method provided in Subparagraph 9.1.3.

9.1.3 If none of the methods set forth in Clauses 9.1.2.1 through 9.1.2.3 is agreed upon, the Contractor, provided he receives a written order signed by the owner, shall promptly proceed with the work involved. The cost of such work shall then be determined on the basis of the reasonable expenditures and savings of those performing the work attributed to the change, including, in the case of an increase in the Guaranteed Maximum Price, a reasonable increase in the Contractor's Fee. In such case, and also under Clauses 9.1.2.3 and 9.1.2.4 above, the Contractor shall keep and present, in such form as the Owner may prescribe, an itemized accounting together with appropriate supporting data of the increase in the Cost of the Project as outlined in Article 8. The amount of decrease in the Guaranteed Maximum Price to be allowed by the Contractor to the Owner for any deletion or change which results in a net decrease in cost will be the amount of the actual net decrease. When both additions and credits are involved in any one change, the increase in Fee shall be figured on the basis of net increase, if any.

9.1.4 If unit prices are stated in this Agreement or subsequently agreed upon, and if the quantities originally contemplated are so changed in a proposed Change Order that application of the agreed unit prices to the quantities of Work proposed will cause substantial inequity to the Owner or the Contractor, the applicable unit prices shall be equitably adjusted.

9.1.5 Should concealed conditions encountered in the performance of the Work below the surface of the ground or should concealed or unknown conditions in an existing structure be at variance with the conditions indicated by the Drawings, Specifications, or Owner-furnished information or should unknown physical conditions below the surface of the ground or should concealed or unknown conditions in an existing structure of an unusual nature, differing materially from those ordinarily encountered and generally recognized as inherent in work of the character provided for in this Agreement, be encountered, the Guaranteed Maximum Price and the Contract Time Schedule shall be equitably adjusted by Change Order upon claim by either party made within a reasonable time after the first observance of the conditions.

9.2 Claims for Additional Cost or Time

9.2.1 If the Contractor wishes to make a claim for an increase in the Guaranteed Maximum Price, or increase in his Fee or an extension in the Contract Time Schedule, he shall give the Owner written notice thereof within a reasonble time after the occurrence of the event giving rise to such claim. This notice shall be given by the Contractor before proceeding to execute the Work, except in an emergency endangering life or property in which case the Contractor shall act, at his discretion, to prevent threatened damage, injury or loss. Claims arising from delay shall be made within a reasonable time after the delay. Increases based upon design and estimating costs with respect to possible changes requested by the Owner, shall be made within a reasonable time after the decision is made not to proceed with the change. No such claim shall be valid unless so made. If the Owner and the Contractor cannot agree on the amount of the adjustment in the Guaranteed Maximum Price, the Contractor's Fee or Contract Time Schedule, it shall be determined pursuant to the provisions of Article 16. Any change in the Guaranteed Maximum Price, the Contractor's Fee or Contract Time Schedule resulting from such claim shall be authorized by Change Order.

9.3 Minor Changes in the Project

9.3.1 The Owner will have authority to order minor Changes in the Work not involving an adjustment in the Guaranteed Maximum Price or an extension of the Contract Time Schedule and not inconsistent with the intent of the Drawings and Specifications. Such Changes may be effected by written order and shall be binding on the Owner and the Contractor.

9.4 Emergencies

9.4.1 In any emergency affecting the safety of persons or property, the Contractor shall act, at his discretion, to prevent threatened damage, injury or loss. Any increase in the Guaranteed Maximum Price or extension of time claimed by the Contractor on account of emergency work shall be determined as provided in this Article.

FIG. 3.4 *(continued)*

ARTICLE 10

Discounts

All discounts for prompt payment shall accrue to the Owner to the extent the Cost of the Project is paid directly by the Owner or from a fund made available by the Owner to the Contractor for such payments. To the extent the Cost of the Project is paid with funds of the Contractor, all cash discounts shall accrue to the Contractor. All trade discounts, rebates and refunds, and all returns from sale of surplus materials and equipment, shall be credited to the Cost of the Project.

ARTICLE 11

Payments to the Contractor

11.1 Payments shall be made by Owner to Contractor according to the following procedure:

11.1.1 On or before the day of each month after work has commenced, the Contractor shall submit to the Owner an Application for Payment in such detail as may be required by the Owner based on the Work completed and materials stored on the site and/or at locations approved by the Owner along with a proportionate amount of the Contractor's Fee for the period ending on the day of the month.

11.1.2 Within ten (10) days after his receipt of each monthly Application for Payment, the Owner shall pay directly to the Contractor the appropriate amounts for which Application for Payment is made therein. This payment request shall deduct the aggregate of amounts previously paid by the Owner.

11.1.3 If the Owner should fail to pay the Contractor at the time the payment of any amount becomes due, then the Contractor may, at any time thereafter, upon serving written notice that he will stop work within five (5) days after receipt of the notice by the Owner, and after such five (5) day period, stop the Project until payment of the amount owing has been received. Written notice shall be deemed to have been duly served if sent by certified mail to the last business address known to him who gives the notice.

11.1.4 Payments due but unpaid shall bear interest at the rate the Owner is paying on his construction loan or at the legal rate, which ever is higher.

11.2 The Contractor warrants and guarantees that title to all Work, materials and equipment covered by an Application for Payment whether incorporated in the Project or not, will pass to the Owner upon receipt of such payment by Contractor free and clear of all liens, claims, security interests or encumbrances hereinafter referred to as Liens.

11.3 No Progress Payment nor any partial or entire use or occupancy of the Project by the Owner shall constitute an acceptance of any Work not in accordance with the Drawings and Specifications.

11.4 Final payment constituting the unpaid balance of the Cost of the Project and the Contractor's Fee shall be due and payable when the Project is delivered to the Owner, ready for beneficial occupancy, or when the Owner occupies the Project, whichever event first occurs, provided that the Project be then substantially completed and this Agreement substantially performed. If there should remain minor items to be completed, the Contractor and the Owner shall list such items and the Contractor shall deliver, in writing, his guarantee to complete said items within a reasonable time thereafter. The Owner may retain a sum equal to 150% of the estimated cost of completing any unfinished items, provided that said unfinished items are listed separately and the estimated cost of completing any unfinished items is likewise listed separately. Thereafter, the Owner shall pay to Contractor, monthly, the amount retained for incomplete items as each of said items is completed.

11.5 Before issuance of Final Payment, the Contractor shall submit satisfactory evidence that all payrolls, materials bills and other indebtedness connected with the Project have been paid or otherwise satisfied.

11.6 The making of Final Payment shall constitute a waiver of all claims by the Owner except those rising from:

FIG. 3.4 *(continued)*

11.6.1 Unsettled Liens.

11.6.2 Improper workmanship or defective materials appearing within one year after the Date of Substantial Completion.

11.6.3 Failure of the Work to comply with the Drawings and Specifications.

11.6.4 Terms of any special guarantees required by the Drawings and Specifications.

11.7 The acceptance of Final Payment shall constitute a waiver of all claims by the Contractor except those previously made in writing and unsettled.

ARTICLE 12

Insurance, Indemnity and Waiver of Subrogation

12.1 Indemnity

12.1.1 The Contractor agrees to indemnify and hold the Owner harmless from all claims for bodily injury and property damage (other than the Work itself and other property insured under Paragraph 12.4) that may arise from the Contractor's operations under this Agreement.

12.1.2 The Owner shall cause any other contractor who may have a contract with the Owner to perform work in the areas where work will be performed under this Agreement, to agree to indemnify the Owner and the Contractor and hold them harmless from all claims for bodily injury and property damage (other than property insured under Paragraph 12.4) that may arise from that contractor's operations. Such provisions shall be in a form satisfactory to the Contractor.

12.2 Contractor's Liability Insurance

12.2.1 The Contractor shall purchase and maintain such insurance as will protect him from the claims set forth below which may arise out of or result from the Contractor's operations under this Agreement whether such operations be by himself or by any Subcontractor or by anyone directly or indirectly employed by any of them, or by anyone for whose acts any of them may be liable:

12.2.1.1 Claims under workers' compensation, disability benefit and other similar employee benefit acts which are applicable to the work to be performed.

12.2.1.2 Claims for damages because of bodily injury, occupational sickness or disease, or death of his employees under any applicable employer's liability law.

12.2.1.3 Claims for damages because of bodily injury, or death of any person other than his employees.

12.2.1.4 Claims for damages insured by usual personal injury liability coverage which are sustained (1) by any person as a result of an offense directly or indirectly related to the employment of such person by the Contractor or (2) by any other person.

12.2.1.5 Claims for damages, other than to the Work itself, because of injury to or destruction of tangible property, including loss of use therefrom.

12.2.1.6 Claims for damages because of bodily injury or death of any person or property damage arising out of the ownership, maintenance or use of any motor vehicle.

12.2.2 The Comprehensive General Liability Insurance shall include premises-operations (including explosion, collapse and underground coverage) elevators, independent contractors, completed operations, and blanket contractual liability on all written contracts, all including broad form property damage coverage.

FIG. 3.4 *(continued)*

12.2.3 The Contractor's Comprehensive General and Automobile Liability Insurance, as required by Subparagraphs 12.2.1 and 12.2.2 shall be written for not less than limits of liability as follows:

a. Comprehensive General Liability

1. Personal Injury	- $______________	Each Occurrence (Completed Operations)
	$______________	Aggregate
2. Property Damage	- $______________	Each Occurrence
	$______________	Aggregate

b. Comprehensive Automobile Liability

1. Bodily Injury	- $______________	Each Person
	$______________	Each Occurrence
2. Property Damage	- $______________	Each Occurrence

12.2.4 Comprehensive General Liability Insurance may be arranged under a single policy for the full limits required or by a combination of underlying policies with the balance provided by an Excess or Umbrella Liability policy.

12.2.5 The foregoing policies shall contain a provision that coverages afforded under the policies will not be cancelled or not renewed until at least sixty (60) days' prior written notice has been given to the Owner. Certificates of Insurance showing such coverages to be in force shall be filed with the Owner prior to commencement of the Work.

12.3 Owner's Liability Insurance

12.3.1 The Owner shall be responsible for purchasing and maintaining his own liability insurance and, at his option, may purchase and maintain such insurance as will protect him against claims which may arise from operations under this Agreement.

12.4 Insurance to Protect Project

12.4.1 The Owner shall purchase and maintain property insurance in a form acceptable to the Contractor upon the entire Project for the full cost of replacement as of the time of any loss. This insurance shall include as named insureds the Owner, the Contractor, Subcontractors and Subsubcontractors and shall insure against loss from the perils of Fire, Extended Coverage, and shall include "All Risk" insurance for physical loss or damage including without duplication of coverage at least theft, vandalism, malicious mischief, transit, collapse, flood, earthquake, testing, and damage resulting from defective design, workmanship or material. The Owner will increase limits of coverage, if necessary, to reflect estimated replacement cost. The Owner will be responsible for any co-insurance penalties or deductibles. If the Project covers an addition to or is adjacent to an existing building, the Contractor, Subcontractors and Subsubcontractors shall be named as additional insureds under the Owner's Property Insurance covering such building and its contents.

12.4.1.1 If the Owner finds it necessary to occupy or use a portion or portions of the Project prior to Substantial Completion thereof, such occupancy shall not commence prior to a time mutually agreed to by the Owner and Contractor and to which the insurance company or companies providing the property insurance have consented by endorsement to the policy or policies. This insurance shall not be cancelled or lapsed on account of such partial occupancy. Consent of the Contractor and of the insurance company or companies to such occupancy or use shall not be unreasonably withheld.

12.4.2 The Owner shall purchase and maintain such boiler and machinery insurance as may be required or necessary. This insurance shall include the interests of the Owner, the Contractor, Subcontractors and Subsubcontractors in the Work.

12.4.3 The Owner shall purchase and maintain such insurance as will protect the Owner and Contractor against loss of use of Owner's property due to those perils insured pursuant to Subparagraph 12.4.1. Such policy will provide coverage for expediting expenses of materials, continuing overhead of the Owner and Contractor, necessary labor expense including

FIG. 3.4 *(continued)*

overtime, loss of income by the Owner and other determined exposures. Exposures of the Owner and the Contractor shall be determined by mutual agreement and separate limits of coverage fixed for each item.

12.4.4 The Owner shall file a copy of all policies with the Contractor before an exposure to loss may occur. Copies of any subsequent endorsements will be furnished to the Contractor. The Contractor will be given sixty (60) days notice of cancellation, non-renewal, or any endorsements restricting or reducing coverage. If the Owner does not intend to purchase such insurance, he shall inform the Contractor in writing prior to the commencement of the Work. The Contractor may then effect insurance which will protect the interest of himself, the Subcontractors and their Subsubcontractors in the Project, the cost of which shall be a Cost of the Project pursuant to Article 8, and the Guaranteed Maximum Price shall be increased by Change Order. If the Contractor is damaged by failure of the Owner to purchase or maintain such insurance or to so notify the Contractor, the Owner shall bear all reasonable costs properly attributable thereto.

12.5 Property Insurance Loss Adjustment

12.5.1 Any insured loss shall be adjusted with the Owner and the Contractor and made payable to the Owner and Contractor as trustees for the insureds, as their interests may appear, subject to any applicable mortgagee clause.

12.5.2 Upon the occurrence of an insured loss, monies received will be deposited in a separate account and the trustees shall make distribution in accordance with the agreement of the parties in interest, or in the absence of such agreement, in accordance with an arbitration award pursuant to Article 16. If the trustees are unable to agree between themselves on the settlement of the loss, such dispute shall also be submitted to arbitration pursuant to Article 16.

12.6 Waiver of Subrogation

12.6.1 The Owner and Contractor waive all rights against each other, the Architect/Engineer, Subcontractors, and Subsubcontractors for damages caused by perils covered by insurance provided under Paragraph 12.4, except such rights as they may have to the proceeds of such insurance held by the Owner and Contractor as trustees. The Contractor shall require similar waivers from all Subcontractors and Subsubcontractors.

12.6.2 The Owner and Contractor waive all rights against each other and the Architect/Engineer, Subcontractors and Subsubcontractors for loss or damage to any equipment used in connection with the Project which loss is covered by any property insurance. The Contractor shall require similar waivers from all Subcontractors and Subsubcontractors.

12.6.3 The Owner waives subrogation against the Contractor, Architect/Engineer, Subcontractors, and Subsubcontractors on all property and consequential loss policies carried by the Owner on adjacent properties and under property and consequential loss policies purchased for the Project after its completion.

12.6.4 If the policies of insurance referred to in this Paragraph require an endorsement to provide for continued coverage where there is a waiver of subrogation, the owners of such policies will cause them to be so endorsed.

ARTICLE 13

Termination of the Agreement And Owner's Right to Perform Contractor's Obligations

13.1 Termination by the Contractor

13.1.1 If the Project is stopped for a period of thirty (30) days under an order of any court or other public authority having jurisdiction, or as a result of an act of government, such as a declaration of a national emergency making materials unavailable, through no act or fault of the Contractor or if the Project should be stopped for a period of thirty (30) days by the Contractor for the Owner's failure to make payment thereon, then the Contractor may, upon seven days' written notice to the Owner, terminate this Agreement and recover from the Owner payment for all work executed, the Contractor's Fee earned to date, and for any proven loss sustained upon any materials, equipment, tools, construction equipment and machinery, including reasonable profit and damages.

13.2 Owner's Right to Perform Contractor's Obligations and Termination by the Owner for Cause

FIG. 3.4 *(continued)*

13.2.1 If the Contractor fails to perform any of his obligations under this Agreement, including any obligation he assumes to perform work with his own forces, the Owner may, after seven days' written notice, during which period the Contractor fails to perform such obligation, make good such deficiencies. The Guaranteed Maximum Price, if any, shall be reduced by the cost to the Owner of making good such deficiencies.

13.2.2 If the Contractor is adjudged a bankrupt, or if he makes a general assignment for the benefit of his creditors, or if a receiver is appointed on account of his insolvency, or if he persistently or repeatedly refuses or fails, except in cases for which extension of time is provided, to supply enough properly skilled workmen or proper materials, or if he fails to make prompt payment to Subcontractors or for materials or labor, or persistently disregards laws, ordinances, rules, regulations or orders of any public authority having jurisdiction, or otherwise is guilty of a substantial violation of a provision of this Agreement, then the Owner may, without prejudice to any right or remedy and after giving the Contractor and his surety, if any, seven (7) days' written notice, during which period Contractor fails to cure the violation, terminate the employment of the Contractor and take possession of the site and of all materials, equipment, tools, construction equipment and machinery thereon owned by the Contractor and may finish the Work by whatever method he may deem expedient. In such case, the Contractor shall not be entitled to receive any further payment until the Work is finished nor shall he be relieved from his obligations assumed under Article 6.

13.3 Termination by Owner Without Cause

13.3.1 If the Owner terminates the Agreement other than pursuant to Article 13.2.2, he shall reimburse the Contractor for any unpaid Cost of the Project due him under Article 8, plus (1) the unpaid balance of the Fee computed upon the Cost of the Work to the date of termination at the rate of the percentage named in Article 7.2.1 or if the Contractor's Fee be stated as a fixed sum, such an amount as will increase the payment on account of his Fee to a sum which bears the same ratio to the said fixed sum as the Cost of the Project at the time of termination bears to the adjusted Guaranteed Maximum Cost, if any, otherwise to a reasonable estimated Cost of the Project when completed. The Owner shall also pay to the Contractor fair compensation, either by purchase or rental at the election of the Owner, for any equipment retained. In case of such termination of this Agreement the Owner shall further assume and become liable for obligations, commitments and unsettled claims that the Contractor has previously undertaken or incurred in good faith in connection with said Work. The Contractor shall, as a condition of receiving the payments referred to in this Article 13, execute and deliver all such papers and take all such steps, including the legal assignment of his contractual rights, as the Owner may require for the purpose of fully vesting in the Owner the rights and benefits of the Contractor under such obligations or commitments.

ARTICLE 14

Assignment and Governing Law

14.1 Neither the Owner nor the Contractor shall assign his interest in this Agreement without the written consent of the other except as to the assignment of proceeds.

14.2 This Agreement shall be governed by the law in effect at the location of this Project.

ARTICLE 15

Miscellaneous Provisions

ARTICLE 16

Arbitration

16.1 All claims, disputes and other matters in question arising out of, or relating to, this Agreement or the breach thereof, except with respect to the Architect/Engineer's decision on matters relating to artistic effect, and except for

FIG. 3.4 *(continued)*

claims which have been waived by the making or acceptance of Final Payment shall be decided by arbitration in accordance with the Construction Industry Arbitration Rules of the American Arbitration Association then obtaining unless the parties mutually agree otherwise. This agreement to arbitrate shall be specifically enforceable under the prevailing arbitration law.

16.2 Notice of the demand for arbitration shall be filed in writing with the other party to this Agreement and with the American Arbitration Association. The demand for arbitration shall be made within a reasonable time after the claim, dispute or other matter in question has arisen, and in no event shall it be made after the date when institution of legal or equitable proceedings based on such claim, dispute or other matter in question would be barred by the applicable statute of limitations.

16.3 The award rendered by the arbitrators shall be final and judgment may be entered upon it in accordance with applicable law in any court having jurisdiction thereof.

16.4 Unless otherwise agreed in writing, the Contractor shall carry on the Work and maintain the Contract Time Schedule during any arbitration proceedings and the Owner shall continue to make payments in accordance with this Agreement.

16.5 All claims which are related to or dependent upon each other shall be heard by the same arbitrator or arbitrators, even though the parties are not the same, unless a specific contract prohibits such consolidation.

This Agreement entered into as of the day and year first written above.

ATTEST: OWNER:

ATTEST: CONTRACTOR:

3.7.2 ADVANTAGES AND DISADVANTAGES

According to many owners, single responsibility is one of the major attractions of design-build contracts. The advantage of dealing with only one administrative entity as opposed to a number of entities under some other forms of contract should not be discounted. Such an arrangement helps protect an owner against errors of omission that can occur with multiple contracts. Owners sometimes refer to these errors of omission as items that "fall through the cracks." Furthermore, under a design-build contract there is usually good coordination between the designers and constructors for both the planning and design stages. An added advantage is the possibility of construction beginning before all design documents are completed. This can result in savings of time and money for the owner.

A possible disadvantage to the design-build method is the absence of a member of the team who expressly looks out for the owner's interests. Under the more traditional forms of contract, this duty is usually identified with the design professional. If the design-build approach is used, the owner must place confidence in the prime contractor, in the areas of both expertise and ethical conduct. An earlier charge leveled against design-build was that the quality of design would be sacrificed to costs and time. Experience with design-build construction has tended to negate this criticism.

Many contractors enjoy their position under a design-build contract because it provides them with an opportunity to participate in the design phase. However, the same reason may be the cause of some resentment on the part of design professionals who feel their authority is being diminished. The truly successful situation is one where neither the contractor nor the design professional dictates to the other, and both are able to participate in a harmonious manner.

3.8 GOVERNMENTAL FORMS

There are several governmental agencies and attendant documents relating to the design phase. The agencies may be at the city, county, state, or federal levels. A multiplicity of requirements and regulations exists and compliance with these is checked at various levels. There are instances where conflicting requirements exist. In most cases conflicts can be resolved by conforming to the more stringent or higher level requirement. However, when opposing regulations apply, the problem becomes vastly more complex. Different requirements affect the project at various times. Most must be satisfied before the commencement of construction.

3.8.1 BUILDING PERMITS

Cities and states, as well as many countries, require a valid building permit before construction can begin. The purpose is to ensure compliance with applicable codes and ordinances. Such a check may include structural adequacy, minimum space provisions, setbacks, and lighting standards. The authority for this action stems from the police power of the state. The fees charged are based upon the size and estimated cost of the project. In some areas permits must be secured from more than one level of authority. An example might be a building of public assembly. In some states permits must be secured from the city or county where the project is located and also from the state board of health. An example of a building permit form as well as the companion zoning permit is shown in Fig. 3.5.

FIG. 3.5 Building permit and zoning permit

CITY OF AMES, IOWA
FORM NO. 120-

BUILDING PERMIT B No. 14964

CITY OF AMES, IOWA

Ames, Iowa, ________________, 19____

Authority is hereby granted to ____

____

to _____________ a _____________ story building of ______________________________

construction, use group _______________ to be used for ___________________________

____

On Lot ____

_______________________ Block ________________ Addition __________________________

At ____

Estimated Cost _________________________ Permit fee _________________________

All work done under this permit shall be in compliance with the laws of the State of Iowa and the Ordinances of the City of Ames.

____
Building Official

This permit **DOES NOT AUTHORIZE** the use of any street, alley, or sidewalk for depositing building materials or the placing of any obstruction within the street or alley limits. Application for all such uses shall be made at the Office of the City Manager.

CITY OF AMES, IOWA
FORM NO. 122-

ZONING PERMIT Z No. 10486

CITY OF AMES, IOWA

Permission is hereby granted by the City of Ames, Iowa, through its Zoning Enforcing Officer to use the premises described herein and to erect the improvements thereon in accordance with the application as approved therefor and all Ordinances of the City of Ames, and the laws of the state of Iowa applicable thereto.

Location ____

____

Address ______________________________ Owner ______________________________

Permitted Use ___________________________________ Zoning District ___________

Reservations ____

____

Issued ______________________, 19____

____
Zoning Enforcing Officer

Prior to the issuance of a building permit the project will be reviewed for possible violations of applicable zoning restrictions. Zoning is primarily concerned with the use and occupancy of a project. Zoning restrictions are enacted to protect the public from dangerous and hazardous conditions. They may also have the purpose of protecting property values, isolating traffic noises, and attempting to ensure an orderly growth pattern. See Fig. 3.5 for an example of a zoning permit.

3.8.2 FEDERAL REGULATIONS

The federal government is increasingly inserting itself into the processes of design and construction. There are myriad federal regulations that must be observed. Ignorance of a particular regulation is never an acceptable excuse and can cause much delay and added expense for the project.

A document that must be filed on some projects is the Environmental Impact Statement. Many designers, owners, and constructors claim that this document can add several years and many millions of dollars to a project. This may be particularly true in the case of a large power plant or dam project. Part of the trouble stems from a lack of clarity in the regulations as well as conflicting decisions from the courts and various federal agencies. Proponents of the project must explore and explain the effects of the project on the environment and show that these effects will not be detrimental. The definition of environment is expanded to include many considerations, such as wildlife, endangered species, air quality, noise, visual impact, and public safety.

As the concern for energy conservation grows, such agencies as the Energy Research and Development Administration (ERDA) are becoming increasingly important to the project approval process. All members of the construction team need to maintain a familiarity with current requirements if they are to satisfactorily meet their obligations.

DIVISION 

BIDDING PHASE DOCUMENTS

4 PRE-BID DOCUMENTS

FIGURES

4.1 COMPETITIVE BIDDING

The federal government as well as most state and municipal bodies have enacted legislation establishing guidelines for the procurement of goods and services. Among these guidelines is the requirement that the competitive bidding process shall be used for all contracts over a specified, minimum amount. The purpose of this requirement is to guard against favoritism as well as collusion between potential suppliers and contracting officers. This is to ensure that public funds are spent wisely and fairly and that the public receives appropriate benefit for its money. Most public funded construction contracts are awarded as a result of competitive bidding. This is especially true in engineered projects where approximately 99 percent of highway work is bid competitively.

Many private owners also prefer to utilize the competitive bid process. Owners may elect the bid approach in order to ensure the most favorable price for their project and to allow a larger number of firms the opportunity to win the contract. This is particularly true when a private owner has a business operation where a show of favoritism could have a negative effect. Use of the competitive bid process establishes a system of evaluating the proposals of the various firms so that each has a fair chance of securing the contract.

4.2 ADVERTISEMENT PUBLICATION

To achieve this goal, a document called an "Advertisement" is used for public funded projects. The objective of the advertisement is to place information regarding the proposed project in the hands of all potential, qualified bidders. (See Fig. 4.1.) Accordingly, the size and scope of the proposed project will indicate the geographic circulation of the advertisement. For a relatively small project within a small community, posting the advertisement on the town's notice board and publication in the local newspapers would be sufficient. This would ensure that the potential bidders received the information since most firms would not be interested in traveling a great distance for a small job. In the case of the multi-million or billion dollar project by the federal government, however, publication in major newspapers and industry publications with nationwide circulation would be indicated. In attempting to determine how widespread the circulation of the advertisement for a particular project should be, the rule of reasonableness should apply.

4.2.1 INFORMATION TO BE INCLUDED

The advertisement should contain sufficient information to allow the potential bidder to make a preliminary decision regarding bidding. The information should not be presented in a manner similar to that used in the advertisement of a department store. The purpose of the document is not to persuade but rather to present factual information so that an informed decision can be made by a potential bidder. The following items of information should be considered essential:

> Type of work proposed. This should state what the proposed project is: bridge, building, dam, etc. Some construction companies are not interested in all types of work and may instead specialize in certain types of projects. Thus the information regarding the type of project will be a determining factor in the decision to bid.

FIG. 4.1 Advertisement

OFFICIAL PUBLICATION

NOTICE OF PUBLIC HEARING ON DRAWINGS, SPECIFICATIONS, ESTIMATE OF COST, AND FORM OF CONTRACT FOR THE CONSTRUCTION OF THE SHAW COMMUNITY CENTER AND FOR THE TAKING OF BIDS THEREIN:

SEALED PROPOSALS will be received by the City of ________________ in the office of the City Clerk until 11:00 o'clock A.M. on the 12th day of May, 1981 for the following improvement as per specifications now on file in the office of the Civil Engineer of the Department of Engineering and City Clerk:

The City Council will hold a public hearing on the matter of the final adoption of drawings, specifications, form of contract, and estimate of cost for the making of said improvements hereinafter described in accordance with Chapter 384, Code of ________________, at 6:30 o'clock P.M. on the 18th day of May, 1981, in the Council Chambers, City Hall, ________________, ________________, and will consider objections thereto or to the cost of the improvements.

Proposals received will be opened and tabulated at a public meeting presided over by the City Manager in the Council Chambers in the City Hall at 11:00 o'clock A.M. on the 12th day of May, 1981, after which the bids, together with a report and recommendation of the Chief Civil Engineer, will be presented to the City Council at its next meeting for action thereon for said improvement as follows:

ONE-STORY SHAW COMMUNITY CENTER, APPROXIMATELY 194′ × 124′
LOCATED IN THE 1900 BLOCK OF STINGER STREET OPPOSITE GEORGE DRIVE

By virtue of statutory authority, a preference will be given to products and provisions grown, and coal produced within the State of ________________, and to ________________ domestic labor.

In accepting this contract the contractor agrees not to commit any of the following unlawful employment practices and to include this nondiscrimination clause in any sub-contracts connected with the performance of this agreement: (1) To refuse to hire employ, or to bar or to discharge from employment any individual because of their race, color, religion, sex, or national origin (2) To discriminate against any individual in terms, conditions or privileges of employment because of their race, color, religion, sex, or national origin (3) To use any form of application for employment, or to make any inquiry in recruiting or advertising for employees which requests the race, color, religion, sex, or national origin of any individual (4) To discharge, expel, or otherwise discriminate against any individual because they have opposed any practices forbidden under this section or because they have filed a complaint, testified or assisted in any proceeding under Section of the Contract-ORD #8920.

Drawings and specifications governing the construction of the proposed improvements have been prepared by TeeJay Associates, 123 Maple, ________________, ________________, which drawings and specifications and the proceedings of the City Council referring to and defining said proposed improvements are hereby made a part of this Notice and the proposed contract by reference, and the proposed contract shall be executed in compliance therewith. Copies of the drawings and specifications for this project may be obtained from the Architect. $75.00 deposit required for one set. Additional sets or partial sets may be obtained for the cost of printing and handling.

Work to be commenced immediately upon approval of the contract by the City Council of the City of ________________ and fully completed not later than September 10, 1982.

Payment to the contractor for said improvement will be made from cash on hand on the basis of monthly estimates in amounts equal to ninety percent (90%) of the work completed, final payment to be made approximately thirty (30) days after the completion and acceptance of the improvement. If satisfactory progress is being made after fifty percent (50%) of the work has been completed, subsequent monthly payments may be made in full.

Separate bids for General, Mechanical and Electrical Construction will be received. Each bid shall be accompanied by a certified check, cashier check, or bid bond as indicated below and as specified in the bidding documents, enclosed in a separate labeled and sealed envelope. Said check to be drawn on some known and responsible bank in the State of ________________, or a bank chartered under the laws of the United States, and made payable to the order of the City Treasurer of the City of ________________.

FIG. 4.1 *(continued)*

CONSTRUCTION	AMOUNT	CHECK OR BOND
General	10% of bid	Certified check, casher check, bid bond
Mechanical	10% of bid	Certified check, cashier check, bid bond
Electrical	10% of bid	Certified check, cashier check ONLY

Such check must not contain any condition in the body of the check or endorsement thereon. Said bid bond must be executed by a corporation authorized to contract as surety in the State of ________________, be for the sum of 10 percent of the amount bid, and be on a form prescribed by the City of ________________. All envelopes must be addressed to the City Clerk, and endorsed with the name of the bidder and the improvement said check or bond and proposal are for. Proposal form, drawings and specifications are on file at the offices of the City Engineer and the City Clerk, City Hall Building, ________________, ________________ and may be examined there.

Certified checks, or cashiers checks or bid bonds accompanying proposals will be returned to the unsuccessful bidders and also the bidder to whom the contract is awarded when he shall have entered into a contract, in the form prescribed by the City Council and given bond in the sum required which shall not be less than $1,000, with a Corporate Surety, to execute said work. In case the successful bidder shall fail to enter into contract or furnish bond as required by law within ten (10) days, said check or the bid bond in its full amount shall be forfeited to the City of ________________ as agreed and liquidated damages.

The right is reserved to reject any and all bids.

John B. Hayrack
Mayor

ATTEST: Helen Rowan
City Clerk

Published in The-----------Herald-Tribune
March 4, 1981

Size of project. This item should include information regarding both the monetary and physical scope of the project. A given construction company may have constructed buildings before, even multistory structures of up to six floors, but would probably not be qualified or interested in submitting a proposal on an 80-story project. Similarly, a firm that has constructed smaller earth-fill dams may not feel qualified to undertake a $400,000,000 concrete dam structure.

Location of project. Many firms prefer to operate within a specific geographic area. In some instances a firm that normally operates nationally or internationally may wish to avoid a particular area because of unfavorable experiences in the past. These may have resulted from labor problems, shortages of skilled personnel, political situations, or conditions of climate. Whatever the reason, the location of the project, along with type and size, is usually a major factor in the bidding decision.

Source of financing. This is a requirement for all public funded projects. Examples would be revenue bonds, general obligation bonds, etc.

Bonding requirements. Information should be presented regarding both bid and contract bond requirements. This would include amount or percentage of bid security required as well as the form in which it is to be furnished. Information pertaining to contract bonds should state whether both payment and performance bonds are required.

Type of bid being requested. This should state whether single or separate contracts are to be awarded and whether the proposal is on a lump sum or a unit price basis. Any alternates to the bid should also be mentioned under this item of information.

Conditions of payment under the proposed contract. How often the contractor is to be paid and the percentage of the monthly pay request that will be retained are important items that should be included.

Time requirements for the project. If specific starting and/or completion dates are required, this information should be included. The existing work load of a particular firm may be such that it would preclude that firm from meeting certain requirements in that regard. Along with time requirements, this section should also present any information related to proposed bonus, penalty, or liquidated damages provisions.

Conditions affecting the award of the contract. A statement should be made as to how the successful bidder will be determined, whether the contract will go to the lowest bidder, how alternates will be considered. In addition, a statement should be made that the owner reserves the right to reject all bids and the reasons for doing this.

Information regarding bidding documents. This should include the date the documents will be available and where they may be obtained, whether a deposit is required and, if it is, the amount of deposit, the conditions for its return, and whether some of it will be retained. It is common practice to return the documents deposit to unsuccessful bidders providing they are returned in good condition. "Good condition" usually means they can be used during the construction of the project. Some design firms follow a practice of refunding only a part of the deposit, a practice that may discourage some potential bidders.

Name and address of the owner. This is a legal requirement. In some instances a decision not to bid may be made because of a previous unpleasant experience with a particular owner. By the same token, a previous favorable experience may encourage a prospective bidder.

The location, date, and time for receiving bids. This information must be specific. An example would be whether the time given is standard or daylight savings time. This section should also give information regarding the procedure that will be followed in the opening and reading of bids. Generally, a statement should be made that the bids will be opened and read publicly. If this is not to be done immediately following the receipt of bids, then the time and location (if different) should be furnished.

Unusual requirements regarding bidding. This may include prequalification requirements, pre-bidding conferences, and other matters of similar nature.

Name and address of design professional. As in the case of the owner, this item sometimes influences the decision of whether or not to bid. The reputation of a particular firm regarding such matters as quality of documents, field inspection, approval of shop drawings, and other items of project administration are quite often taken into consideration by potential bidders.

4.2.2 PREPARATION AND TIME OF PUBLICATION

The advertisement is usually written by the design professional retained by the owner. Close consultation with the owner is advised to make certain that all items of particular interest and concern are included. The size and scope of the project will control the amount of lead time that is necessary in the publication of the advertisement prior to the receipt of the bids. Sufficient time should be allowed for potential bidders to accurately and thoroughly prepare their proposals. Generally, two weeks is considered the minimum amount of time required, even for a minor project. On larger projects, the time should be increased accordingly. It is generally safer to err on the side of allowing too much rather than not enough time.

The time of year and the number of other projects being bid in the same area at about the same time are also factors that should be considered. The spring is traditionally a busy time for firms bidding on building projects. A firm's estimating staff can only prepare so many quotations. Poor timing and lack of recognition of work loads may reduce the number of bidders. In many areas the local contractors' associations will furnish information without charge to owners and design professionals so that bids may be scheduled with a minimum of conflicts. Following this procedure may increase the number of bidders and result in a more favorable price for the owner.

4.3 INVITATION TO BID

If a project is not funded with public monies, the legal requirements regarding advertising procedures do not apply. The owner may elect to give no advance public notice regarding the proposed project. In many cases the owner may negotiate a contract with a particular construction firm without affording any other firms the opportunity to be considered. This is strictly the owner's option and prerogative. However, the owner is often interested in obtaining reasonable, competitive bids in an effort to minimize the cost. In accordance with this desire, the owner may advertise the project for bids but will not use an advertisement. The document used under these circumstances is called an "Invitation to Bid." It is basically identical to the advertisement except that it is not required to contain information relative to the method of financing for the project. (See Fig. 4.2.)

The invitation to bid also differs from the advertisement with respect to circulation procedures. Although the owner may elect to have the invitation to bid published in either newspapers or industry magazines, this would not be considered normal procedure. More often, the owner will mail copies of the invitation to bid to a selected list of construction firms. This list is normally prepared in cooperation with the design professional. It should be noted that some public owners label the required public notice as an invitation to bid. Regardless of the title used, it must meet the requirements of an advertisement.

4.3.1 ETHICAL CONSIDERATIONS

The owner does not have a legal obligation to award the contract to the lowest responsible bidder on privately funded projects. It should be kept in mind, however, that each firm submitting a proposal incurs a cost of bidding. Therefore the owner would be well advised to prepare the bidding list with due care and not send an invitation to bid to any firm to whom the contract would not be awarded. There have been occasions where an owner has preselected a contractor but still goes through the bidding process to make certain the price is

FIG. 4.2 Invitation to bid

INVITATION TO BID

ADDITION TO SCIENCE CENTER
VALLEYTOWN COLLEGE
__________, __________

College Project No. 3211-07

Bid Openings: 2:00 P.M., June 10, 1981

Sealed bids for the project designated above will be received for and in behalf of the Board of Trustees of Valleytown College in Room 180, Shaw Building, 765 College Drive, ________________, before the time and date indicated above, at which time they will be opened publicly and read aloud.

Drawings and specifications have been prepared by TeeJay Associates, 123 Maple, ________________.

A prebid conference and tour will be held on May 6, 1981, at 9:00 A.M.; refer to Instructions to Bidders.

In general, the work consists of interior remodeling involving approximately 11,300 square feet and new additions of approximately 25,300 square feet. Remodeling work varies from a complete renovation of some spaces to only a minor requirement at other spaces. The new construction and remodeling work also includes plumbing, heating, ventilating, air conditioning and electrical systems.

Lump sum bids will be required for each of the following: 1) General Construction, 2) Plumbing, 3) Heating, Ventilating and Air Conditioning, 4) Electrical, 5) Laboratory Casework and Fume Hoods.

Drawings, specifications and other bidding documents may be obtained from the Director of Physical Plant, Room 78 Wallace, Valleytown College, ________________, ________________ 99989, telephone: 800/555-1212; or may be seen at the Master Builders of ________________, 7864 Irwin, ________________, ________________.

A check in the amount of $125.00 will be required as a deposit for bidding documents. The check should be made payable to the Board of Trustees, Valleytown College. Checks will be refunded upon the return of the drawings, specifications and addenda in good condition.

Bid guarantee in the amount of 10% of the Bid must accompany each bid submitted.

Drawings and specifications will be available after April 24, 1981.

BOARD OF TRUSTEES
VALLEYTOWN COLLEGE

April 10, 1981

reasonable and fair. The preferred construction firm will then be offered an opportunity to meet the low bid price and receive the contract. This practice ignores the cost of bidding incurred by the other construction firms and takes unfair advantage of their services. The owner receives the benefit of price confirmation without having paid for it. At the least, this practice is unethical and should be discouraged.

4.4 INSTRUCTIONS TO BIDDERS

Each construction company that has taken out a set of drawings and specifications with the stated intention of submitting a proposal shall receive a copy of the "Instructions to Bidders." (See Fig. 4.3.) This document is sometimes called the "Information for Bidders." Regardless of the title used, the purpose of the document is to furnish prospective bidders with detailed information on a uni-

FIG. 4.3 Instructions to bidders

INSTRUCTIONS TO BIDDERS
(Example)

Date: _____

1. Sealed Bids will be received by the _________________, hereinafter referred to as the "Owner," for performing the work as set forth in the plans and specifications attached hereto, on or before _____ o'clock _____.M., C._____.T., in the _________________ at which time said bids will be publicly opened and read.

2. All bids must be made on the blank form of proposal attached hereto.

3. Each bid must be accompanied by a bid bond or certified check payable to the Owner for _____ per cent (_____%) of the total amount of the bid. As soon as the bid prices have been compared, the Owner will return the check of all except the three lowest responsible Bidders. When the contract is awarded, any checks of the two remaining unsuccessful Bidders will be returned. Any check of the successful Bidder will be retained until the contract and surety bond have been executed and approved, after which it will be returned. The bids will hold firm for a ninety day period in order to allow a borrower to complete its financing arrangements.

4. A performance and payment bond in the amount of one hundred per cent (100%) of the contract price with a corporate surety approved by the Owner and the Engineer, will be required for the faithful performance of the contract, and the Bidder shall state in the proposal the name and address of the surety or sureties who will sign this bond in case the contract is awarded to him. The bond guaranteeing the repair of all damage due to improper materials or workmanship for a period of _________________ after the acceptance of the work by the Owner will also be required.

5. The party to whom the contract is awarded will be required forthwith to execute the contract and a performance and payment bond within ten (10) calendar days from the date when the written "Notice of Award" of the contract is mailed to the Bidder at the address given by him. In case of failure to do so, the Owner may at his option consider that the Bidder has abandoned the contract, in which case the certified check accompanying the proposal shall become the property of the Owner.

6. The Owner reserves the right to reject any and all bids or to accept any bid. Contractors' proposals shall hold firm for ninety (90) days to allow the Owner to complete its financing arrangements. Mutually agreed upon extensions of time may be made, if necessary.

7. Before the award of the contract, any Bidder may be required to furnish evidence satisfactory to the Owner and to the Engineer of the necessary facilities, ability, and pecuniary resources to fulfill the conditions of the said contract.

8. Prior to signing the contract, the Contractor shall submit on a form acceptable to the Owner and Engineer, an overall construction schedule for the project. This construction schedule shall start with the proposed date of signing the contract, and the completion date shall be the date specified in the contract.

9. Bidders must satisfy themselves by personal examination of the location of the proposed work, by examination of the plans and specifications and requirements of the work and the accuracy of the estimate of the quantities of the work to be done, and shall not at any time after the submission of a bid, dispute or complain of such estimate nor assert that there was any misunderstanding in regard to the nature or amount of work to be done.

10. The construction contract and the detailed specifications contain the provisions required for the construction of the project. No information obtained from any officer, agent, or employee of the Owner on any such matters shall in any way affect the risk or obligation assumed by the Contractor or relieve him from fulfilling any of the conditions of the contract.

11. Proposals which are incomplete, unbalanced, conditional or obscure or which contain additions not called for, erasures, alterations or irregularities of any kind or which do not comply with the "Notice and Instructions to Bidders" may be rejected at the option of the Owner.

12. A Bidder may withdraw any proposal he has submitted at any time prior to the hour set for the closing of the bids provided the request for withdrawal is signed in a manner identical with the proposal being withdrawn. No withdrawal or modification will be permitted after the hour designated for closing the bids.

13. Computation of quantities that will be the basis for payment estimates, both monthly and final, will be made by the Engineer.

FIG. 4.3 *(continued)*

14. The word "Owner" means __________________. The Owner will be responsible for payment in accordance with the terms of the contract.
The word "Contractor" means the person, firm or corporation to whom the award is made. Subcontractors as such will not be recognized.
The word "Engineer" refers to __________________, designated by the Owner as its engineering representative during the course of construction to make appropriate inspections and computations of payments.

15. "Bidder's Proposals," plans and specifications may be obtained from the __________________. A deposit of $______ will be required for the plans and specifications, which will be refunded to those returning them in good condition within ten (10) days after the date of the bid opening.

16. Inspection trips for Contractors will leave from the office of the __________________ at ______ o'clock, ______.M., C.______.T. on __________________ and __________________.

____
Owner

By______________________________

Title_____________________________

form basis. Most of the information included in the advertisement or the invitation to bid will be repeated in the instructions to bidders and will be sometimes expanded when it is deemed advisable. The purpose of the instructions to bidders is not only to provide potential bidders with the detailed information necessary for the preparation of their proposal, but also to try to ensure that the basis of preparation and the form of submittal are based on common, uniform information that will result in a responsive bid. This will enable the owner and the design professional to perform a more fair and equitable comparison of the bids submitted, with improved chances of securing the maximum return for the owner's investment.

4.4.1 INFORMATION TO BE INCLUDED

As previously stated, items of information listed in the advertisement or the invitation to bid should be repeated in the instructions to bidders. In addition to those items, the following information should be given:

Procedure for preparation of the proposal. This would include instructions indicating if a furnished proposal form is to be used or if the bid is to be submitted in another manner, such as on the company stationery. It should also include procedures to be followed in acknowledging receipt of any addenda, how the bid figure is to be presented (usually in words and figures), and any special requirements as to signatures.

Procedure for submission of the proposal. This would include instructions relative to the size, type, and color of envelope to be used as well as how the envelope is to be addressed. It may also be required that the name of the bidding firm along with the part of the contract on which the bid is being submitted be shown in a particular way on the outside of the envelope containing the bid.

Bid security. This item should include information regarding the form of bid security that is acceptable (bond or certified check) and the percentage or fixed amount of such security. It may also include provisions relative to the return of bid security for unsuccessful bidders, as well as which bid security the owner will retain and for how long.

Withdrawal of bids. Presented in this section would be the procedure along with the applicable restrictions for bidders to withdraw their bids. This information is of particular importance in public funded projects because of the necessity to guard against fraud and collusion. Definite time limits should be established for this eventuality.

Opening of bids. On public funded projects, the time, place, and date of the public opening and reading of bids would be given in this section. If the project is one using private funds, this process may also be used but is not mandatory. In such a case, however, it is usually desirable to give information relative to how the successful bidder will be selected and when and how the notification will be made.

Rejection of bids. The owner usually needs a statement specifying that if none of the bids are satisfactory because of exceeding the budget or some similar reason, then all bids may be rejected. This is of particular importance on public funded projects where the source of financing may be a bond issue where only a fixed amount of funds is available. It is also recommended for privately funded projects that the owner be provided with this option to ensure a minimum of misunderstandings.

Procedure for discrepancies. A completely error-free set of bidding documents is the exception rather than the rule. In recognition of this fact, it is necessary to indicate what procedure is to be followed when a bidder discovers an error in the documents or a conflict between two or more of them. The purpose of establishing this procedure is again to ensure that all bidders receive the necessary information and that their bids are based upon equal information.

Return of drawings and specifications. Most design offices establish limits relative to the return of the deposit on drawings and specifications. This may include statements about what constitutes acceptable conditions and what time limit is to be observed. Information as to whether the refund will be entire or partial should also be included in this section.

Time limitations. Some projects carry requirements concerning starting and completion times. These requirements should be accompanied by information on how these times will be interpreted, such as working days or calendar days, as well as specifics regarding applicable bonus or penalty provisions.

Requirements unique to the project. This may include such items as prequalification of bidders, specific hiring requirements, special purchasing provisions, unusual methods requirements, or any number of other items called for by legislation or special owner interests. An example might be how the "or equal" phrase is to be applied.

4.4.2 PREPARED BY WHOM

The instructions to bidders is usually prepared by the design professional retained by the owner. The same person or firm will normally accept the responsibility for its distribution. A common practice is for the document to be given to the potential bidders at the time they pick up the drawings and specifications, although the document can be mailed at a later time. It is important to remember that it should be placed in the hands of the bidders with sufficient time allowed for its thorough review and incorporation of the requirements in the bidders' proposal.

4.5 ADDENDA

If a change is indicated as necessary in a drawing during the bidding phase, the normal procedure is to issue a revised drawing to all potential bidders. If a change is required in the specifications or one of the other documents during the bidding phase, an addendum is used. (See Fig. 4.4.) The purpose of an addendum is to present any change in bidding conditions except those involving only a revision in the drawings. The use of addenda should not be limited to changes in the documents that will affect price. Any change in requirements or bidding conditions should be covered by an addendum.

4.5.1 REASONS FOR ISSUING

There are several reasons for issuing addenda. Among them are the following:

To correct errors. This may include errors in the specifications or discrepancies between the drawings and the specifications. If an error is discovered in the drawings only, it is better to issue a revised drawing.

To make owner-initiated changes. It is not uncommon for an owner to have second thoughts relating to a particular item of the project. If the owner desires to change the quality of a material or to cancel or add an alternate, an addendum should be issued.

To furnish clarification. There may be occasions when a given requirement is not clearly understood by the potential bidders. In response to their questions, an addendum explaining and clarifying the provision will be issued.

To add or delete products. If an "or equal" clause is used in specifying certain products, bidders may raise questions regarding the possible addition or deletion of a particular brand name to the list. If the design professional is in agreement, an addendum will be issued modifying the list of acceptable products.

To change stated requirements. In addition to correcting or clarifying statements regarding product requirements, such revisions as a change in bid due date may be the cause for issuing an addendum. In other words, any change in bidding conditions would be reflected by an addendum.

4.5.2 RESTRICTIONS ON ISSUING

There are certain safeguards that should be observed in the issuance of addenda. Just as it is the intention of the original bidding documents to ensure fair and equal competition between the bidders with resulting benefit to the owner, so it should be the intention of the addenda to support and reinforce this purpose. Any procedure is only as effective as its weakest link or member, and in some cases the addenda have been the "weak link." To avoid such an occurrence the following guidelines should be strictly adhered to:

FIG. 4.4 Addendum

ADDENDUM NO. 4

NEW CONSTRUCTION
OF THE
_______________ HOSPITAL

_______________, _______________

August 18, 1981

The specifications prepared by TeeJay Associates, Architects, for the new construction of _______________ Hospital, _______________, _______________, dated June 17, 1981, are hereby amended in the following particulars:

Item No. 1: Specifications, Page 15M.4, I. **HEATING COILS IN DUCTWORK.**

8. Specified ratings shall be developed with water entering at 180°F. and leaving at 160°F. This information was omitted from schedule on drawings.

Item No. 2: Specifications, Page 15M.5, J. **CONDENSATE PUMPING UNIT.** Add Paragraph 5 as follows:

5. Capacity of each pump shall be 9.0 GPM at 20 PSI. Motor shall be ½ HP.

Item No. 3: This will clarify details of welded pipe construction called for in Section 15H, 15L and 15M. All steel pipe lines 2½" and larger shall be welded. Lines 2" and smaller shall have screwed connections. If desired, welding may also be used in 2" size pipe.

Item No. 4: Copper pipe specified in Section 15I shall be Type "L" for water lines and Type "M" for waste and vent lines.

Item No. 5: Article N on Page 15J.6 of specification describes shower mixing valve incorrectly. This valve must be thermostatic type and not balanced pressure type.

No addendum should be issued later than five days prior to the bidding due date. Violation of this guideline may result in an insufficient amount of time for the bidders to reflect the changes in their proposal, and thus constitutes a disservice to the owner. If it is imperative that changes be made after this deadline, the due date should be extended. If this procedure is not possible or practical, changes may be made before signing the contract, or afterwards by means of change orders.

Addenda should be issued only in proper written form. Issuance of addenda by telephone or similar means may result in a lack of complete understanding, with resulting confusion in the bidding process.

A procedure should be established that requires acknowledgment of the receipt of all addenda. This may consist of a signed receipt for each addendum returned to the office of the design professional and/or acknowledgment of the addenda on the proposal form. Only by using such a system can the owner be assured that the bids were prepared on a common basis.

4.5.3 INFORMATION TO BE INCLUDED

Addenda are most commonly prepared and issued by the office of the design professional. Each addendum should include the following information:

Name of the project to which it applies
Number of the addendum
Date of issuance
Branch, or branches, of the work affected
Provisions for receipt procedure
Name and address of the issuing office
Reason for issuing
Changes in the work (or conditions) under the addendum

5 BID SUBMISSION DOCUMENTS

FIGURES

5.1 PROPOSAL

The estimating process is normally expected to culminate in the submission of a proposal. A proposal is an offer by the construction company to the owner to perform the work in accordance with the documents furnished for a stated remuneration. The proposal constitutes the first half of an "offer and acceptance" contract.

The offer is usually submitted on a proposal form that is furnished to each bidder. This form is supplied as part of the bidding documents and enables each bidder to submit a proposal in a common manner. When a proposal form is not furnished, the instructions to bidders may dictate that the proposal be submitted on the company letterhead. Such an approach clearly identifies the firm submitting the proposal but does not provide a convenient means of comparison. This is especially true if prices other than the base bid are requested.

5.1.1 PURPOSE

The purpose of using a proposal form is to permit the owner to more easily compare the bids that are submitted. The prepared form consists of the owner's name and address, along with a statement that the bidder agrees to furnish all the labor, material, and equipment needed to complete the required work for the remuneration that the bidder lists in the proposal. Instructions given to the bidders relative to the completion of the proposal form will normally instruct them not to make any changes in the form that is furnished. Changes to the form—either additions or deletions—are generally considered sufficient to adjudge the bid nonresponsive and may result in its disqualification. By requiring all bidders to submit their proposals on furnished forms, the owner is able to compare "apples to apples" rather than "apples to oranges."

5.1.2 INFORMATION TO BE INCLUDED

The proposal form is prepared by the design professional as part of the total bid documents package. The content of the form should reflect the needs and wishes of the owner as well as the particular requirements of the project. For this reason, although few proposal forms will be identical, most will be similar in content. Some information to be included on the proposal form is the following:

Name and address of the owner. In some cases the name of the individual authorized to receive bids may be added or listed in addition to the owner.

Title of the project. If bids are being taken on government projects, a project identification number is also given.

An opening statement. As a minimum, this statement will specify that the bidder agrees to furnish all labor, material, and equipment to complete the project. In addition, there may be provisions stating that the bidder has examined the site, that special hiring requirements will be adhered to, that a performance bond will be furnished, or that a disclaimer of any fraud or collusion in connection with the preparation of the bid will be furnished.

Blank spaces for bid amount. The amount of the bid must be given in both words and figures, with spaces provided for this purpose.

Date of proposal.

Name and address of firm submitting the bid along with the required signatures. Provisions may also be made to identify the type of firm bidding, i.e., corporation, partnership.

Receipt of addenda. This would consist of a statement that the bidder acknowledges receipt of addenda, with blanks provided for the numbers of addenda received and possibly the dates when each was delivered.

Alternate bids. If alternates are used in the proposal, provisions should be made for the additive or deductive price for each alternate. Although the prices for alternates are often given in figures only, it is preferable to require them in both words and figures.

Time of completion. If a completion time will be required in the contract, space should be provided for the bidder to state the number of days required or the date when this can be accomplished. If it is according to the number of days, reference should be made to calendar days rather than working days. In some instances the completion time is designated by the owner, in others by the contractor.

Subcontractor listing. If this is required, the proposal form should provide spaces for the listing of subcontractors and major material suppliers.

Unit prices. Many owners desire unit prices for any additive or deductive work in the contract due to unforeseen circumstances. These will most often be requested in connection with excavation or earth work. Provisions should be made for the bidder to present this information in a common form.

On some contracts the owner may already have taken bids on a particular branch of the work, such as site clearance to provide for an early beginning for the project, or structural steel if timely delivery is important. Under these circumstances, the contracts may be assigned to the general contractor. The proposal form for this situation may include provisions for stating the percentage of markup to be applied to the assigned contracts. It may also request that the percentage of markup be applied to "extra" work under the contract.

5.1.3 SIGNATURE REQUIREMENTS

The name and signature of the bidder is an item deserving special notice. If the bidding firm is a single proprietorship, the name and address of the firm should be given along with the typed name and title of the person whose signature is on the bid. If the bidding firm is a partnership, the full name of the partnership should be given, not just the name of the signing partner. In the case of a corporation, the signature must be that of a person authorized to sign on behalf of the corporation. If the bid is signed by an officer of the corporation other than the president, it is usually required that the signature be accompanied by a power of attorney. An additional, desirable safeguard for corporate signatures is to require a copy of the resolution by the corporation's board of directors approving the bid.

5.2 LUMP SUM PROPOSAL

Proposals under the competitive bidding system are usually submitted as a lump sum or a unit price bid. The lump sum bid is used for most building construction projects but may be used on other types of works where the scope and requirements of the project are well defined. Most civil projects, such as street and highway work, utilize the unit price system.

The five elements of construction costs are labor, equipment, material, overhead, and profit. All of these elements are included within a lump sum bid. Therefore when the owner elects the lump sum bid method, the proposal form must provide spaces for only one amount to be given in words and figures. This is referred to as the base bid. (See Fig. 5.1.)

FIG. 5.1 Lump sum proposal

BID FORM C-1

VALLEYTOWN COLLEGE
CENTER FOR SCIENCES
____________, ____________

Project No. 3211-07

Bid Opening: 2:00 P.M., June 10, 1981

To: The Board of Trustees of the Valleytown College

(a corporation)
(a partnership)
We____(an individual)
(Cross out inapplicable)

of____
Street City County State Zip

hereby agree to execute the proposed contract and to furnish a satisfactory surety bond in the amount specified within 10 days of offering, and to provide all labor and material required for the construction of the project designated above, for the prices hereinafter set forth, in strict accordance with the Contract Documents prepared by TeeJay Associates, 123 Maple, ____________, ____________ 99989, for the Valleytown College Board of Trustees, and dated March 18, 1981.

GENERAL

BASE BID—General Work as per specifications and related drawings:
For the sum of____
______________________________________Dollars ($______________)

ALTERNATE BID NO. 1—3rd floor for Annex "B":
Add to Base Bid the sum of____________________________________
______________________________________Dollars ($______________)

ALTERNATE BID NO. 2—omit floortile in Labs 107, 112, 204:
Deduct from Base Bid the sum of________________________________
______________________________________Dollars ($______________)

ADDENDUM RECEIPT
We acknowledge receipt of the following Addenda:

NUMBER AND DATE	NUMBER AND DATE

COMMENCEMENT AND COMPLETION OF CONTRACT WORK

The undersigned agrees, if awarded the contract, to commence the contract work on or before a date to be specified in a written notice to proceed, and to complete the work within the time slated in the Instructions to Bidders.

____
(Firm Name)

____
(Area Code & Telephone Number)

By____________________________
(Signature of Bidder)

Title__________________________

(Seal, if bid is by a corporation)
Date______________

READ NOTES BEFORE SIGNING

NOTE: 1) This bid will be rejected if the Bid Form has been altered or changed in such a way that it incorporates unsolicited material, either directly or by reference, which would alter any essential provision of the contract documents or require consideration of unsolicited material in determining the award of contract.
2) If this bid is not accompanied by a bid guarantee, it will be considered NO BID and will not be read at the bid opening.
3) This bid will be rejected if receipt of an addendum applicable to the award of Contract has not been acknowledged on the bid form.

5.3 UNIT PRICE PROPOSAL

A unit price bid should be requested when the scope of the work cannot be well defined and when the project will mainly consist of a relatively small number of standard activities. In order to use this type of bidding system, an estimate must be made of the quantities of units of work in the various branches. While this may appear inconsistent with the fact that the scope of the work cannot be well defined, it is necessary so that a comparison of the bids can be made. These estimated quantities are furnished to the bidders and listed on the proposal form. Each bidder is then required to provide a price per unit for each type of work represented. Each unit price will include all five elements of construction costs.

In order to compare the bids, each quantity listed is multiplied by the unit price submitted by the contractor on that branch of the work. These extensions are then added to arrive at a single figure that is called the equivalent lump sum. The firm with the lowest equivalent lump sum would be designated the successful bidder and awarded the contract. The equivalent lump sum is used for the purpose of determining the low bidder only and is not the basis for contract or payment amount. The proposal form used for this type of bid must include provisions for this information. (See Fig. 5.2.)

5.4 BID ALTERNATES

Alternates and substitutions are changes to the base bid. An alternate is a change that has been requested by the bidding documents and can be either additive or deductive. It is often called for when the owner wants to compare costs for different products or methods, or when the owner is not certain of finances at the time the bids are called for. Examples might include the requirement for precast exterior wall panels with an alternate price requested for cut stone panels. In another case, a multistage bond issue may be placed before the voters. As a result of uncertainty regarding the outcome of the election, the base bid may call for a specified amount of four-lane highway, with the work pertaining to shoulders and median strip included within an additive alternate. For alternate bids refer to Fig. 5.1.

Alternates should not be used to replace a lack of knowledge on the part of the design professional regarding relative costs of construction. Design offices should keep well informed regarding costs so that the owner may be properly advised in connection with various options for the project. Depending upon the size of the project, the use of four or five alternates may be considered reasonable, but the use of a dozen alternates may indicate poor preparation on the part of the design professional. Using too many alternates has a tendency to increase the chances for error in the bidding, and it may indicate collusion and cause a good contractor to decide not to bid.

It is a difficult task for a design office to maintain current knowledge of various construction costs. This often requires breaking down projects that have been bid into various components and working with the succesful bidder to distribute costs. Some bidders are reluctant to furnish this information or they may not devote sufficient time to develop accurate information. Owners should not be surprised if the estimate furnished by the design professional differs from the low bid by 15 to 20 percent since the spread between the high and low bid bidders will often exceed this. It should also be remembered that although design professionals are concerned with construction costs, their involvement with these costs is less direct than that of the constructor.

FIG. 5.2 Unit price proposal

PROPOSAL

FOR THE CONSTRUCTION OF
1981 STREET PROGRAM NO. 2

Name of Bidder____________________
Address of Bidder ____________________

To the City Council
__________, __________

Division 1A—Assessable (17th Street & Stinger Road)

The undersigned bidder submits herewith bid security in the amount of $21,000.00 in accordance with the terms set forth in the Instructions to Bidders.

The undersigned bidder having examined and determined the scope of the Contract Documents, hereby proposes to provide the required labor, services, materials, and equipment and to perform the work as described in the Contract Documents, including Addenda __________, __________ and __________ as follows:

Item	Description	Estimated Quantity	Unit	Unit Price	Amount
1.	Excavation	2,377	c.y.	$____	$____
2.	Borrow Fill	7,492	c.y.	$____	$____
3.	Concrete Removal	2,486	s.y.	$____	$____
4.	Asphalt Removal	343	s.y.	$____	$____
5.	7″ P.C.C. w/integral curbs	12,033	s.y.	$____	$____
6.	12″ R.C.P.	743	l.f.	$____	$____
7.	15″ R.C.P.	189	l.f.	$____	$____
8.	18″ R.C.P.	222	l.f.	$____	$____
9.	Type "D" Intakes	5	ea.	$____	$____
10.	Type "E" Intakes	4	ea.	$____	$____
11.	Rock Embedment	330	ton	$____	$____
12.	Road & Drive Rock	16	ton	$____	$____
13.	Asphalt Base	354	ton	$____	$____
14.	Asphalt Mat	175	ton	$____	$____
15.	Tack Coat	327	gal.	$____	$____
TOTAL	Division 1A—Assessable				$____

The undersigned bidder certifies that this proposal is made in good faith without collusion or connection with any other person or persons bidding on the work.

The undersigned bidder states that this proposal is made in conformity with the Contract Documents and agrees that, in the event of any descrepancies or differences between any conditions of his proposal and the Contract Documents prepared by the City of __________, __________, the provisions of the latter shall prevail.

Bidder____________________
By____________________
Title____________________

Date__________

5.5 BID SUBSTITUTIONS

A substitution is a suggested change in price to the base bid, which has been volunteered by the bidding firm—not the owner or the design professional. It is usually an attempt on the part of the bidder to interest the owner in the bid of that particular firm and to thereby gain the contract. Some owners react favorably to such action, as do some design firms. One disadvantage to substitutions is that there are no comparative prices available. Although a bidder may offer a savings to the owner of "x" number of dollars if a certain substitution is allowed, it is possible that if comparative bids were taken on the suggested change, "x" plus "y" dollars in savings could be realized. Firms that make it a practice to offer substitute quotations have stated that it is representative of the bidding firm's initiative in saving the owner money on the project. Some design firms react unfavorably to substitute quotations, feeling that they imply a cirticism of the design and may endanger the intent of the total design.

5.6 LOW BIDDER DETERMINATION

Determination of the low bidder under either the lump sum or unit price method is a fairly simple matter. Under the lump sum process a direct comparison of the base bids is all that is required. Under the unit price proposal system, the method used is that of equivalent lump sum, as described earlier in this chapter. However, the process becomes more complex when alternates and substitutions to the base bid are taken into consideration.

If the bidding documents for a project call for a lump sum base bid along with three additive alternates and three deductive alternates, a variety of "low" bidders might be established. One firm may be low on the base bid with another firm being low bidder if the base bid is combined with one or more of the additive alternates. Still another firm may be low if the base bid is combined with some of the deductive alternates. Although it would be preferable for the owner to state in the bidding documents which combination will be used to determine the successful bidder, this is generally not practical nor is it in the owner's best interest.

There have been some situations when the use of alternates in determining the low bidder has been used to apply favoritism in public contracts, with award being made on base bid and then alternates being added to the contract at a later date by means of change orders. Although unsuccessful bidders will no doubt be disappointed not to receive a contract because of alternates, they generally will bear the owner no ill will if they are convinced the action taken was impartial and that there was no favoritism.

The process of determining the successful bidder when consideration is given to substitute quotations is even more difficult than the process of using alternates. With alternate bids, each bidder quotes on the same changed conditions. As a result, there are comparative prices available, and the owner is still able to compare "apples to apples." On the other hand, a bidder who offers a substitute bid may be the only one doing so. As a result, the scope of the change may not be clear and the owner will not have an opportunity to compare it with any other price. This means that the owner must compare "apples to oranges." Although substitute bids may present the owner with a strong incentive to save money by awarding the contract to the innovative bidder, it would be better practice to make the award according to base bid and alternates and then to negotiate any possible substitutions with the successful bidder.

5.7 BID SECURITY

When construction firms submit bids on a project, they are making offers to perform work for the owner. It rationally follows that the owner should have the right to accept these offers. To ensure that the successful bidder will enter into a contract in conformance with the accepted offer, bid security may be required. This requirement is optional on privately funded projects and mandatory on publicly funded works.

The use of bid security in connection with the construction bidding process may be compared to the "earnest money" that accompanies the offer to purchase a piece of real estate. The bid security ensures that the offer will not be withdrawn and that the offering party will sign an agreement in accordance with the bid if the offer is accepted. Without any bid security, the low bidder could raise the quoted price until it was slightly below the next bid, and the owner would be powerless to prevent it.

5.7.1 FORFEITURE OF BID SECURITY

When a bid security accompanies an offer, it is understood that if the offer is accepted and the submitting firm refuses to enter into a contract, the bid security is forfeited to the owner. This is to reimburse the owner for any possible loss as a result of losing the bargain. In the event the successful bidder fails to enter into a contract, he may be liable for the difference between his low bid and the second bid up to the face amount of the bid security. Prior to the mid-1950s most construction firms expected this loss if they failed to honor their bids. In more recent years, however, an increasing number of bidders have successfully secured release from the loss of bid security by claiming "honest clerical errors" or similar reasons. The courts appear to be increasingly lenient in this regard, the result being that the quality of protection afforded the owner by bid security has been diminished.

5.7.2 AMOUNT OF BID SECURITY

One of the forms under which bid security can be furnished is a bid bond. Other possible forms include a certified check or negotiable securities. The bidding documents will normally specify which form or forms are acceptable. The amount of bid security may be stated as a fixed amount or as a given percentage of the firm's bid amount.

There is a distinct advantage to the bidders when the design professional specifies a fixed amount for the bid security. To arrive at a percentage figure, the approximate total of the estimate must first be determined. This may be quite difficult if the bid is not being put together locally or if there are strict time constraints. This in turn can lead to errors in the amount of bid security, the most common being that an insufficient amount is furnished. The usual result of such an error is disqualification of the proposal. In the case of a fixed amount, the bid security, in whatever form, can be secured ahead of time with a resulting decrease in the chances for error.

5.7.3 RETENTION OF BID SECURITY

The owner usually retains the bid security for the low bidder until a contract has been signed. If the designation of the low bidder is in doubt or difficult to determine immediately, the owner may retain the bid security of the apparent second and third bidders as well. Bid security for all other bidders is usually returned shortly after the opening of the bids. This releases bonding monies into working capital, increases the bonding capacity of the unsuccessful bidders, and makes it easier for them to pursue other contracts.

Bidding documents should state the time limit for which the owner may retain the bid security before a contract is offered. Normal acceptance time within the construction industry is 30 days, but 60 or 90 days may be stipulated if warranted by unusual circumstances.

5.8 BID BOND PREMIUMS

A bid bond is issued by a surety company to the owner. A surety may be defined as one who insures the actions of another, for a consideration. In this case, the consideration is a fee. A surety may charge a premium (or fee) for the bid bond; however, normal practice is to furnish the bid bond without a fee on the anticipation of collecting a fee for the performance bond. When a surety does furnish a bid bond for a company, there is an implied obligation to furnish the contract bond for the same company in the amount of the firm's bid.

It should be emphasized that the bid bond does not constitute any form of insurance for the contractor but rather serves to guarantee to the owner that the bidder will enter into a contract. Failure to do so will make the surety liable to the owner up to the face amount of the bond. The construction firm submitting the bid pays any premium attached to the furnishng of the bid bond but passes this cost on to the owner through the bid as a cost of doing business.

5.9 SUBCONTRACTOR BID BONDS

Although it is more common for owners to require bid bonds from firms submitting bids to them, there may be occasions when the general contractor places a requirement for bid bonds upon the subcontractors. This would normally be done only when the general contractor is working in an unfamiliar area or when a particular subcontracting firm has not bid with them or done work for them before. The purpose of having subcontractors post bid bonds is the same as that for the general contractor—to ensure that they will enter into a contract if their proposal is accepted. There may also be occasions when the general contractor's surety will require subcontractor bid bonds as a condition to furnishing a bid bond for the general contractor. This would normally be limited to cases where the general contractor's financial capacity is approaching the top limit.

5.10 FORM OF BID BOND

On some projects the form of bid bond is furnished by the owner, while on other projects the surety's form of bid bond is accepted. (See Fig. 5.3.) Regardless of what bond form is used, the following information should be included:

The name and address of the principal submitting the bid, that is, the one whose action is being guaranteed.

The name of the surety company, that is, the party furnishing the guarantee.

The name of the owner, also known as the obligee.

The title of the project and any applicable project identification numbers, as well as the branch or branches of work for which the bid is being submitted.

A statement as to the purpose and limits of obligation of the bond. This may include information to the effect that obligation under the bond shall not exceed the difference between the lowest and second lowest bids, or the face amount of the bond, whichever is less.

The date of the bond. This should be the same as the date of the proposal.

(text continues on page 96)

FIG. 5.3 Two examples of bid bonds

BID BOND
(Example)

KNOW ALL MEN BY THESE PRESENTS, that we, ________________
________________, as Principal, and ________________
________________, as Surety are held and firmly bound unto the ________________
________________, hereinafter called "Owner", in the penal sum of ________________ Dollars ($______) lawful money of the United States, for the payment of which sum will and truly be made, we bind ourselves, our heirs, executors, administrators, and successors, jointly and severally, firmly by these presents. The condition of this obligation is such that whereas the Principal has submitted the accompanying bid, dated ________________, 19______, for ________________ Project.

NOW, THEREFORE,

(a) If said Bid shall be rejected, or in the alternate,
(b) If said Bid shall be accepted and the Principal shall execute and deliver a contract in the form specified and shall furnish a bond for his faithful performance of said contract, and for the payment of all persons performing labor or furnishing materials in connection therewith, and shall in all other respects perform the agreement created by the acceptance of said Bid,

then this obligation shall be void, otherwise the same shall remain in force and effect; it being expressly understood and agreed that the liability of the Surety for any and all claims hereunder shall, in no event, exceed the penal amount of this obligation as herein stated.

By virtue of statutory authority, the full amount of this bid bond shall be forfeited to the Owner in liquidation of damages sustained in the event that the Principal fails to execute the contract and provide the bond as provided in the specifications or by law.

The Surety, for value received, hereby stipulates and agrees that the obligations of said Surety and its bond shall be in no way impaired or affected by any extension of the time within which the Owner may accept such Bid or execute such contract; and said Surety does hereby waive notice of any such extension.

IN WITNESS WHEREOF, the Principal and the Surety have hereunto set their hands and seals, and such of them as are corporations, have caused their corporate seals to be hereto affixed and these presents to be signed by their proper officers this ______ day of ________________, A.D. 19______.

________________________ Witness

________________________ (Seal)
Principal
By________________________
(Title)

________________________ (Seal)
Surety
By ________________________
(Attorney-in-fact)

________________________ Witness

Attach Power-of-Attorney

FIG. 5.3 (*continued*)

THE AMERICAN INSTITUTE OF ARCHITECTS

AIA Document A310

Bid Bond

KNOW ALL MEN BY THESE PRESENTS, that we
(Here insert full name and address or legal title of Contractor)

as Principal, hereinafter called the Principal, and
(Here insert full name and address or legal title of Surety)

a corporation duly organized under the laws of the State of
as Surety, hereinafter called the Surety, are held and firmly bound unto
(Here insert full name and address or legal title of Owner)

as Obligee, hereinafter called the Obligee, in the sum of

Dollars ($),
for the payment of which sum well and truly to be made, the said Principal and the said Surety, bind ourselves, our heirs, executors, administrators, successors and assigns, jointly and severally, firmly by these presents.

WHEREAS, the Principal has submitted a bid for
(Here insert full name, address and description of project)

NOW, THEREFORE, if the Obligee shall accept the bid of the Principal and the Principal shall enter into a Contract with the Obligee in accordance with the terms of such bid, and give such bond or bonds as may be specified in the bidding or Contract Documents with good and sufficient surety for the faithful performance of such Contract and for the prompt payment of labor and material furnished in the prosecution thereof, or in the event of the failure of the Principal to enter such Contract and give such bond or bonds, if the Principal shall pay to the Obligee the difference not to exceed the penalty hereof between the amount specified in said bid and such larger amount for which the Obligee may in good faith contract with another party to perform the Work covered by said bid, then this obligation shall be null and void, otherwise to remain in full force and effect.

Signed and sealed this day of 19

	(Principal) *(Seal)*
(Witness)	
	(Title)
	(Surety) *(Seal)*
(Witness)	
	(Title)

AIA DOCUMENT A310 • BID BOND • AIA ® • FEBRUARY 1970 ED • THE AMERICAN INSTITUTE OF ARCHITECTS, 1735 N.Y. AVE., N.W., WASHINGTON, D. C. 20006 **1**

This document has been reproduced with the permission of the American Institute of Architects under application number 79082. Further reproduction, in part or in whole, is not authorized. Because AIA documents are revised from time to time, users should ascertain from AIA the current edition of the document reproduced above.

The signature(s) of the principal (contractor), along with signatures of any required witnesses.

The signature of the surety.

The amount of the bond. Normal practice is to require the bid bond to be in an amount equal to 5 to 10 percent of the bid. There have been instances on some private projects, as well as on some governmental projects, where unusual circumstances have necessitated bid bonds of up to 20 percent of the bid.

5.11 ATTACHED DOCUMENTS

A number of other documents are often required to be attached to the proposal. One that is often used on both private and public funded projects is the "Prequalification Statement." The owner may elect to limit the bidding to those firms meeting certain criteria in areas such as financial capacity, experience in the type of work involved, or possession of a stated minimum amount of equipment. As can be seen from the example illustrated in Fig. 5.4, the information requested by this form is quite extensive.

The increasing involvement of the federal government in the construction process has spawned several other requirements that generate their own brand of documents. Many of these must be attached to the proposal. This may apply to private projects even if no federal funds are involved. These forms may pertain to equal employment practices regarding minorities, employment of women, minority business enterprise requirements, employment of the handicapped and disadvantaged, etc. Typical requirements are illustrated by Fig. 5.5.

Many private owners as well as some federal agencies encourage or require the listing of subcontractors with the proposal. The purpose of this requirement is to discourage bid shopping and bid peddling. This practice has resulted in fairly strong feelings on both sides of the controversy and will be discussed in more detail in Chapter 6. An example of the manner in which the list of major subcontractors and material suppliers is requested will also be illustrated there. (See Fig. 6.4).

FIG. 5.4 Prequalification statement

STANDARD FORM OF

CONTRACTORS' EXPERIENCE QUESTIONNAIRE AND FINANCIAL STATEMENT

Approved and recommended by

THE JOINT CONFERENCE ON CONSTRUCTION PRACTICES

CLEARING HOUSE SECTION OF THE AMERICAN BANKING ASSOCIATION

THE ASSOCIATED GENERAL CONTRACTORS OF AMERICA

Issued by

THE ASSOCIATED GENERAL CONTRACTORS OF AMERICA, INC.

WASHINGTON, D. C.

A. G. C. STANDARD FORM 28

Reproduced by permission of The Associated General Contractors of America

FIG. 5.4 *(continued)*

EXPERIENCE QUESTIONNAIRE

Submitted to ______________________________

By ______________________________ ☐ A Corporation ☐ A Co-partnership ☐ An Individual

Principal Office ______________________________

The signatory of this Questionnaire and Financial Statement guarantees the truth and accuracy of all statements and of all answers to interrogatories hereinafter made.

1. How many years has your organization been in business as a general contractor under your present business name? ______________________________

2. How many years experience in construction work has your organization had: (a) As a General Contractor ______________ (b) as a Sub-Contractor ______________?

3. Have you ever failed to complete any work awarded to you? ______________ If so, where and why?

____

____

____

4. Has any officer or partner of your organization ever been an officer or partner of some other organization that failed to complete a construction contract? ______________ If so, state name of individual, other organization and reason therefor ______________________________

____

____

5. Has any officer or partner of your organization ever failed to complete a construction contract handled in his own name? ______________ If so, state name of individual, name of Owner and reason therefor.

____

____

____

6. In what other lines of business are you financially interested? ______________________________

____

____

____

7. For what corporations or individuals have you performed work, and to whom do you refer? ______________

____

____

____

____

8. For what cities have you performed work and to whom do you refer? ______________________________

____

____

____

____

9. For what counties have you performed work and to whom do you refer? ______________________________

____

____

____

____

FIG. 5.4 *(continued)*

10. For what State Bureaus or Departments have you performed work and to whom do you refer? ________________

11. Have you ever performed any work for the U. S. Government? ________________

If so, when and to whom do you refer? ________________

12. What projects has your organization completed?

Contract Amt.	Class of Work	When Completed	Name and Address of Owner

FIG. 5.4 *(continued)*

13. What is the construction experience of the principal individuals of your organization?

Individual's Name	Present Position or Office	Years of Construction Experience	Magnitude and Type of Work	In What Capacity

14. Give the names and addresses of all banks with whom you have done business during the past five years.

15. Give the names and addresses of all surety companies with whom you have done business during the last five years.

FIG. 5.4 *(continued)*

FINANCIAL STATEMENT

Condition at close of business __________ *19*____

ASSETS	Dollars	Cts.
1. Cash: (a) On hand $_____, (b) In bank $_____, (c) Elsewhere $_____		
2. Notes receivable (a) Due within 90 days		
(b) Due after 90 days		
(c) Past due		
3. Accounts receivable from completed contracts, exclusive of claims not approved for payment		
4. Sums earned on uncompleted contracts as shown by Engineer's or Architect's estimate		
(a) Amount receivable after deducting retainage		
(b) Retainage to date, due upon completion of contracts		
5. Accounts receivable from sources other than construction contracts		
6. Deposits for bids or other guarantees: (a) Recoverable within 90 days		
(b) Recoverable after 90 days		
7. Interest accrued on loans, securities, etc.		
8. Real estate: (a) Used for business purposes		
(b) Not used for business purposes		
9. Stocks and bonds: (a) Listed—present market value		
(b) Unlisted—present value		
10. Materials in stock not included in Item 4 (a) For uncompleted contracts (present val.)		
(b) Other materials (present value)		
11. Equipment, book value		
12. Furniture and fixtures, book value		
13. Other assets		
Total assets		
LIABILITIES		
1. Notes payable: (a) To banks regular		
(b) To banks for certified checks		
(c) To others for equipment obligations		
(d) To others exclusive of equipment obligations		
2. Accounts Payable: (a) Not past due		
(b) Past due		
3. Real estate encumbrances		
4. Other liabilities		
5. Reserves		
6. Capital stock paid up:		
(a) Common		
(b) Common		
(c) Preferred		
(d) Preferred		
7. Surplus (net worth) Earned $_____ Unearned $_____		
Total liabilities		
CONTINGENT LIABILITIES*		
1. Liability on notes receivable, discounted or sold		
2. Liability on accounts receivable, pledged, assigned or sold		
3. Liability as bondsman		
4. Liability as guarantor on contracts or on accounts of others		
5. Other contingent liabilities		
* For co-partnerships a separate list of contingent liabilities of individual members is required. Total contingent liabilities		

FIG. 5.4 *(continued)*

DETAILS RELATIVE TO ASSETS

1 **Cash** (a) on hand .... $........
(b) deposited in banks named below ....
(c) elsewhere—(state where) ....

NAME OF BANK	LOCATION	DEPOSIT IN NAME OF	AMOUNT

2* **Notes receivable** (a) due within 90 days .... $........
(b) due after 90 days ....
(c) past due ....

RECEIVABLE FROM: NAME AND ADDRESS	FOR WHAT	DATE OF MATURITY	HOW SECURED	AMOUNT

Have any of the above been discounted or sold?........ If so, state amount, to whom, and reason........

3* **Accounts receivable from completed contracts exclusive of claims not approved for payment** .... $........

NAME AND ADDRESS OF OWNER	NATURE OF CONTRACT	AMOUNT OF CONTRACT	AMOUNT RECEIVABLE

Have any of the above been assigned, sold, or pledged?........ If so, state amount, to whom, and reason........

4* **Sums earned on uncompleted contracts, as shown by engineer's or architect's estimate:**
(a) Amount receivable after deducting retainage .... $........
(b) Retainage to date due upon completion of contract........ ....

DESIGNATION OF CONTRACT AND NAME AND ADDRESS OF OWNER	AMOUNT OF CONTRACT	AMOUNT EARNED	AMOUNT RECEIVED	RETAINAGE		AMOUNT EXCLUSIVE OF RETAINAGE
				WHEN DUE	AMOUNT	

Have any of the above been sold, assigned, or pledged?........ If so, state amount, to whom, and reason........

* List separately each item amounting to 10 per cent or more of the total and combine the remainder.

FIG. 5.4 *(continued)*

DETAILS RELATIVE TO ASSETS (continued)

5* **Accounts receivable not from construction contracts** $

RECEIVABLE FROM: NAME AND ADDRESS	FOR WHAT	WHEN DUE	AMOUNT

What amount, if any, is past due $

6 **Deposits with bids or otherwise as guarantees** $

DEPOSITED WITH: NAME AND ADDRESS	FOR WHAT	WHEN RECOVERABLE	AMOUNT

What amount, if any, has been assigned, sold or pledged $

7 **Interest accrued on loans, securities, etc.** $

ON WHAT ACCRUED	TO BE PAID WHEN	AMOUNT

What amount of the above, if any, has been assigned or pledged $

8 **Real Estate** / **Book value** { (a) Used for business purposes $
(b) Not used for business purposes

DESCRIPTION OF PROPERTY	IMPROVEMENTS		TOTAL BOOK VALUE
	NATURE OF IMPROVEMENTS	BOOK VALUE	
1.			
2.			
3.			
4.			
5.			
6.			
7.			

LOCATION	HELD IN WHOSE NAME	ASSESSED VALUE	AMOUNT OF ENCUMBRANCES
1.			
2.			
3.			
4.			
5.			
6.			
7.			

* List separately each item amounting to 10 per cent or more of the total and combine the remainder.

FIG. 5.4 *(continued)*

DETAILS RELATIVE TO ASSETS (continued)

9 **Stocks and Bonds**
(a) Listed—present market value .... $........
(b) Unlisted—present value ....

	DESCRIPTION	ISSUING COMPANY	LAST INT. OR DIV. PAID		PAR VALUE	PRESENT MARKET VALUE	QUAN-TITY	AMOUNT
			DATE	%				
1.								
2.								
3.								
4.								
5.								
6.								
7.								

	WHO HAS POSSESSION	IF ANY ARE PLEDGED OR IN ESCROW, STATE FOR WHOM AND REASON	AMOUNT PLEDGED OR IN ESCROW
1.			
2.			
3.			
4.			
5.			
6.			
7.			

10 **Materials in stock and not included in Item 4, Assets:**
(a) For use on uncompleted contracts (present value) .... $........
(b) Other materials (present value) ....

DESCRIPTION OF MATERIAL	QUANTITY	PRESENT VALUE	
		FOR UNCOMPLETED CONTRACTS	OTHER MATERIALS

11* **Equipment at book value** .... $........

QUAN-TITY	DESCRIPTION AND CAPACITY OF ITEMS	AGE OF ITEMS	PURCHASE PRICE	DEPRECIATION CHARGED OFF	BOOK VALUE

Are there any liens against the above? .... If so, state total amount .... $........

* If two or more items are lumped above, give the sum of their ages.

FIG. 5.4 *(continued)*

DETAILS RELATIVE TO ASSETS (continued)

12 **Furniture and fixtures at book value** $

13 **Other assets** $

DESCRIPTION	AMOUNT

TOTAL ASSETS $

DETAILS RELATIVE TO LIABILITIES

1 **Notes payable**
(a) To banks, regular $
(b) To banks for certified checks
(c) To others for equipment obligations
(d) To others exclusive of equipment obligations

TO WHOM: NAME AND ADDRESS	WHAT SECURITY	WHEN DUE	AMOUNT

2 **Accounts payable**
(a) Not past due $
(b) Past due

TO WHOM: NAME AND ADDRESS	FOR WHAT	DATE PAYABLE	AMOUNT

3 **Real estate encumbrances** (See Item 8, Assets) $

4 **Other liabilities** $

DESCRIPTION	AMOUNT

5 **Reserves** $

INTEREST	INSURANCE	BLDGS. AND FIXT.	PLANT DEPR.	TAXES	BAD DEBTS		
$	$	$	$	$	$	$	$

6 **Capital stock paid up**
(a) Common $
(b) Preferred

7 **Surplus: $** $

TOTAL LIABILITIES $

FIG. 5.4 *(continued)*

Condensed Operating Statement

For period beginning .... 19...... and ending .... 19......

Income

Gross receipts from contracts	
Value of work performed on contracts but not yet paid for	
Return from investments	
From other sources (specify)	
Total income	$

Expense

Costs of construction, labor, material, etc., not listed below	
Salary of officials, partners or proprietor	
Depreciation: Plant $.......... Buildings $..........	
Interest paid	
Federal taxes paid during fiscal period	
Reserved for interest, taxes, etc., excepting depreciation	
Travel, estimating, rent and other expense of doing business	
Other expenditures (specify)	
Total Expense	
Net Profit or Loss	$

Reconcilement or Surplus

Undivided surplus (net worth) at close of previous fiscal year .... $....................

Items not applicable to current year:

Add .... $....................

Add ....

Deduct ....

Deduct ....

Net addition or reduction ....

Balance ....

Net profit or loss as above ....

$....................

Less dividends or withdrawals by partners or proprietor, except as salary above ....

Undivided surplus (See fifth page, item 7, liabilities) .... $....................

Contracts on Hand (1)

	Name and Address of Owner	Name and Address of Architect or Engineer	Name of Surety
1			
2			
3			
4			
5			
6			
7			

	Character of Work	Probable Date of Completion	Amount of Contract	Amount Receivable on Contract	Amt. Disbursed on Contract	Amt. Owning on Contracts (2)	Estimate of Cost to complete work
1							
2							
3							
4							
5							
6							
7							
8	Total of all contracts not listed above						

Total for all contracts on hand .... $....................

(1) List the seven largest contracts and lump the remainder. (2) Owed by Contractor.

FIG. 5.4 *(continued)*

Insurance Carried:	Fire on work under Contract $	Plant and Equipment $	Total Insurance carried is $
	Other Ins. on work under Contract $	Pay Roll $	
	Life Ins. payable into business $	Miscellaneous $	

Are any suits or unsatisfied judgments pending against you? ________ Amount $ ________

If your books have been audited by a C. P. A., give date, name and address ________

If a corporation, state: capital paid in cash, $ ________	If a co-partnership, state: date organized ________
Where and date incorporated ________	Whether general, limited or association ________
President's name ________	Name and address of parents:
Secretary's name ________	________
Treasurer's name ________	________

The undersigned hereby declares: that the foregoing is a true statement of the financial condition of the individual, co-partnership or corporation herein first named, as of the date herein first given; that this statement is for the express purpose of securing a loan from the party to whom it is submitted; and that any depository, vendor or other agency herein named is hereby authorized to supply such party with any information necessary to verify this statement.

Affidavit for Individual

STATE OF ________ }
COUNTY OF ________ } ss.:

________ being duly sworn, deposes and says that the foregoing financial statement, taken from his books, is a true and accurate statement of his financial condition as of the date thereof and that the answers to the foregoing interrogatories are true.

Sworn to before me this
________ day of ________ 19

____
(Applicant must also sign here)

____
Notary Public

Affidavit for Co-partnership

STATE OF ________ }
COUNTY OF ________ } ss.:

________ being duly sworn, deposes and says that he is a member of the firm of ________; that he is familiar with the books of the said firm showing its financial condition; that the foregoing financial statement, taken from the books of the said firm, is a true and accurate statement of the financial condition of the said firm as of the date thereof and that the answers to the foregoing interrogatories are true.

Sworn to before me this
________ day of ________ 19

____
(Members of firm must also sign here)

____
Notary Public

Affidavit for Corporation

STATE OF ________ }
COUNTY OF ________ } ss.:

________ being duly sworn, deposes and says that he is ________ of the ________ the corporation described in and which executed the foregoing statement; that he is familiar with the books of the said corporation showing its financial condition; that the foregoing financial statement, taken from the books of the said corporation, is a true and accurate statement of the financial condition of said corporation as of the date thereof and that the answers to the foregoing interrogatories are true.

Sworn to before me this
________ day of ________ 19

____
(Officer must also sign here)

____
Notary Public

FIG. 5.5 Affirmative action plan

AFFIRMATIVE ACTION PLAN INSTRUCTIONS
(Example)

These instructions and plan example are provided to assist the Contractor in conforming to the Statutes and the General Condition Article entitled Equal Employment Opportunities, regarding affirmative action plans.

1. The main objective of an affirmative action plan is a balanced work force which is achieved when a company (a) fully promotes equal employment opportunities for women, minorities and persons with disabilities, (b) identifies possible causes or effects of discrimination to eliminate them in its employment practices, (c) makes every effort to hire and train female, minority and disabled candidates through apprenticeship programs whenever possible and (d) uses the services of minority and small businesses as subcontractors through positive and continuing programs.

2. The affirmative action plan shall include, but is not limited to, the following parts: (a) the Contractor's equal employment opportunity policy in all personnel actions, (b) formal internal and external dissemination of the Contractor's policy, (c) appointment of a company official as responsible for the implementation of the affirmative action plan, (d) analysis of the Contractor's workforce by job classifications and organizational units, (e) objectives and time frames to further balance representation of women, minorities and disabled persons in the structural organization of the company, and (f) an internal system to monitor and evaluate regularly the affirmative action plan.

3. The estimated percentage of minority, women and disabled employees available in the labor market, and other technical assistance regarding affirmative action plan is available from the Department of Administration, Equal Employment Contract Compliance Office.

AFFIRMATIVE ACTION PLAN
(Example)

(The following is a sample of what an affirmative action plan should contain and will serve as a checklist both for preparation of the plan and for review of a plan.)

1. Consistent with the requirements set forth by Section______of the statutes, ____________________ states its policy
(Name of Company)
regarding equal employment opportunities to ensure that all recruitment and placement of employees shall be done regardless of age, race, religion, color, handicap, sex, physical condition, developmental disability, or national origin, and that all employees shall be treated equally with respect to compensation, training, layoff and recall as well as opportunities for advancement including upgrading, promotion and transfer.
In furtherance of this policy, this company will take affirmative actions to achieve the following objectives: to reach a balance of workforce which will reflect the representation of women, minorities and persons with disabilities in the labor market; to increase the subcontracting opportunities of companies owned by women, minorities and disabled individuals; and to encourage similar efforts from those companies with which we do business on state contracts.

2. The company's chief administrators are responsible for the implementation and enforcement of this plan. They have appointed __________________________________, __________________________________,
(Appointee's Name) (Appointee's Title)
Equal Employment Opportunity Officer to be accountable for implementing the company's affirmative action plan, for maintaining audit and report systems to measure the plan's effectiveness, for identifying problems, objectives and time frames, and for facilitating compliance reviews and the submission of reports and other pertinent documents when so requested by the Department of Administration.

3. Copies of this Affirmative Action Plan are available to employees and job applicants upon request and notices of the nondiscrimination provisions are posted in conspicuous places. In addition, the company's policy is to notify and advise subcontractors, employment agencies, schools, community groups and labor organizations of its equal employment and affirmative action commitment and to request their assistance in the implementation of its affirmative action plan.

FIG. 5.5 *(continued)*

4. The company will make good faith efforts to undertake the following actions.

OBJECTIVE TIME FRAME

A. Recruitment* (see example below)
B. Hiring* (see example below)
C. Promotion
D. Training
E. Apprenticeship
F. Other Employment Practices

*EXAMPLE

1.	Recruitment	To review recruitment process, application forms, interview process to delete potential discriminatory practices.	Within 4 months of contract.
2.	Hiring	To hire one minority and one disabled person to assist administration.	Within 6 months.

5. We expect that during our contract with ________________, our company's good faith efforts to implement the affirmative action plan will result in women, minorities and disabled individuals being employed a proportionate number of work hours in comparison with the work hours accumulated by the Contractors' aggregate work force, as is represented by the percentage of such persons present in the relevant county labor market. The estimated percentages in the labor market are:

Women _____% Minorities _____% Persons with Disabilities _____%

6. The company agrees to assert leadership within the community and to put forth the maximum effort to achieve employment and utilization of the capabilities and productivity of all our citizens.

We submit this plan to assure compliance with the nondiscrimination and affirmative action provisions.

The undersigned on behalf of the company and with its authorization acknowledges to have read and reviewed this affirmative action plan and upon approval of the Department of Administration agrees to be bound by it.

____
(Printed Name of Executive Officer)

____
(Signature of Executive Officer) (Date)

____
(Printed Name of EEO)

____
(Signature of Equal Opportunity Officer) (Date)

PROCEDURAL CONSIDERATIONS OF BIDDING

6

FIGURES

6.1 COMPETITIVE BIDDING

The competitive bidding process has long been one of the major strengths of the American construction industry. Most heavy highway contracts as well as the majority of building construction works are awarded as a result of competitive bids. The purpose of competitive bids is to deliver to the owner an ever-improving product at the lowest possible cost. It is often difficult to remember this goal during times of economic inflation; the goal, however, remains. Competition encourages construction firms to increase their efficiency so that they may be able to bid at a lower price and, therefore, secure a greater amount of work with a corresponding increase in profits.

From time to time, various owners or contract awarding agencies have attempted to find viable substitutes for this process, generally without success. Some European companies take competitive bids and then award the contract to the bidder whose bid is closest to the average of all bids submitted. Others award to the second or third bidder in an attempt to secure the price that is most beneficial to the owner. Although there has been a decrease in the percentage of building construction projects put out for competitive bids and a corresponding increase in negotiated contracts, competitive bidding still remains one of the strong, identifying characteristics of the American construction industry.

6.1.1 BIDDING TIME PERIOD

The amount of time established for the bidding process to take place for a given project is of vital interest to all parties concerned. Quite naturally, the owner is interested in securing the completed project at the earliest possible time. For this reason, the design professional may be pressured to set the shortest possible bidding period. It should be remembered, however, that the general contractor and the many subcontractors may be preparing quotations on a number of projects at the same time. In fact, it would be rare for a firm to be working on the estimate for only one job at any given time. Therefore sufficient time should be allowed for a thorough estimate to be prepared.

Even for smaller projects, two weeks is usually considered to be the minimum amount of bidding time necessary. This time should be increased in accordance with the size and complexity of the project being bid. Some extremely large projects may have a bidding period as long as three months or more.

Failure to allow a sufficient amount of time for the bidding process will usually result in a higher group of bids. When faced with a time crunch, most bidders have a tendency to increase their price to cover any items they may have missed because of the lack of time. Some firms may decide not to prepare a bid on a project if they feel the amount of time allowed is inadequate. The resulting decrease in the number of proposals submitted may act to the owner's detriment. Other firms may increase their price upon learning of the decrease in the number of competitors. At the least, the owner will have reduced chances of securing the optimum quotation. Recommended estimating times for various types and sizes of projects are shown in Fig. 6.1.

6.1.2 LOW BIDDER

Under the competitive bid process for public funded projects, the contract award is made to the "lowest responsible bidder." On privately funded projects, the owner has the legal right to award to any bidder, whether that bidder is low or not. However, such practice can cause ill feelings toward an owner and create problems in the bidding for future projects. Most of the discussion that follows pertains to the competitive bid process on public works.

FIG. 6.1 Recommended estimating times in calendar days

SIZE OF PROJECT	TYPE OF PROJECT		
	Simple	*Average*	*Complex*
Less than $50,000	14	14	15
$50,000 to $500,000	18	21	25
$500,000 to $1,000,000	24	27	32
$1,000,000 to $3,000,000	30	32	36
Minimum time for projects over $3,000,000	36	38	44

The term "responsible" is interpreted to mean that the firm is responsive to the request for bids and is also qualified to perform the work. Being responsive to the request for bids requires that the bidder submit the proposal in acceptable form and that the bid be on the work as outlined in the bidding documents. Changes in the proposal form, either additive or deductive, may result in the bid being disqualified. If the documents call for road paving of Portland cement concrete and the bidder states that the proposal is based on asphaltic concrete, the bid is not responsive to the request. A bidder may be disqualified if a deficiency exists in financial capacity, experience, reputation, etc.

The matter of disqualifying a bidder for any of these reasons is fairly serious. Owners will generally do so only after serious reflection. Since such decisons may in some instances be challenged in the courts, owners should be quite sure of their grounds for rejection before taking this step.

6.1.3 TIME LIMITATION FOR ACCEPTANCE

The standard time period in the construction industry for an owner to accept or reject a proposal is 30 days. Private owners and some governmental agencies may on occasion desire a longer acceptance period, often to arrange financing. If such is the case, the specific acceptance period should be stated in the bidding documents. During periods of economic inflation, wages and material prices may be changing with sufficient rapidity to make a bidder somewhat reluctant to guarantee an estimate for more than 30 days.

If the owner desires or needs an extension of time for acceptance of the offer, agreement from the contractor should be obatained in writing. Before agreeing to a request for extension from the owner, the general contractor should check with the subcontractors and material suppliers whose quotations have been used in the bid to make sure of their agreement to extend the time period. Failure to do so may leave the general contractor in the unfortunate position of having guaranteed a price to the owner without having a similar guarantee from the subcontractors and material suppliers. Under such circumstances, the general contractor may be forced to pay higher prices with no corresponding increase in contract price from the owner.

6.1.4 BID CHANGES

In most instances, changes can be made in a bid or the bid can be withdrawn up to the time when bids are due. Return of the bid for purposes of withdrawal or change should be accomplished only through a formal, written process. Failure to follow such procedures could conceivably make it possible for one bidder to withdraw the bid of another in order to remove competition. Court decisions have emphasized the importance of following a formal or stated procedure for withdrawal. Instances where this has not been done have resulted in the contractor being bound by the proposal.

6.1.5 WITHDRAWAL OF BIDS

Once the bids have been opened and read, the owner has the power of acceptance for the specified acceptance period. Under no circumstances should a bidder be permitted to make anything but minor changes in a bid after the other prices are known. Violation of this principle is a disservice to the other bidders and tends to weaken the competitive bidding system.

The courts have generally held that withdrawal of the bid after the time when bids are due should be permitted only for the reason of an "honest error." The following guidelines are used in the application of this principle:

The mistake is of sufficient magnitude that holding the bidder to a contract in that amount would be unconscionable. The interpretation of what constitutes "sufficient magnitude" is often a knotty problem for the court, with some decisions seeming to favor the bidder and others favoring the owner.

The mistake is not the result of negligence on the part of the bidder. Although it can be held that all mistakes are the result of negligence, the courts use the principle of culpable negligence. This principle holds that a person or firm must be guilty of a wrongful act (either criminal or civil) to be blamable for the error. If this is not the finding, the courts may conclude that an "honest error" has been made.

The owner suffers no great harm except the loss of the bargain that contained the error.

A private owner has the legal right to permit the withdrawal of bids at any time, even after they have been opened and read. This same right does not pertain to public works. Public owners should follow only formal, court-approved procedures for the withdrawal of bids since any other action may be suspected of fraud and collusion. Violations of this practice have sometimes resulted in a taxpayer's suit being filed and the court issuing an injunction restraining the awarding authority.

Permitting changes in bids after the time when bids are due has generally not been allowed in public projects. If changes were allowed, it would permit the low bidder to increase a bid to just under that of the second bidder, it would encourage unrealistic bids in order to be designated low bidder, and it would remove any real competition from the bidding process. However, a recent court decision seems to have changed even this practice. Accordingly, future cases should be followed closely to determine future trends. Although private owners are not prohibited by law from permitting bid changes after bids are received, it is not a recommended procedure. It could easily lead to charges of favoritism, with a possible reduction in the number of bidders on subsequent projects.

6.1.6 LATE BIDS

The courts have almost unanimously held that due time for bids is absolute on public projects. If the time is given as 10:00 a.m. EST, bids turned in even one minute late must be rejected. The easiest way for an owner to reject a bid so that everyone is aware that it is being done and to avoid the appearance of vacillating is to return the bid to the bidder unopened. The courts have generally stated that although a bidder may derive no benefit from a one-minute extension, this may logically lead to an extension of an hour, a day, or even a week. This would completely defeat the safeguards established to protect the public interests. Private owners would also be well advised to reject any bids submitted after the stated time in order to avoid giving the impression of unfair action.

6.1.7 REJECTION OF BIDS

Each set of bidding documents should contain provisions reserving for the owner the right to reject any and all bids and the right to waive any informalities. The first provision is to preclude the owner from having to enter into a contract for an amount that exceeds the available financing. On public works projects, many contracts are financed by a bond issue that contains a stated maximum. If all bids came in over the bond issue, the project could not proceed financially, and yet the owner might be required to enter into a contract for the work. To avoid this eventuality, the provision to reject bids is included in the documents.

An owner may also wish to reject an individual bid if there is a failure to meet qualifications, if there is an unbalanced bid for a unit price contract, or if financial incapacity, fraud, or collusion is suspected. Although the private owner is legally free to reject a bid for any of these reasons, the public owner will often be required to show justifiable cause for such action.

6.1.8 BID INFORMALITIES

An informality in the bid is created whenever a bidder fails to comply with one of the many requirements pertaining to the submission of a proposal. Examples of informalities would be failure of the bidder to sign the bid bond, failure of the date on the proposal to agree with the bidding due date, failure to initial a correction in the bid, or any one of a host of other possibilities. Any one informality may be the basis for a bid being termed "not responsive" because all requirements have not been met. Each bidder should develop a checklist to guard against such an occurrence.

It is quite often to the owner's advantage to be able to waive informalities in a bid. While the private owner has great latitude of action in this regard, the public owner needs authorization within the bidding documents for such action. To create this option, the bidding documents should always contain a provision permitting the owner to waive informalities in the bids.

6.2 SUBCONTRACTOR RELATIONS

The task of putting a bid together requires the effort of several individuals from a number of different firms. In the past, general contracting firms employed craft workers in sufficient numbers and types to enable the firm to complete a project with its own forces. Under those circumstances the individual firm estimated all branches of the work and was solely responsible for the preparation of the bid. The years following the end of World War II saw a change in this practice when a large number of specialty contracting firms came into existence. Today general contractors tend to operate more as brokers, performing only a portion of the work with their own forces and awarding subcontracts for the balance of a project. In a building construction project, the general contractor may perform only 10 to 20 percent of the work, with this percentage being considerably higher for the general contractor on a civil project.

6.2.1 CONTRIBUTION TO LOW BID

In light of the above situation, it would not be accurate for a general contractor to say, "I was low bidder on the job." It would be more accurate to say that the low bid was a result of the general contractor's efforts along with those of a number of subcontractor and material supply firms. While it is mandatory for the subcontractor and material supply firms to maintain the good will of the general contractor, it is also advisable for the general contractor to maintain excellent working relations with them since they may be primarily responsible for the firm being able to secure the low bid.

6.2.2 DOMINANCE OF GENERAL CONTRACTOR

The two types of firms (general and specialty contractors) do not operate from positions of equal power. It is a fact of life within the industry that the general contracting firm enjoys the position of dominance in the relationship. This has been confirmed by many court decisions, which have held that although the subcontractor can be held to a bid if the general has relied upon it, the mere fact that the general has used the subs quotation in the preparation of the bid does not entitle the sub to a contract.

Even though the general contractor is not legally bound to award the subcontract to the firm that is low bidder for a particular branch of the work, ethical considerations and good business practice indicate that this should be done. Much time and effort on the part of many people within the industry is constantly being devoted to creating a system that will improve the relationship between the parties involved. While these efforts are commendable, they will probably never substitute honest contractors dealing with one another in a fair and equitable manner.

6.3 QUOTATION REQUESTS

One of the first actions that a general contractor takes after the decision to bid has been made is to send out quotation requests to various subcontractor and material supply firms. (See Fig. 6.2.) These requests for quotations are usually presented in written form, but if the bidding period is short, they may be made by telephone. Regardless of what form is used, the request should contain the following information:

Name of the project for which quotations are requested.

Branch of the work on which that particular firm is being asked to quote. References to specific sections of the specifications are often helpful in defining the branch of the work.

The place where plans and specifications and other bidding documents may be examined. This would include a list of plan rooms that have the documents and whether or not they are available in the offices of the general contractor and design professional.

How quotations are to be submitted. Some firms may elect to consider only written quotations while others may accept either written or telephoned prices.

Any unusual requirements in the particular branch of the work that the general contractor is already aware of.

Date by which quotations are desired. This will not be the same as the bidding due date for the general contractor since time must be allowed for analysis of subcontractor and material supplier quotations and their assimilation into the final bid.

Some firms may use the quotation request form to make a statement regarding their bidding philosophy. It is useless to make such a statement unless the firm intends to adhere to the principles stated.

FIG. 6.2 Quotation request

Invitation to Bid

To:..

....

....

Date:..............................

Gentlemen:

We propose to submit a bid for the construction of..............................

....
(*Name and location of project*)

Name of Owner..............................

Name of Architect or Engineer..............................

Date and hour for submission of our bid..............................

You are invited to submit a bid for the following work:..............................

....

....

Plans and specifications covering the work you are invited to bid on are available at..............................

....

You are advised that we subscribe to the Code of Ethical Conduct of The Associated General Contractors of America, Inc., appropriate parts of which are reprinted on the back of this page, and we intend to conform to the letter and spirit contained therein in the handling of your proposal, provided:

1. That your bid is in our hands in time for us to tabulate and evaluate it before making up our own general bid, in accordance with date set forth below.
2. That your bid is in accordance with the plans and specifications, and all items or alternates are bid as required.
3. That your bid truly represents a reasonable and fair value of the work to be performed.
4. That if we are awarded the work, and your bid is accepted by us you will enter into a contract for the performance of the work covered at the price quoted and that if requested you agree to furnish a bond, at our expense.

We request that you have your bid in our office not later than..............................

that it be in writing, that it clearly indicates the work covered and in case of alternates, each to be listed together with prices therefor.

Signed.............................. *Company*..............................

Title.............................. *Address*..............................

(Over)

Reproduced by permission of The Associated General Contractors of America.

FIG. 6.2 *(continued)*

SECTION 3, RULES OF ETHICAL PRACTICE OF THE ASSOCIATED GENERAL CONTRACTORS OF AMERICA, INC.

"The operations of the contractor are made possible through the functioning of those agencies which furnish him with service or products, and in contracting with them he is rightfully obligated by the same principles of honor and fair dealing that he desires should govern the actions toward himself of architects, engineers and client owners.

"Ethical conduct with respect to sub-contractors and those who supply materials requires that:

1. Proposals should not be invited from anyone who is known to be unqualified to perform the proposed work or to render the proper service.
2. The figures of one competitor shall not be made known to another before the award of the subcontract, nor should they be used by the contractor to secure a lower proposal from another bidder.
3. The contract should preferably be awarded to the lowest bidder if he is qualified to perform the contract, but if the award is made to another bidder, it should be at the amount of the latter's bid.
4. In no case should the low bidder be led to believe that a lower bid than his has been received.
5. When the contractor has been paid by a client owner for work or material, he should make payment promptly, and in just proportion, to subcontractors and others."

Requests should only be sent to those firms of such quality and reputation that the general contractor will be willing to award them the contract if they are the low bidder. Unsolicited quotations pose a problem. Upon receiving such a quotation, the general contractor should make every effort to gain information about the firm and its reputation. If it appears not to be a firm to which a contract would be given, the quotation should be rejected and returned. Such action will help to establish the general contractor's reputation for fair and honest dealings.

6.3.1 ANALYSIS OF QUOTATIONS

Each subcontractor and material supplier quotation must be thoroughly reviewed and analyzed to determine how it may be incorporated into the final bid. Among the items that should be checked are:

Is the quotation from an acceptable firm?

Does the quotation include all items within the branch of the work for which the quotation was requested, or does it overlap other quotations? This is one of the most troublesome items to check. Failure to do so, however, may result in a serious error in the final bid.

Does the quotation include sales tax (if applicable), and does it include delivery, storage, and protection of material?

Does the material being proposed under the quotation meet the specification requirements? This would include review of approved brand names or an evaluation of performance specifications.

Can the quoting firm supply the material and/or labor within a time limit that is acceptable under the contract and the general contractor's schedule?

Are there any exclusions or limitations to the time of acceptance that may create a hardship for the general contractor?

Has an obvious error been made in the price quoted? The general contractor gains no advantage by accepting and using a price that is ridiculously low. In the first place, the firm would probably not accept the contract; in the second place, if the contract were accepted, the quality of work furnished might be lower than acceptable as the firm tried to minimize its losses. It is often difficult to determine immediately whether a firm has made a mistake in the bid or is just bidding very tightly. Only through experience and judgment can one arrive at an answer.

Is the bidding firm union or open shop, and how will this coordinate with the rest of the project? If the general contractor is signatory to collective bargaining agreements, an award to an open shop subcontractor may be asking for trouble.

Does the price compare favorably with other quotations for that branch of the work?

It may be helpful to compare the above items with the example of a subcontractor's proposal form shown in Fig. 6.3.

FIG. 6.3 Subbid proposal form

STANDARD SUBBID PROPOSAL

(Developed as a guide by The Associated General Contractors of America, The National Electrical Contractors Association, The Mechanical Contractors Association of America. The Sheet Metal and Air Conditioning Contractors National Association and The National Association of Plumbing-Heating-Cooling Contractors.)

SUBCONTRACTOR__________ Project __________

Address __________ Location __________

Location __________

GENERAL CONTRACTOR __________ A&E __________

Bid Time & Date __________

Address __________ Subbid Time & Date* __________

Type of work (including specification sections)__________

(List the category(ies) this proposal will cover, such as plumbing, heating, air conditioning and ventilation, electrical and elevators.)

This proposal includes furnishing all materials and performing all work in the category(ies) listed above, as required by the plans, specifications, general and special conditions and addenda__________,
(Here list addenda by numbers)

Identify work to be **excluded** by specification paragraph otherwise the subcontractor will be responsible for all work in the above category(ies) required by the specifications and plans.

If this proposal, including prices, is accepted, the subcontractor agrees to enter into a subcontract and, if required, furnish performance and payment bonds from__________
(Name of surety company or agency)

guaranteeing full performance of the work and payment of all costs incident thereto, and the cost of the bond is **not** included in this proposal.

This proposal will remain in effect and will not be withdrawn by the subcontractor for a period of 30 days or for the same period of time required by the contract documents for the general contractor in regard to the prime bid, plus 15 days, whichever period is longer.

____
Subcontractor

____
By (Title)

BASE BID__________

Alternates

	Add	Deduct
1.	$	$
2.	$	$
3.	$	$
4.	$	$
5.	$	$
6.	$	$
7.	$	$
8.	$	$

Reproduced by permission of The Associated General Contractors of America

FIG. 6.3 *(continued)*

UNIT PRICES

(Insert Unit Prices if Requested)

	Unit	Add	Deduct
1.		$	$
2.		$	$
3.		$	$
4.		$	$
5.		$	$
6.		$	$
7.		$	$
8.		$	$

(Any Additional Information)

*General Services Administration procurement regulations and the CMSCI and the AGC urge that this proposal be submitted to the general contractor at least 48 hours before the opening of the prime bids. As contained in Title 41—Public Contracts & Property Management, Chapter 5B—Public Buildings Service, General Services Administration—Subpart 5B-2.2, paragraph (j), these regulations specifically state: "In order to effectively implement the objectives of the foregoing provisions and to assure the timely receipt of accurate bids, the bidder is requested to urge all subcontractors intending to submit a proposal for work involved in the project to submit to all bidders to whom they intend to bid, a written proposal (or written abstract) with or without price, **outlining in detail the specific sections of the specifications to be included in their work as well as any exceptions or exclusions therefrom.** It is suggested that such written proposal be submitted to the bidder at least 48 hours in advance of the bid opening."

6.3.2 BID SHOPPING AND BID PEDDLING

Two practices that are the cause of a lot of ill will within the construction industry are bid peddling and bid shopping. They are the same except for the time when each occurs. Bid peddling takes place during the bidding phase, and bid shopping after the general contract has been awarded. The practice of bid peddling involves the disclosure of the quotations of other bidders within a certain branch to a favored subcontractor or material supplier in the same branch of work. The purpose of this practice is for the general contractor to secure a lower price from a preferred firm, thus enabling the general contractor to improve the chances of being the low bidder. In actuality, however, the subcontractors and material suppliers soon learn of this practice and respond by inflating their original quotation or refusing to bid with the firm altogether.

In practicing bid shopping, general contractors take the low bid in a branch of the work (which was used in the preparation of the bid) and go "shopping" for lower prices. By this method, they hope to increase their profit on the project. Again, once word of this gets around the industry, the subcontractors and material suppliers either inflate their original bids or cease bidding with such firms.

The net result of both of these practices is to reduce the quality of the owner's project. When subcontractors or material suppliers are forced to reduce their quoted price in order to secure the contract, it is highly unlikely that they will reduce their profit. Rather, the tendency is to cheapen either the quality of material and/or workmanship and to try to maintain the same level of anticipated profit. As previously stated, if the general contractor expects to receive the contract as a result of being the low bidder, the same expectation on the part of the subcontractors and material suppliers should be fulfilled.

6.3.3 LISTING OF SUBCONTRACTORS

One attempt on the part of the owners and design professionals to reduce the practice of bid peddling and bid shopping has been to require the listing of subcontractors and major material suppliers on the proposal form. This has been strongly resisted by many general contractors in the industry. They argue that there is not sufficient time to permit a thorough analysis of the quotations prior to the bid submission. This argument is partly supported by the practice of most subcontractors, who wait until the last minute to turn in their bids. However, the subcontractors do this primarily to guard against their bid being peddled by the general contractor.

If listing is required, provisions are made on the proposal form for the name and amount of the bid for various subcontractors and material suppliers. (See Fig. 6.4.) Indications are that it is at least partially effective in reducing bid peddling.

6.3.4 BID DEPOSITORIES

Another attempt to reduce bid shopping and bid peddling has been the introduction of bid depositories. When a bid depository is used, all subcontractors turn in their bids to a central location. Copies are included for each general contracting firm with which the sub wishes to bid, along with a copy to be retained by the depository. Bids must be submitted by an established deadline before the general contractors' bids are due. This time allowance varies, the minimum usually being set at four hours. The depository may be a local bank, a plan room, or some other independent organization so designated. Following the

FIG. 6.4 Subcontractor listing

LIST OF SUBCONTRACTORS

Project (title, number):

Architect/Engineer:

Contractor:

Date:

The following is a list of the major subcontractors (and material suppliers) whose quotations have been used in the preparation of our proposal:

Branch	Firm Name	Price (Optional)

opening of bids, subbidders may then contact the bid depository to determine their standing within that particular branch. All firms and factions must agree to full participation and conformance with award to low bidders for such a system to be successful. While bid depositories have been put into effect in several localities, most have remained in operation only a short time. Failure can be attributed to a lack of full cooperation by both general and subcontractor firms.

DIVISION 

CONTRACT PHASE

THE AGREEMENT 7

7.1 PURPOSE OF AGREEMENT

As has been previously discussed, there are several methods by which an owner may select the construction firm to perform the proposed work. The method may be chosen because of conditions or personal preference, or it may be limited by statutory requirements. Whatever method is used, however, a contract should be entered into once selection has been made. When the contract is in written form, it is called the "Agreement." As used in connection with a construction project, the purpose of the agreement is to record in written form those items agreed to by the owner and the contractor. This document then constitutes legal evidence that a contract exists and forms the basis for it.

7.2 CONDITIONS FOR CONTRACT

There are four conditions that must be met in order for a valid contract to exist. The absence of any one of these conditions may lead to a dispute and render the document invalid. These conditions are:

The contract must have a legal nature. Parties may agree to conditions and price, but if the subject of the agreement is contrary to law, no contract can exist. For example, if an agreement between a contractor and a labor union included discriminatory hiring practices that were contrary to law, the contract would be ruled invalid.

The agreement must be between two or more parties who possess the power to make a contract. In most states minors do not have the power to make a contract. An individual who falsely represents either a private or a public body does not have the power to make a contract.

There must be provisions recognizing that which is to be received. Although in most contracts this is in the form of a monetary figure, anything of value can be used as long as the parties involved are in agreement. A contractor might agree to construct a building for an owner in return for an interest in the company. While no dollar figure is mentioned, the two parties have agreed to this arrangement for compensation and the contract is valid.

The parties to the contract must have a "meeting of the minds"; that is, there must be an offer and an acceptance of that offer. This particular contract condition is one of some complexity and often leads to litigation. Although the courts generally hold that the parties to the contract must have agreed to agree, they are also concerned with protecting the reasonable expectations of the innocent party. This may easily occur in the case of verbal contracts where one party may mislead the other. The claim may be made that a party was "only joking," but if the other party could not reasonably be expected to have known that, the "joking" party will be bound by the implied contract. In construction contracts where written agreements are used, this situation is rare. However, it can easily happen in connection with subcontracts and material orders, where much of the preliminary negotiations are conducted by telephone.

7.3 AGREEMENT INFORMATION

There are many forms of agreement that can be used. Regardless of the form, however, the agreement should contain the following information:

The date of the agreement.

The names of those who are party to the agreement. In addition to names, all parties are usually identified by their role under the contract. For a construction contract, one party would be identified as the owner and the other party as the contractor.

The scope of the work. This section may include the name, title, or number of the project, and a brief description of the work or the branches of work that are to be included. It may also refer to the drawings and specifications of the project along with the name of the preparing design professional.

References to time limitations. If a starting and/or completion time is included in the contract, it will be listed here along with any provisions for bonuses, penalties, or liquidated damages, and a statement that time is of the essence under the contract.

The contract consideration. This normally consists of the dollar amount that is to be paid the contractor for the work. The form of stating this dollar amount will depend upon the type of contract that is being used (lump sum, unit price, etc.).

Other conditions relating to payment. This may include information concerning progress payments, amount of retainage to be withheld, conditions of final payment, or any other payment conditions that are peculiar to the contract.

Inclusion of other documents by reference. Most construction contracts include by reference such documents as the drawings, specifications, general conditions, and special conditions. An enumeration of the drawings and specifications is usually given.

Signatures. Each party should sign and date the contract. Signatures should be accompanied by seals and witnessed as necessary to further validate the signatures of the parties.

7.4 COMPARISON WITH OTHER CONTRACTS

Although contracts are used in almost all types of business ventures, the construction contract differs from most other contracts in several respects. Construction contracts are usually longer and more complex than those used in other businesses or industries because the work that is to be undertaken is often more involved and requires a more lengthy description and consequently a longer document. Furthermore, the construction contract is seldom an instrument that can stand alone. Many other documents, such as the drawings and the specifications, are included by reference and thereby become an actual part of the contract. The fact that other documents are included by reference may lead to conflicts with the agreement. If conflicts exist, the provisions of the agreement will usually govern.

7.5 STANDARDIZED FORMS

Many industry and professional associations, as well as many private and public owners, have prepared and distributed for use standardized forms of construction agreements. Such forms have been issued by the Associated General Contractors of America, the American Institute of Architects, the National Society of Professional Engineers, the Corps of Engineers, the Department of Defense, and many others. In some instances the use of one of these standard forms may be recommended, while in others it may be required.

7.5.1 ADVANTAGES

The major advantage to using one of these standardized forms is that the provisions that they contain have been tested in the courts. Although a "meeting of the minds" is regarded as one of the essentials for a valid contract, when a contract dispute results in litigation, the courts will interpret what the document actually says—not what the parties intended it to say. The courts have defended this approach as the best way to protect the reasonable expectations of all parties concerned. Because of this, a contract is probably the worst place to be innovative and creative about the choice of words and phrases. What was intended by the preparers of the contract may not agree with the interpretation placed upon it by the court. For this reason, the standard forms of agreement try to make use primarily of clauses and phrases that have stood the test of the courts and are the least likely to be misunderstood.

It is often necessary to modify even such standard forms by either additions or deletions. In most cases such changes should be reviewed by a lawyer to avoid future legal disputes. Any non-standard form of agreement should be prepared by an attorney, or at least reviewed by one, so there is reasonable assurance that it is legally correct.

7.6 TIME CONSIDERATIONS

Time is generally considered important in a construction contract. Most contracts will contain a phrase stating that the contractor shall pursue the work in an expeditious and economical manner. However, many contracts are more explicit and will specify a period of time or a date by which the project is to be completed. As a means of emphasizing this requirement, some form of monetary incentive—such as bonuses, penalty clauses, or liquidated damages—is often provided.

7.6.1 BONUS AND PENALTY CLAUSES

A bonus clause contains provisions for the contractor to earn extra money in the event that the project is completed earlier than the normal schedule would indicate. Such a clause should only be used when there is actual benefit to the owner for early completion. The amount per unit of time saved should be clearly stated, along with any other conditions upon which the bonus is contingent.

A penalty clause is an attempt to provide the contractor with incentive, in a negative way, to complete the project on schedule. It requires the contractor to pay the owner a stated amount of money per unit of time that completion is delayed. As with the bonus clause, completion must be clearly defined. Substantial completion is the usual term rather than total completion. This is sometimes interpreted as meaning the project is sufficiently completed to provide beneficial use or occupancy to the owner. Most court decisions have disallowed penalty clause provisions unless they are accompanied by a corresponding bonus clause and are based on real and measurable losses to the owner. If the penalty clause provisions appear to be excessive, and are contained within an adhesion type of contract (where a more powerful party makes the offer on a "take it or leave it" basis), the courts will often reduce the amount of penalty to more closely approximate actual losses suffered by the owner due to the late completion.

7.6.2 LIQUIDATED DAMAGES

If time of completion is made a condition of the contract, failure to complete the project by the assigned time constitutes a breach of contract. As a result the owner may suffer losses and be entitled to damages. It is often difficult to determine precisely the amount of damages suffered, and this difficulty may lead to litigation. In an attempt to avoid this problem, the contract may include a preagreed unit-of-time amount that the contractor shall pay the owner in case of delayed completion. The amount named in the damages provision should be as realistic as the two parties can make it. The fact that the contractor has signed the agreement, and has therefore accepted the amount, is not a guarantee that the damages amount will be upheld. Many court decisions attest to the fact that an unreasonable amount of damages will be overturned by the court, even when both parties have previously agreed to that amount. Penalty clauses in construction contracts have been largely replaced by liquidated damage clauses because they are more easily enforced.

7.7 RETAINAGE

It is standard practice within the construction industry for the contractor to be paid only a portion of the amount requested on the monthly statement. The part withheld is called a "retainage" and can vary from 5 to 20 percent. The purpose of the retainage is to provide the owner with an economic lever to be exerted upon the contractor to ensure the correction and/or completion of improper or uncompleted parts of the project. General contractors do not bear the total impact of the retainage since they in turn retain a similar percentage from payments made to the subcontractors.

This practice has been the subject of considerable discussion and debate within the industry for many years. An example often cited is that of the excavation subcontractor who may finish the assigned part of the project within the first two months of a two-year project. The question then asked is if it is fair to require this subcontractor to wait an additional 22 months for the remaining portion of payment under the subcontract. Many systems have been tried to improve this situation for both the subcontractors and the general contractor. The system that seems to have been somewhat successful is that of cutting the retainage percentage in half when the project is fifty percent completed. This allows subcontractors who perform work in the early part of the contract to be paid in full at this point, with sufficient time elapsing for obvious defects to surface. Such a system still retains a portion of the economic lever that the owner feels to be essential.

8 FORMS OF AGREEMENT

FIGURES

8.1 COMPETITIVELY BID VS. NEGOTIATED

Although most publicly funded projects are required by statute to be competitively bid, the private owner must decide between negotiating a contract or taking competitive bids. When a project is submitted to the competitive bid process, there is an implied obligation to award the contract to the lowest responsible bidder. This is an absolute requirement for public projects and one that is highly advised on private projects. The private owner who fails to award to the lowest responsible bidder after taking competitive bids runs a great risk of eliminating potential bidders on future projects. Few construction firms will bid with a client a second time if they failed to secure a contract after being low bidder on a previous project. The cost of bidding is too great for it to be wasted by bidding on projects where there is low potential of securing the contract.

Many private owners prefer to negotiate a contract with a selected construction firm rather than taking competitive bids. It may be that the owner has had previous experience with a contractor on an earlier job, is well aquainted with the reputation of the firm, or is a close personal friend of one of the principals. By negotiating the contract, the owner will save a considerable amount of time. This in turn can be translated into a given sum of dollars. The owner also reduces the risk of being in a position of having to award a contract to a questionable or unknown firm, although this can be greatly reduced by allowing only invited firms to bid under the competitive system. Some owners follow the practice of taking competitive bids and then negotiating with the two or three low bidders. This is patently unfair to the construction firms, since it puts them in a "cutthroat" situation. If they turned in a reasonable bid in the first place, all they can do is cheapen the job during the negotiation stage.

8.1.1 REASONS FOR SELECTION

The decision to use a competitively bid or a negotiated contract is influenced by several factors. If the project is being financed by public funds, the competitive bid system is required by statute and there is basically no decision to be made. For the private owner, however, the option still remains. Competitive bids, especially if based upon a lump sum proposal, provide the owner with advance knowledge of what the total construction costs will be. A similar advantage can be realized under a unit price proposal if an accurate estimate of quantities is performed. The assurance of projected costs is often of great importance to an owner who is operating under a tight or restricted budget.

The competitive bid system does present some disadvantages to the owner. Unless a selected list of bidders is prepared, the bid may be won by a contractor who is not of the caliber or reputation satisfactory to the owner. In addition, the bidding process uses valuable time that might be eliminated, or at least reduced, by going to a negotiated contract.

Many negotiated contracts are of the cost plus type and may present a situation where the owner is not assured of a total construction cost before the start of the project. Under such a contract the owner must assume a greater risk than under the competitive bid contract. With a competitively bid contract the contractor assumes the greater risk since the owner's cost is guaranteed. On the other hand, under such a contract an unscrupulous contractor may try to "cut corners" in an attempt to protect or increase the anticipated profit.

Owners must make an individual decision in each case. The decision is often made after consultation with a design professional who can offer valuable advice. Such advice may also be sought from the bonding company or financial institution used by the owner.

8.1.2 DIFFERING ROLE OF CONTRACTOR

The position of the construction firm as well as its relationship to the owner will vary, depending upon whether the contract was secured as a result of negotiation or the competitive bidding system. If the contract was secured through a competitive bid, the firm may well feel that it has earned the contract. It has beaten out a certain number of competitors, and the contract is a justly deserved reward. If the contract was secured through negotiation, however, there exists some possibility that they were favored by the owner. Even though there may have been modified competition during the negotiation process, it is more likely that the contractor will feel some degree of obligation to the owner for granting the contract. Such feelings will influence the contractor's conduct during the construction stage.

Under a competitive bid contract, the contractor has entered into a straight business agreement. The owner has selected the contractor because of the merit of the low bid and will conduct inspections to see that the material and workmanship called for in the bidding documents are delivered. The contractor will in turn receive the monies as stipulated in the agreement. Under such an agreement, the contractor is required to assume a certain degree of risk to ensure that the firm's potential profit can be protected. The risk of the owner is greatly reduced, since the cost and what is to be received is known ahead of time.

Under a negotiated contract, the contractor is in a more professional position, operating as a member of the "team" with the owner and the design professional. The owner assumes a greater share of the risk under such a contract because most negotiated contracts are of an "open end" type, with the total cost not being determined until the project has been completed.

8.2 LUMP SUM CONTRACTS

A lump sum contract is one in which the contractor agrees to perform the work called for in the contract documents for a given sum of money. An example is illustrated in Fig. 8.1. The owner knows in advance the approximate total cost of the project. The only additions to the contract sum would be additive change orders authorized by the owner and any contingency costs. Contingency costs are those costs resulting from unforeseen circumstances, usually changes from predicted underground conditions. The wise and well-informed owner will budget for contingency costs. A general discussion regarding information that should be included in the contract is covered in Chapter 7.

8.2.1 WHEN TO USE

There are certain conditions that would indicate that it is to the owner's advantage to use a lump sum contract. Included among these would be the following:

> The project is fairly standard in nature. That is, it does not involve new or innovative construction methods or materials, and it is of a type that has been constructed by several firms in the past.

(text continues on page 139)

FIG. 8.1 Lump sum contract

THE AMERICAN INSTITUTE OF ARCHITECTS

AIA Document A101

Standard Form of Agreement Between Owner and Contractor

where the basis of payment is a

STIPULATED SUM

1977 EDITION

THIS DOCUMENT HAS IMPORTANT LEGAL CONSEQUENCES; CONSULTATION WITH AN ATTORNEY IS ENCOURAGED WITH RESPECT TO ITS COMPLETION OR MODIFICATION

Use only with the 1976 Edition of AIA Document A201, General Conditions of the Contract for Construction.

This document has been approved and endorsed by The Associated General Contractors of America.

AGREEMENT

made as of the day of in the year of Nineteen Hundred and

BETWEEN the Owner:

and the Contractor:

The Project:

The Architect:

The Owner and the Contractor agree as set forth below.

Copyright 1915, 1918, 1925, 1937, 1951, 1958, 1961, 1963, 1967, 1974, © 1977 by the American Institute of Architects, 1735 New York Avenue, N.W., Washington, D. C. 20006. Reproduction of the material herein or substantial quotation of its provisions without permission of the AIA violates the copyright laws of the United States and will be subject to legal prosecution.

This document has been reproduced with the permission of the American Institute of Architects under application number 79082. Further reproduction, in part or in whole, is not authorized. Because AIA documents are revised from time to time, users should ascertain from AIA the current edition of the document reproduced above.

FIG. 8.1 *(continued)*

ARTICLE 1

THE CONTRACT DOCUMENTS

The Contract Documents consist of this Agreement, the Conditions of the Contract (General, Supplementary and other Conditions), the Drawings, the Specifications, all Addenda issued prior to and all Modifications issued after execution of this Agreement. These form the Contract, and all are as fully a part of the Contract as if attached to this Agreement or repeated herein. An enumeration of the Contract Documents appears in Article 7.

ARTICLE 2

THE WORK

The Contractor shall perform all the Work required by the Contract Documents for
(Here insert the caption descriptive of the Work as used on other Contract Documents.)

ARTICLE 3

TIME OF COMMENCEMENT AND SUBSTANTIAL COMPLETION

The Work to be performed under this Contract shall be commenced

and, subject to authorized adjustments, Substantial Completion shall be achieved not later than

(Here insert any special provisions for liquidated damages relating to failure to complete on time.)

FIG. 8.1 *(continued)*

ARTICLE 4
CONTRACT SUM

The Owner shall pay the Contractor in current funds for the performance of the Work, subject to additions and deductions by Change Order as provided in the Contract Documents, the Contract Sum of

The Contract Sum is determined as follows:
(State here the base bid or other lump sum amount, accepted alternates, and unit prices, as applicable.)

ARTICLE 5
PROGRESS PAYMENTS

Based upon Applications for Payment submitted to the Architect by the Contractor and Certificates for Payment issued by the Architect, the Owner shall make progress payments on account of the Contract Sum to the Contractor as provided in the Contract Documents for the period ending the day of the month as follows:

Not later than days following the end of the period covered by the Application for Payment percent (%) of the portion of the Contract Sum properly allocable to labor, materials and equipment incorporated in the Work and percent (%) of the portion of the Contract Sum properly allocable to materials and equipment suitably stored at the site or at some other location agreed upon in writing, for the period covered by the Application for Payment, less the aggregate of previous payments made by the Owner; and upon Substantial Completion of the entire Work, a sum sufficient to increase the total payments to percent (%) of the Contract Sum, less such amounts as the Architect shall determine for all incomplete Work and unsettled claims as provided in the Contract Documents.

(If not covered elsewhere in the Contract Documents, here insert any provision for limiting or reducing the amount retained after the Work reaches a certain stage of completion.)

Payments due and unpaid under the Contract Documents shall bear interest from the date payment is due at the rate entered below, or in the absence thereof, at the legal rate prevailing at the place of the Project.
(Here insert any rate of interest agreed upon.)

(Usury laws and requirements under the Federal Truth in Lending Act, similar state and local consumer credit laws and other regulations at the Owner's and Contractor's principal places of business, the location of the Project and elsewhere may affect the validity of this provision. Specific legal advice should be obtained with respect to deletion, modification, or other requirements such as written disclosures or waivers.)

FIG. 8.1 *(continued)*

ARTICLE 6

FINAL PAYMENT

Final payment, constituting the entire unpaid balance of the Contract Sum, shall be paid by the Owner to the Contractor when the Work has been completed, the Contract fully performed, and a final Certificate for Payment has been issued by the Architect.

ARTICLE 7

MISCELLANEOUS PROVISIONS

7.1 Terms used in this Agreement which are defined in the Conditions of the Contract shall have the meanings designated in those Conditions.

7.2 The Contract Documents, which constitute the entire agreement between the Owner and the Contractor, are listed in Article 1 and, except for Modifications issued after execution of this Agreement, are enumerated as follows:
(List below the Agreement, the Conditions of the Contract (General, Supplementary, and other Conditions), the Drawings, the Specifications, and any Addenda and accepted alternates, showing page or sheet numbers in all cases and dates where applicable.)

This Agreement entered into as of the day and year first written above.

OWNER	CONTRACTOR
______________________	______________________
______________________	______________________
______________________	______________________

©1977 • THE AMERICAN INSTITUTE OF ARCHITECTS, 1735 NEW YORK AVE., N.W., WASHINGTON, D. C. 20006

FIG. 8.1 *(continued)*

INSTRUCTION SHEET *AIA DOCUMENT A101a*

FOR AIA DOCUMENT A101, STANDARD FORM OF AGREEMENT BETWEEN OWNER AND CONTRACTOR — JUNE 1977 EDITION

AIA Document A101, Standard Form of Agreement Between Owner and Contractor, is for use where the basis of payment is a stipulated sum (fixed price). The 1977 Edition has been prepared for use with the 1976 Edition of AIA Document A201, General Conditions of the Contract for Construction. It is suitable for any arrangement between the Owner and the Contractor where the cost has been set in advance either by bidding or by negotiation. Although the Owner has the advantage of advance knowledge of the cost of the Work, increased efforts to assure Contract compliance may be required, in view of the fact that the price is fixed and the Contractor has a financial interest in minimizing the cost of carrying out the Work. A more complete explanation of A101 is provided in Architect's Handbook of Professional Practice, Chapter 17: Owner-Contractor and Contractor-Subcontractor Agreements.

Below is a listing of pertinent provisions revised or added to the 1977 Edition of the Stipulated Sum Owner-Contractor Agreement Form:

Article 3 — Modified to read, "Time of Commencement and *Substantial* Completion." The General Conditions, AIA Document A201, 1976 Edition, make it clear that the Contract Time runs until the Date of Substantial Completion; the Owner should be aware that an additional period of time will be required to reach final completion.

Article 4 — Revised to include reference to the Contract Documents for determination of amounts of Change Orders. Parenthetical instruction describing basis of payment now includes *base bid* and *accepted alternates.*

Article 5 — A sentence has been added at the end of the first paragraph to stipulate a specific day of the month as the end of the period for which progress payments will be made. The Agreement requires that the Owner make progress payments not later than *an agreed-upon number of days* following the end of that period covered by the Application for Payment. (Note that the General Conditions, AIA Document A201, 1976 Edition, require in Subparagraph 9.3.1 that the Contractor apply for payment at least 10 days in advance of the date payment is due.)

The provision for interest on payments due and unpaid has been revised to provide for the entry of a specific rate of interest in accordance with the changes in the interest provision of A201, Paragraph 7.8. A parenthetical statement has been added drawing attention to Truth-in-Lending and other laws which may govern the use and form of an interest provision under certain circumstances.

Article 6 — Modified to provide that final payment is due when the Work has been completed (the reference to an agreed-upon number of days after Substantial Completion of the Work has been deleted). The Certificate of Substantial Completion will provide the time period within which the Contractor will bring the Work to final completion.

Completing the form:

(NOTE: Prospective bidders should be aware of any additional provisions which may be included in A101, such as liquidated damages, retainage, or payment for stored materials, by an appropriate notice in the Bidding Documents.)

Cover Page—The names of the Owner and the Architect should be shown in the same form as in the other Project documents; include the full legal or corporate names under which the Owner and Contractor are entering the Agreement.

Article 1 — The Contract Documents

The Contract Documents must be enumerated in detail under Article 7. If unit prices are incorporated in the Contractor's bid, the bid itself may be incorporated into the Contract; similarly, other bidding documents, bonds, etc. may be incorporated, particularly in public work.

Article 2 — The Work

The general scope of the Work should be carefully defined here, since changes by Change Order, under Paragraph 12.1 of A201, must be within the general scope of the Work contemplated by the Contract. This Article should be used to describe the portions of the Project for which the Contractor is responsible, if separate contracts are used.

Article 3 — Time of Commencement and Substantial Completion

The following items should be included as appropriate:

- Date of commencement of the Work
- Provision for notice to proceed, if any
- Date of Substantial Completion of the Work
- Provision, if any, for liquidated damages if not included in the Supplementary Conditions (see AIA Document A511)

Date of commencement of the Work should not be earlier than the date of execution of the Contract. When time of performance is to be strictly enforced, the statement of starting time should be carefully considered.

A sample provision where a notice to proceed will be used is as follows:

The Work shall commence on the date stipulated in the notice to proceed and shall be substantially completed on ______________________.

The Date of Substantial Completion of the Work may be expressed as a number of days (preferably calendar days) or as a specific date. The time requirements will ordinarily have been fulfilled when the Work is Substantially Complete, as defined in A201, Subparagraph 8.1.3, even if a few minor items may remain to be completed or corrected.

FIG. 8.1 *(continued)*

If liquidated damages are to be assessed because delayed construction will result in the Owner actually suffering loss, the amount per day should be entered in the Supplementary Conditions or the Agreement. Factors such as confidentiality will help determine the choice of location. Liquidated Damages are not a penalty to be inflicted on the Contractor, but must bear an actual and reasonably estimated relationship to the loss to the Owner if the building is not completed on time; for example, the cost per day of renting space to house students if a dormitory cannot be occupied when needed, additional financing costs, loss of profits, etc. This provision, which should be carefully reviewed, if not drafted, by the Owner's attorney, may be as follows:

The Owner will suffer financial damage if the Project is not Substantially Completed on the date set forth in the Contract Documents. The Contractor (and his Surety) shall pay to the Owner the sums hereinafter stipulated as fixed, agreed and liquidated damages for each calendar day of delay until the Work is Substantially Completed: ______________ dollars ($).

A provision for penalty and *bonus*, where such is appropriate, is suggested as follows:

The Contractor agrees to pay to the Owner a sum of ______________ dollars ($) for each calendar day beyond the established completion date that the Work remains uncompleted, in consideration of which the Owner agrees to pay the Contractor a sum of ______________ dollars ($) for each calendar day ahead of the established completion date that the Work is determined to be Substantially Completed.

Note that a liquidated damages provision may be placed in the Supplementary Conditions in order to put Subcontractors on notice of this condition.

Article 4 — Contract Sum

The following items should be included as appropriate:

- The Contract Sum
- Unit prices, cash allowances, or cash contingency allowances, if any

If not covered elsewhere in the Contract Documents in more detail, the following provision for unit prices is suggested:

The unit prices listed below shall determine the value of extra Work or changes, as applicable. They shall be considered complete including all material and equipment, labor, installation costs, overhead and profit, and shall be used uniformly for either additions or deductions.

Specific allowances for overhead and for profit on Change Orders may also be included here.

Article 5 — Progress Payments

The following items should be included as appropriate:

- Due dates for payments
- Retained percentage
- Payment for materials stored off the site

The due date for payment is often arbitrarily set. It should be a date mutually acceptable to both the Owner and the Contractor in consideration of the time required for the Contractor to prepare an Application for Payment, for the Architect to check and certify payment, and for the Owner to make payment, within the time limits set in Subparagraph 9.4.1, of A201, and in this Article of A101.

The last date upon which Work may be included in an Application should be normally not less than fourteen days prior to the payment date to allow seven days for the Architect to evaluate the Application and issue a Certificate for Payment and seven days for the Owner to make payment as provided in Article 9 of AIA Document A201. The Contractor may prefer an additional few days to allow time for preparation of his Application.

Retained percentage: It is a frequent practice to pay the Contractor 90 percent of the earned sum when payments fall due, retaining 10 percent to assure faithful performance of the Contract. These percentages may vary with circumstances and localities. AIA endorses the concept of reducing retainage as rapidly as possible consistent with the continued protection of all affected interests. See AIA Document A511, Guide for Supplementary Conditions, for a complete discussion.

A provision for reducing retainage should provide that the reduction will be made only if, in the judgment of the Architect, satisfactory progress is being made and maintained in the Work. If the Contractor has furnished a bond, he should be required to provide a Consent of Surety to Reduction In or Partial Release of Retainage (AIA Document G707A), before the retainage is reduced.

Payment for materials stored *off* the site should be provided for in a specific agreement and included in Article 7. Provisions regarding transportation to the site and insurance to protect the Owner's interests should be included.

Article 6 — Final Payment

At the time final payment is requested, the Architect should be particularly meticulous in ascertaining that all claims have been settled, in defining any claims that may still be unsettled, in obtaining from the Contractor the certification required in Article 9 of AIA Document A201 that no indebtedness against the Project remains, and in being assured that to the best of his knowledge and belief, based on the final inspection, the Contract requirements have been fulfilled.

Article 7 — Miscellaneous Provisions

An accurate, detailed enumeration of all Documents included in the Contract must be made in this Article.

Signatures — Subparagraph 1.2.1 of AIA Document A201, states that the Contract Documents shall be executed in not less than triplicate by the Owner and the Contractor. The Agreement should be executed by the parties in their capacities as individuals, partners, officers, etc., as appropriate.

The scope of the project can be well defined. The drawings and the specifications can be used to give a description of the work complete enough to allow the potential bidders to accurately determine quantities for each branch of the work. As a result of an accurate quantity survey, firm prices can be established.

It is important that the owner know a fairly exact cost before entering into a contract. Because of budgetary reasons, the owner often has little or no leeway in the cost of a project and therefore elects the lump sum form of contract.

8.2.2 FACTORS OF COST

The figure given in a lump sum contract will include all elements of construction costs. These elements are:

Labor. This includes both on-site and off-site manpower used in construction of the project. All factors related to labor costs—such as workman's compensation, unemployment compensation, social security contribution, and fringe benefits such as medical plans, vacation, and retirement contributions—would be included in this category.

Material. Any material that is incorporated into the finished project as well as temporary material that is used to accomplish the construction must be included. Additional items may include sales taxes, delivery and storage costs, waste allowances, and demurrage charges.

Equipment. Equipment is usually charged to the job on a unit of time or a unit of production basis. The successful contractor follows the principle that equipment is not paid for, it must instead pay for itself. Accordingly, equipment charges include allowances for depreciation, maintenance, operation, fuel, delivery and set-up, and several other factors such as special job conditions.

Overhead. This element of construction costs is usually subdivided into the categories of general and job overhead. All non-productive costs are considered to be overhead. Those costs that can be identified with a particular project are charged to that project as job overhead, all others are charged as general overhead. The expenses of the home office and the telephones located there would fall under general overhead, while the expenses of the field office and its telephones would be classified as job overhead. Special contract conditions may also affect this item.

Profit. Profit is perhaps the only valid reason for the existence of any business enterprise. In any venture, the investor should be entitled to a fair and reasonable return on the investment. The contractor in the construction industry can separate profit into two categories. Under the first category, the contractor should be entitled to a return equal to that which would result from investing the funds used for the project in a safe commercial investment, such as certificates of deposit. If the anticipated profit on a project is less than this, the contract should not be undertaken. After all, it would be foolish for a contractor to assume a risk for less return than could be realized from depositing the funds in a bank or savings and loan association.

Under the second category, the contractor should be entitled to a return in recognition of the risk being taken. This percentage of return varies according to the size and complexity of the proposed project. Many contractors feel that the risk rate of return should be no less than 10 percent, with proportionate increases as the risk factor increases. It is important to point out that these percentages would be figured not on the total contract amount but only on the amount of monies that the contractor has invested to construct the project.

8.2.3 COMPLETENESS OF DOCUMENTS

When a lump sum contract is contemplated, the drawings and specifications must be as complete as possible. This requirement is only reasonable if a fixed total price is expected. Underground conditions should be thoroughly explored through such means as borings and load tests. This will also reduce the possibility of additive change orders or contingency costs. Under a lump sum contract, the necessity for the drawings and specifications to be extensive and complete imposes an added time restraint upon the entire process.

8.2.4 ASSIGNMENT OF RISK

The major share of the risk is borne by the contractor under a lump sum contract. Although the owner is informed in advance what the project costs will be, the contractor will not be able to accurately determine the total costs until the project is completed. Instead, the contractor attempts to make as accurate a prediction of probable costs as is possible through the construction estimate. It is then the contractor's responsibility to construct the project in accordance with the estimate in order to protect the firm's potential profit. The project must also be constructed in accordance with the contract documents, and the design professional or the owner's inspector will observe the progress of construction to ensure that this is done. If the total cost of constructing the project exceeds the contractor's estimate, the additional cost must be paid by the contractor. The only place the monies can come from is out of the contractor's overhead and profit figures. This explains in part the relatively high rate of business failures among construction contracting firms.

8.2.5 COST CONTINGENCIES

It is virtually impossible for every conceivable cost connected with a construction project to be anticipated. Those costs covered by the contract documents must be paid by the contractor, whether or not they have been included in the estimate. It is probable, however, that some costs will be incurred that are not covered by the contract documents. As mentioned previously, these quite often involve unknown and undiscovered underground conditions. Such unforeseen costs are referred to as "contingencies" and must be paid for by the owner. The test that determines a contingency is whether the cost has been covered in the contract documents and/or whether a reasonable person could reasonably have anticipated it. If the answer to both points is negative, it is regarded as a contingency cost. Naturally, the quality of the drawings and specifications can have a marked effect upon the contingency charges for a project. Even with an excellent set of documents, however, an owner would be well advised to budget 2 to 3 percent of the contract price for contingencies.

8.3 UNIT PRICE CONTRACTS

A unit price contract is one where a bid price is given for each unit of work within each branch of the contract. For example, a price may be quoted per cubic yard of excavation, per square yard of paving, and per lineal foot of curb and gutter. Each unit price quoted must include the same five elements of construction costs (labor, equipment, material, overhead, and profit) that must be included in the lump sum contract.

Under a lump sum bid, the price quoted by a contractor is compared with the prices quoted by all other bidders. The contract is then awarded to the lowest responsible bidder. The process is a bit more complicated under a unit price bid. The owner, through the design professional, furnishes as part of the bidding documents a list of estimated quantities for each branch of the work. These quantities are used by each bidder in determining unit prices. Upon receipt of the bids, the unit price quoted is multiplied by the estimated quantity for that branch of the work. The resulting subtotals are then added together to arrive at an "equivalent lump sum" bid for each bidder. The firm having the lowest equivalent lump sum bid will be awarded the contract. An example of a unit price contract is shown in Fig. 8.2, and an example of calculations to determine the equivalent lump sum is shown in Fig. 8.3.

8.3.1 BASIS FOR PAYMENT

Payment to the contractor for work performed is determined by multiplying the number of units put in place by the unit price that had been bid by that firm. As a result of this procedure, the total cost of the project to the owner will seldom be exactly equal to the equivalent lump sum that was used to determine the successful bidder. It may be more or less than this amount, depending upon the number of units put in place. Relative accuracy in estimating the number of units in each branch of the work is of vital concern to both the owner and the contractor. Aside from the fact that the owner must be concerned because of budgetary considerations such as bond issue limits, both the owner and the contractor must be concerned to avoid being penalized financially.

8.3.2 UNBALANCED UNIT PRICE CONTRACTS

Of major concern to all parties involved is an unbalanced unit price contract. This can come about when there is a marked discrepancy between the estimated quantities for a branch of the work and the actual number of quantities needed for completion. Let us assume that the cost of transporting and setting up a particular piece of equipment needed for one branch of the work is $2,000. If the estimated quantity is 2,000 units, the contractor will allow $1.00 per unit for this expense. If there is a drastic difference between this estimated quantity and the quantity of units actually put in place, one of the parties will be penalized. Assume the actual quantity is 500. This would allow the contractor to recover only 25 percent of the set-up expenses. Now assume that the actual quantity is 5,000. This would allow the contractor to collect two and one-half times the actual amount of expenses incurred. For this reason, unit price contracts should contain provisions for price adjustments when actual quantities vary from estimated quantities by more than an agreed-upon percentage. Five percent is a figure commonly used for this situation.

(text continues on page 149)

FIG. 8.2 Unit price contract

FORM OF AGREEMENT
FOR
ENGINEERING CONSTRUCTION

THIS AGREEMENT, made on the ________________ of ____________________ 19______

by and between ____

____

party of the first part, hereinafter called the OWNER, and_______________________________

____

____

party of the second part, hereinafter called the CONTRACTOR.

WITNESSETH, that the Contractor and the Owner, for the considerations hereinafter named, agree as follows:

ARTICLE I—Scope of the Work

The Contractor hereby agrees to furnish all of the materials and all of the equipment and labor necessary, and to perform all of the work shown on the drawings and described in the specifications for the project entitled ____

____

all in accordance with the requirements and provisions of the following Documents as well as the Contract Documents as hereinafter defined in the General Conditions which are hereby made a part of this Agreement:

(*a*) Drawings prepared for same by____

numbered ____

and dated ________________________ , 19____.

(*b*) Specifications consisting of:

1. "Standard General Specifications" issued by_________________________________

____ , _______________Edition

2. "Special Conditions" as prepared by____________________________________

dated ______________________________ .

Reproduced by permission of The American Public Works Association and The Associated General Contractors of America.

FIG. 8.2 *(continued)*

3. The "General Conditions of Contract for Engineering Construction"—19____ Edition.
4. Addenda

   No.________________________ Date________________________

ARTICLE II — Time of Completion

(*a*) The work to be performed under this Contract shall be commenced within ________ calendar days after receipt of written notice to proceed. The work shall be completed within ________ (calendar) (working) days after receipt of notice to proceed.

(*b*) Failure to complete the work within the number of (calendar) (working) days stated in this Article, including extension granted thereto as determined by Section 19 of the General Conditions, shall entitle the Owner to deduct from the moneys due to the Contractor as "Liquidated Damages" an amount equal to $____________ for each (calendar) (working) day of delay in the completion of work.*

(*c*) If the Contractor completes the work earlier than the date determined in accordance with Paragraph (b), and the Engineer shall so certify in writing, the Owner shall pay the Contractor an additional amount equal to $____________ for each (calendar) (working) day by which the time of completion so determined has been reduced.*

(*d*) Failure to complete the work within the number of (calendar) (working) days stated in this Article, including extension granted thereto as determined by Section 19 of the General Conditions, shall entitle the Owner to deduct from the moneys due to the Contractor as a penalty, an amount equal to $____________ for each (calendar) (working) day of delay in the completion of work.*

ARTICLE III — Extra Work

If the Engineer orders, in writing, the performance of any work not covered by the Drawings or included in the Specifications, and for which no item in the Contract is provided, and for which no unit price or lump sum basis can be agreed upon, then such extra work shall be done on a Cost-Plus-Percentage basis of payment as follows:

(*a*) The Contractor shall be reimbursed for all costs incurred in doing the work, and shall receive an additional payment of ______ percent of all such cost to cover his indirect overhead costs, plus ______ percent of all costs, including indirect overhead, as his fee.

(*b*) Extra Work is to be performed as agreed upon by the Owner and Contractor on a reasonable basis including but not limited to the following items:

1. Wages: The actual payroll costs plus fringe benefits of all workmen such as laborers, mechanics, craftsmen and foremen.

*(Can be used, where applicable, at the discretion of the Owner.) If paragraph (d) is used then paragraph (c) should be used.

FIG. 8.2 *(continued)*

2. Material: The contractor's or subcontractor's net costs for materials and supplies, including sales and user taxes where applicable.
3. Equipment Rental: The rental charges for vehicles, construction machinery and equipment, including transportation charges if any, and charges for electrical power, fuel, lubricants, water and special services as published by the Associated Equipment Distributors.

   For equipment which the contractor regularly keeps and uses on the job site, the rental charges shall not exceed the applicable monthly rates published by the Associated Equipment Distributors. However, the rental charges for special equipment which must be moved to the job site, because of the nature of the work, shall not exceed the applicable monthly, weekly or daily rates including transportation charges.
4. Taxes and Insurance: The charges for payroll taxes; bond premiums; workmen's compensation, public liability, property damage, special hazard, social security and other insurance premiums.
5. Fee: The contractor will be paid the actual costs listed above plus an allowance of fifteen (15) percent thereof if the work is performed by the contractor's or the subcontractor's forces. This fee is considered as full compensation for the contractor's or the subcontractor's general superintendent, office expense, overhead and profit.

(*c*) The hours and rates of labor and equipment and costs of materials used each day shall be submitted to the Engineer in a satisfactory form on the succeeding day, and shall be approved by him or adjusted at once.

(*d*) Monthly payments are to be handled in the same manner as regular progress payments.

FIG. 8.2 *(continued)*

For Use When Contract Is On A "Unit Price" Basis

ARTICLE IV-B — Payment

(*a*) The Contract Sum. The Owner shall pay to the Contractor for the performance of the Contract the amounts determined for the total number of each of the following units of work completed at the unit price stated thereafter. The number of units contained in this schedule is approximate only, and the final payment shall be made for the actual number of units that are incorporated in or made necessary by the work covered by the Contract.

Item No.	Classification	Estimated No. of Units	Unit	Unit Price Bid	Total for Item
	(LIST ALL ITEMS)				

Should the number of units of completed work of any individual item of the above schedule vary by more than ______ percent from the number of units stated in such schedule of units, either the Owner or the Contractor may request a revision of the unit price for the item so affected, and both parties agree that under such conditions an equitable revision of the price shall be made.

Changes in the work made under Section 18 of the General Conditions, and not included in Article I, that cannot be classified as coming under any of the Contract units may be done at mutually agreed-upon unit prices, or on a lump sum basis, or under the provisions of Article III "Extra Work."

(*b*) Progress Payments.

The Owner shall make payments on account of the Contract as follows:

1. On not later than the fifth day of every month the Contractor shall present to the Owner or his authorized representative an invoice covering the total quantities as determined by the Engineer under each item of work that has been completed from the start of the job up to and including the last day of the preceding month, and the value of the work so completed determined in accordance with the schedule of unit prices for such items, together with such supporting evidence as may be required. This invoice shall also include an allowance for the cost of such materials and equipment required in the permanent work as have been delivered to the site but not as yet incorporated in the work. Measurements of units for payment shall be made in accordance with the Special Conditions of the Contract.*

*In addition to payment for materials delivered to site, wording should indicate that, where applicable, payment may be in order for materials in storage away from the site, for field plant and equipment, access roads, etc.—details to be spelled out in the Special Conditions.

FIG. 8.2 *(continued)*

2. On not later than the 15th day of the month, the Owner shall, after deducting previous payments made, pay to the Contractor 90 percent of the amount of the approved invoice. The 10 percent retained percentage may be held by the Owner until the value of the work completed at the end of any month equals 50 percent of the total amount of the Contract after which, if the Engineer finds that satisfactory progress is being made, he shall recommend that all of the remaining monthly payments be paid in full. Payments for work, under Subcontracts of the General Contractor, shall be subject to the above conditions applying to the general Contract after the work under a Subcontract has been 50 percent completed.
3. Final payment of all moneys due on the contract shall be made within ______ days of the completion and acceptance of the work, or in accordance with local law.
4. If the owner fails to make payment as herein provided, or as provided in Article III (*d*), in addition to those remedies available to the Contractor under Section 25 of the General Conditions, there shall be added to each such payment daily interest at the rate of 6 percent per annum commencing on the first day after said payment is due and continuing until the payment is delivered or mailed to the Contractor.

(Optional, in accordance with applicable local law.)

IN WITNESS WHEREOF the parties hereto have executed this Agreement, the day and year first written above.

______________________________ OWNER

WITNESS:

______________________________ By: ______________________________
Title

______________________________ CONTRACTOR

WITNESS:

______________________________ By: ______________________________
Title

FIG. 8.3 Determination of equivalent lump sum

FORM 650016
6-78 H 15478

TABULATION OF CONSTRUCTION AND MATERIAL BIDS
IOWA DEPARTMENT OF TRANSPORTATION

LOCATION ON I-380 AND PRIMARY ROUTE 520 IN THE CITY OF WATERLOO
MILES OVER THE CEDAR RIVER

BID ORDER NO. 4

COUNTY BLACK HAWK
TYPE OF WORK STRUCTURES
PROJECT NO. I-IG-380-7(39)315--04-07
DATE OF LETTING JAN. 16, 1980

NO.	ITEM	QUANTITY	UNIT	BIDDER "A" UNIT PRICE	BIDDER "A" AMOUNT	BIDDER "B" UNIT PRICE	BIDDER "B" AMOUNT	BIDDER "C" UNIT PRICE	BIDDER "C" AMOUNT
	GRP 1 DESIGN 577								
1	CONCRETE, STRUCTURAL	3461.300	CU. YDS	150.00	519,195.00	170.00	588,421.00	192.00	664,569.60
2	STEEL, REINFORCING	824976	LBS.	.44	362,989.44	.45	371,239.20	.43	354,739.68
3	EXCAVATION, CLASS 20	1886	CU. YDS	20.00	37,720.00	10.00	18,860.00	9.00	16,974.00
4	PILING, FURNISH STEEL BEARING HP 10 X 42	5868	LIN. FT	12.00	70,416.00	12.00	70,416.00	13.50	79,218.00
5	PILING, DRIVE STEEL BEARING HP 10 X 42	5868	LIN. FT	1.00	5,868.00	1.50	8,802.00	.50	2,934.00
6	PILING, FURNISH STEEL BEARING HP 14 X 73	19535	LIN. FT	19.00	371,165.00	22.00	429,770.00	20.00	390,700.00
7	PILING, DRIVE STEEL BEARING HP 14 X 73	19535	LIN. FT	1.00	19,535.00	2.50	48,837.50	.50	9,767.50
8	CONCRETE SLOPE PROTECTION	1044	SQ. YDS	30.00	31,320.00	20.00	20,880.00	18.00	18,792.00
9	SEALANT, CONCRETE SURFACE	43591	SQ. FT.	.50	21,795.50	.75	32,693.25	.45	19,615.95
10	PREBORED HOLES - AS PER PLAN	907	LIN. FT	4.00	3,628.00	5.00	4,535.00	6.00	5,442.00
11	EXCAVATE AND DEWATER, PIER 11	LUMP SUM			87,000.00		30,000.00		56,500.00
12	EXCAVATE AND DEWATER, PIER 5	LUMP SUM			87,000.00		80,000.00		103,000.00
13	EXCAVATE AND DEWATER, PIER 6	LUMP SUM			89,000.00		70,000.00		103,000.00
14	EXCAVATE AND DEWATER, PIER 7	LUMP SUM			89,000.00		70,000.00		103,000.00
15	EXCAVATE AND DEWATER, PIER 8	LUMP SUM			87,000.00		70,000.00		103,000.00
16	EXCAVATE AND DEWATER, PIER 9	LUMP SUM			87,000.00		70,000.00		46,500.00
17	CLEARING & GRUBBING ENGRS. EST. $1,177.98	% OF ESTIMATE		300.00%	3,533.94	500.00%	5,889.90	1000.00%	11,779.80
18	TRAINEE REIMBURSEMENT	1000	HOURS	.80	800.00	.80	800.00	.80	800.00
	GRP 1 TOTAL	. . . .		. . . . .	$1,973,965.88		$1,991,143.85		$2,090,332.53

FIG. 8.3 *(continued)*

FORM 650016
6-78 H 15478

TABULATION OF CONSTRUCTION AND MATERIAL BIDS
IOWA DEPARTMENT OF TRANSPORTATION

LOCATION ON I-380 AND PRIMARY ROUTE 520 IN THE CITY OF WATERLOO
MILES OVER THE CEDAR RIVER

BID ORDER NO. 4

COUNTY BLACK HAWK
TYPE OF WORK STRUCTURES
PROJECT NO. 1-IG-380-7(39)315--04-07
DATE OF LETTING JAN. 16, 1980

NO.	ITEM	QUANTITY	UNIT	BIDDER "D" UNIT PRICE	BIDDER "D" AMOUNT	BIDDER "E" UNIT PRICE	BIDDER "E" AMOUNT	BIDDER "F" UNIT PRICE	BIDDER "F" AMOUNT
	GRP 1 DESIGN 577								
1	CONCRETE, STRUCTURAL	3461 300	CU. YDS	280 00	969,164 00	250 00	865,325 00	195 00	674,953 50
2	STEEL, REINFORCING	824976	LBS.	40	329,990 40	40	329,990 40	40	329,990 40
3	EXCAVATION, CLASS 20	1886	CU. YDS	10 00	18,860 00	10 00	18,860 00	8 00	15,088 00
4	PILING, FURNISH STEEL BEARING HP 10 X 42	5868	LIN. FT	12 50	73,350 00	14 00	82,152 00	15 00	88,020 00
5	PILING, DRIVE STEEL BEARING HP 10 X 42	5868	LIN. FT	2 50	14,670 00	10	586 80	1 00	5,868 00
6	PILING, FURNISH STEEL BEARING HP 14 X 73	19535	LIN. FT	20 00	390,700 00	21 00	410,235 00	22 00	429,770 00
7	PILING, DRIVE STEEL BEARING HP 14 X 73	19535	LIN. FT	3 00	58,605 00	10	1,953 50	1 50	29,302 50
8	CONCRETE SLOPE PROTECTION	1044	SQ. YDS	25 00	26,100 00	22 00	22,968 00	30 00	31,320 00
9	SEALANT, CONCRETE SURFACE	43591	SQ. FT.	50	21,795 50	35	15,256 85	50	21,795 50
10	PREBORED HOLES - AS PER PLAN	907	LIN. FT	10 00	9,070 00	7 00	6,349 00	20 00	18,140 00
11	EXCAVATE AND DEWATER, PIER 11	LUMP SUM			15,000 00		38,461 50		15,000 00
12	EXCAVATE AND DEWATER, PIER 5	LUMP SUM			70,000 00		77,640 00		125,000 00
13	EXCAVATE AND DEWATER, PIER 6	LUMP SUM			70,000 00		115,368 00		125,000 00
14	EXCAVATE AND DEWATER, PIER 7	LUMP SUM			70,000 00		107,019 00		125,000 00
15	EXCAVATE AND DEWATER, PIER 8	LUMP SUM			70,000 00		98,653 50		125,000 00
16	EXCAVATE AND DEWATER, PIER 9	LUMP SUM			50,000 00		73,920 00		125,000 00
17	CLEARING & GRUBBING ENGRS. EST. $1,177.98	% OF ESTIMATE		1000.00%	11,779 80	500.00%	5,889 90	1500.00%	17,669 70
18	TRAINEE REIMBURSEMENT	1000	HOURS	80	800 00	80	800 00	80	800 00
	GRP 1 TOTAL				$2,269,884 70		$2,271,428 45		$2,302,717 60
	TIE STATUS AT END OF PROJECT								

There are some cases on record where collusion has been proved in connection with an unbalanced estimate of quantities for a unit price contract. A provision in the contract requiring adjustment of prices when the quantities vary by more than the agreed-upon percentage will discourage such unethical activities.

8.3.3 WHEN TO USE

The use of a unit price contract is advised when the total scope of the project cannot be determined in advance. Such is often the case for projects involving tunneling, dredging, street repair, piling, etc. Although unit price contracts are more commonly used in conjunction with civil works, it is not unusual for an owner to elect a unit price contract for the substructure of a building and a lump sum contract for the superstructure. It is also advisable to restrict unit price contracts to situations involving a relatively small number of branches of work that are mostly standard in nature.

8.3.4 EFFECT UPON DOCUMENTS

As indicated above, the drawings and specifications for a unit price contract will not normally be able to totally define the scope of the project. They should be complete and accurate, however, with regard to details and procedures. The unit price contract should not be used as an excuse for lower quality work on the part of the design professional. It is also desirable for the documents to make every attempt to define the scope of the project as completely as possible so as to encourage better bids. It is a fact of life within the construction industry that the reputation of an individual design firm relative to the quality of their drawings and specifications has a definite influence upon the quality of bids that the owner receives from the construction firms.

8.3.5 ASSIGNMENT OF RISK

Aside from the normal risk of making a mistake in pricing, the major risk for the contractor under a unit price contract is in encountering an unbalanced set of quantities. As previously discussed, this can be greatly reduced by parameter provisions in the contract. The greater risk under this type of contract is borne by the owner because of the difficulty in establishing ahead of contract time the total number of quantities that will be required for different parts of the work. As a result, the owner is sometimes subject to an unpleasant surprise.

8.4 COST PLUS CONTRACTS

Although lump sum and unit price are used for both competitive bid and negotiated contracts, more common under negotiated contracts is some form of cost plus a fee contract. An example of a cost plus contract is shown in Fig. 8.4. An owner may elect to use such a form of contract in order to save time on the overall project. This is made possible because the contractor can start construction before drawings and specifications have been completed. Such an arrangement also makes it possible for the contractor to make a contribution during the design phase by acting as an adviser or consultant to the design professional. In many instances this can result in considerable savings for the owner since the contractor's expertise is reflected in the design. On some occasions, the drawings and specifications have been completed before the owner negotiates the contract. Under these circumstances the owner may elect a negotiated contract because of a preference for a particular contractor and because of the time saved by not going through the bidding process.

(text continues on page 158)

FIG. 8.4 Cost plus fee contract

THE AMERICAN INSTITUTE OF ARCHITECTS

AIA Document A111

Standard Form of Agreement Between Owner and Contractor

where the basis of payment is the

COST OF THE WORK PLUS A FEE

1978 EDITION

THIS DOCUMENT HAS IMPORTANT LEGAL CONSEQUENCES; CONSULTATION WITH AN ATTORNEY IS ENCOURAGED WITH RESPECT TO ITS COMPLETION OR MODIFICATION

Use only with the 1976 Edition of AIA Document A201, General Conditions of the Contract for Construction.

This document has been approved and endorsed by The Associated General Contactors of America

AGREEMENT

made as of the day of in the year of Nineteen Hundred and

BETWEEN the Owner:

and the Contractor:

the Project:

the Architect:

The Owner and the Contractor agree as set forth below.

Copyright 1920, 1925, 1951, 1958, 1961, 1963, 1967, 1974, © 1978 by The American Institute of Architects, 1735 New York Avenue, N.W., Washington, D.C. 20006. Reproduction of the material herein or substantial quotation of its provisions without permission of the AIA violates the copyright laws of the United States and will be subject to legal prosecution.

This document has been reproduced with the permission of the American Institute of Architects under application number 79082. Further reproduction, in part or in whole, is not authorized. Because AIA documents are revised from time to time, users should ascertain from AIA the current edition of the document reproduced above.

FIG. 8.4 *(continued)*

ARTICLE 1

THE CONTRACT DOCUMENTS

1.1 The Contract Documents consist of this Agreement, the Conditions of the Contract (General, Supplementary and other Conditions), the Drawings, the Specifications, all Addenda issued prior to and all Modifications issued after execution of this Agreement. These form the Contract, and all are as fully a part of the Contract as if attached to this Agreement or repeated herein. An enumeration of the Contract Documents appears in Article 16. If anything in the Contract Documents is inconsistent with this Agreement, the Agreement shall govern.

ARTICLE 2

THE WORK

2.1 The Contractor shall perform all the Work required by the Contract Documents for

(Here insert the caption descriptive of the Work as used on other Contract Documents.)

ARTICLE 3

THE CONTRACTOR'S DUTIES AND STATUS

3.1 The Contractor accepts the relationship of trust and confidence established between him and the Owner by this Agreement. He covenants with the Owner to furnish his best skill and judgment and to cooperate with the Architect in furthering the interests of the Owner. He agrees to furnish efficient business administration and superintendence and to use his best efforts to furnish at all times an adequate supply of workmen and materials, and to perform the Work in the best way and in the most expeditious and economical manner consistent with the interests of the Owner.

ARTICLE 4

TIME OF COMMENCEMENT AND SUBSTANTIAL COMPLETION

4.1 The Work to be performed under this Contract shall be commenced and, subject to authorized adjustments, Substantial Completion shall be achieved not later than

(Here insert any special provisions for liquidated damages relating to failure to complete on time.)

© 1978 • THE AMERICAN INSTITUTE OF ARCHITECTS, 1735 NEW YORK AVE., N.W., WASHINGTON, D.C. 20006

FIG. 8.4 *(continued)*

ARTICLE 5

COST OF THE WORK AND GUARANTEED MAXIMUM COST

5.1 The Owner agrees to reimburse the Contractor for the Cost of the Work as defined in Article 8. Such reimbursement shall be in addition to the Contractor's Fee stipulated in Article 6.

5.2 The maximum cost to the Owner, including the Cost of the Work and the Contractor's Fee, is guaranteed not to exceed the sum of dollars ($); such Guaranteed Maximum Cost shall be increased or decreased for Changes in the Work as provided in Article 7.

(Here insert any provision for distribution of any savings. Delete Paragraph 5.2 if there is no Guaranteed Maximum Cost.)

ARTICLE 6

CONTRACTOR'S FEE

6.1 In consideration of the performance of the Contract, the Owner agrees to pay the Contractor in current funds as compensation for his services a Contractor's Fee as follows:

6.2 For Changes in the Work, the Contractor's Fee shall be adjusted as follows:

6.3 The Contractor shall be paid percent (%) of the proportional amount of his Fee with each progress payment, and the balance of his Fee shall be paid at the time of final payment.

© 1978 • THE AMERICAN INSTITUTE OF ARCHITECTS, 1735 NEW YORK AVE., N.W., WASHINGTON, D.C. 20006

FIG. 8.4 *(continued)*

ARTICLE 7

CHANGES IN THE WORK

7.1 The Owner may make Changes in the Work as provided in the Contract Documents. The Contractor shall be reimbursed for Changes in the Work on the basis of Cost of the Work as defined in Article 8.

7.2 The Contractor's Fee for Changes in the Work shall be as set forth in Paragraph 6.2, or in the absence of specific provisions therein, shall be adjusted by negotiation on the basis of the Fee established for the original Work.

ARTICLE 8

COSTS TO BE REIMBURSED

8.1 The term Cost of the Work shall mean costs necessarily incurred in the proper performance of the Work and paid by the Contractor. Such costs shall be at rates not higher than the standard paid in the locality of the Work except with prior consent of the Owner, and shall include the items set forth below in this Article 8.

8.1.1 Wages paid for labor in the direct employ of the Contractor in the performance of the Work under applicable collective bargaining agreements, or under a salary or wage schedule agreed upon by the Owner and Contractor, and including such welfare or other benefits, if any, as may be payable with respect thereto.

8.1.2 Salaries of Contractor's personnel when stationed at the field office, in whatever capacity employed. Personnel engaged, at shops or on the road, in expediting the production or transportation of materials or equipment, shall be considered as stationed at the field office and their salaries paid for that portion of their time spent on this Work.

8.1.3 Cost of contributions, assessments or taxes incurred during the performance of the Work for such items as unemployment compensation and social security, insofar as such cost is based on wages, salaries, or other remuneration paid to employees of the Contractor and included in the Cost of the Work under Subparagraphs 8.1.1 and 8.1.2.

8.1.4 The portion of reasonable travel and subsistence expenses of the Contractor or of his officers or employees incurred while traveling in discharge of duties connected with the Work.

8.1.5 Cost of all materials, supplies and equipment incorporated in the Work, including costs of transportation thereof.

8.1.6 Payments made by the Contractor to Subcontractors for Work performed pursuant to Subcontracts under this Agreement.

8.1.7 Cost, including transportation and maintenance, of all materials, supplies, equipment, temporary facilities and hand tools not owned by the workers, which are consumed in the performance of the Work, and cost less salvage value on such items used but not consumed which remain the property of the Contractor.

8.1.8 Rental charges of all necessary machinery and equipment, exclusive of hand tools, used at the site of the Work, whether rented from the Contractor or others, including installation, minor repairs and replacements, dismantling, removal, transportation and delivery costs thereof, at rental charges consistent with those prevailing in the area.

8.1.9 Cost of premiums for all bonds and insurance which the Contractor is required by the Contract Documents to purchase and maintain.

8.1.10 Sales, use or similar taxes related to the Work and for which the Contractor is liable imposed by any governmental authority.

8.1.11 Permit fees, royalties, damages for infringement of patents and costs of defending suits therefor, and deposits lost for causes other than the Contractor's negligence.

8.1.12 Losses and expenses, not compensated by insurance or otherwise, sustained by the Contractor in connection with the Work, provided they have resulted from causes other than the fault or neglect of the Contractor. Such losses shall include settlements made with the written consent and approval of the Owner. No such losses and expenses shall be included in the Cost of the Work for the purpose of determining the Contractor's Fee. If, however, such loss requires reconstruction and the Contractor is placed in charge thereof, he shall be paid for his services a Fee proportionate to that stated in Paragraph 6.1.

8.1.13 Minor expenses such as telegrams, long distance telephone calls, telephone service at the site, expressage, and similar petty cash items in connection with the Work.

8.1.14 Cost of removal of all debris.

© 1978 • THE AMERICAN INSTITUTE OF ARCHITECTS, 1735 NEW YORK AVE., N.W., WASHINGTON, D.C. 20006

FIG. 8.4 (*continued*)

8.1.15 Costs incurred due to an emergency affecting the safety of persons and property.

8.1.16 Other costs incurred in the performance of the Work if and to the extent approved in advance in writing by the Owner.

(Here insert modifications or limitations to any of the above Subparagraphs, such as equipment rental charges and small tool charges applicable to the Work.)

ARTICLE 9

COSTS NOT TO BE REIMBURSED

9.1 The term Cost of the Work shall not include any of the items set forth below in this Article 9.

9.1.1 Salaries or other compensation of the Contractor's personnel at the Contractor's principal office and branch offices.

9.1.2 Expenses of the Contractor's principal and branch offices other than the field office.

9.1.3 Any part of the Contractor's capital expenses, including interest on the Contractor's capital employed for the Work.

9.1.4 Except as specifically provided for in Subparagraph 8.1.8 or in modifications thereto, rental costs of machinery and equipment.

9.1.5 Overhead or general expenses of any kind, except as may be expressly included in Article 8.

9.1.6 Costs due to the negligence of the Contractor, any Subcontractor, anyone directly or indirectly employed by any of them, or for whose acts any of them may be liable, including but not limited to the correction of defective or nonconforming Work, disposal of materials and equipment wrongly supplied, or making good any damage to property.

9.1.7 The cost of any item not specifically and expressly included in the items described in Article 8.

9.1.8 Costs in excess of the Guaranteed Maximum Cost, if any, as set forth in Article 5 and adjusted pursuant to Article 7.

© 1978 • THE AMERICAN INSTITUTE OF ARCHITECTS, 1735 NEW YORK AVE., N.W., WASHINGTON, D.C. 20006

FIG. 8.4 *(continued)*

ARTICLE 10

DISCOUNTS, REBATES AND REFUNDS

10.1 All cash discounts shall accrue to the Contractor unless the Owner deposits funds with the Contractor with which to make payments, in which case the cash discounts shall accrue to the Owner. All trade discounts, rebates and refunds, and all returns from sale of surplus materials and equipment shall accrue to the Owner, and the Contractor shall make provisions so that they can be secured.

(Here insert any provisions relating to deposits by the Owner to permit the Contractor to obtain cash discounts.)

ARTICLE 11

SUBCONTRACTS AND OTHER AGREEMENTS

11.1 All portions of the Work that the Contractor's organization does not perform shall be performed under Subcontracts or by other appropriate agreement with the Contractor. The Contractor shall request bids from Subcontractors and shall deliver such bids to the Architect. The Owner will then determine, with the advice of the Contractor and subject to the reasonable objection of the Architect, which bids will be accepted.

11.2 All Subcontracts shall conform to the requirements of the Contract Documents. Subcontracts awarded on the basis of the cost of such work plus a fee shall also be subject to the provisions of this Agreement insofar as applicable.

ARTICLE 12

ACCOUNTING RECORDS

12.1 The Contractor shall check all materials, equipment and labor entering into the Work and shall keep such full and detailed accounts as may be necessary for proper financial management under this Agreement, and the system shall be satisfactory to the Owner. The Owner shall be afforded access to all the Contractor's records, books, correspondence, instructions, drawings, receipts, vouchers, memoranda and similar data relating to this Contract, and the Contractor shall preserve all such records for a period of three years, or for such longer period as may be required by law, after the final payment.

ARTICLE 13

APPLICATIONS FOR PAYMENT

13.1 The Contractor shall, at least ten days before each payment falls due, deliver to the Architect an itemized statement, notarized if required, showing in complete detail all moneys paid out or costs incurred by him on account of the Cost of the Work during the previous month for which he is to be reimbursed under Article 5 and the amount of the Contractor's Fee due as provided in Article 6, together with payrolls for all labor and such other data supporting the Contractor's right to payment for Subcontracts or materials as the Owner or the Architect may require.

© 1978 • THE AMERICAN INSTITUTE OF ARCHITECTS, 1735 NEW YORK AVE., N.W., WASHINGTON, D.C. 20006

FIG. 8.4 (*continued*)

ARTICLE 14

PAYMENTS TO THE CONTRACTOR

14.1 The Architect will review the Contractor's Applications for Payment and will promptly take appropriate action thereon as provided in the Contract Documents. Such amount as he may recommend for payment shall be payable by the Owner not later than the day of the month.

14.1.1 In taking action on the Contractor's Applications for Payment, the Architect shall be entitled to rely on the accuracy and completeness of the information furnished by the Contractor and shall not be deemed to represent that he has made audits of the supporting data, exhaustive or continuous on-site inspections or that he has made any examination to ascertain how or for what purposes the Contractor has used the moneys previously paid on account of the Contract.

14.2 Final payment, constituting the entire unpaid balance of the Cost of the Work and of the Contractor's Fee, shall be paid by the Owner to the Contractor days after Substantial Completion of the Work unless otherwise stipulated in the Certificate of Substantial Completion, provided the Work has been completed, the Contract fully performed, and final payment has been recommended by the Architect.

14.3 Payments due and unpaid under the Contract Documents shall bear interest from the date payment is due at the rate entered below, or in the absence thereof, at the legal rate prevailing at the place of the Project.

(Here insert any rate of interest agreed upon.)

(Usury laws and requirements under the Federal Truth in Lending Act, similar state and local consumer credit laws and other regulations at the Owner's and Contractor's principal places of business, the location of the Project and elsewhere may affect the validity of this provision. Specific legal advice should be obtained with respect to deletion, modification, or other requirements such as written disclosures or waivers.)

ARTICLE 15

TERMINATION OF CONTRACT

15.1 The Contract may be terminated by the Contractor as provided in the Contract Documents.

15.2 If the Owner terminates the Contract as provided in the Contract Documents, he shall reimburse the Contractor for any unpaid Cost of the Work due him under Article 5, plus (1) the unpaid balance of the Fee computed upon the Cost of the Work to the date of termination at the rate of the percentage named in Article 6, or (2) if the Contractor's Fee be stated as a fixed sum, such an amount as will increase the payments on account of his Fee to a sum which bears the same ratio to the said fixed sum as the Cost of the Work at the time of termination bears to the adjusted Guaranteed Maximum Cost, if any, otherwise to a reasonable estimated Cost of the Work when completed. The Owner shall also pay to the Contractor fair compensation, either by purchase or rental at the election of the Owner, for any equipment retained. In case of such termination of the Contract the Owner shall further assume and become liable for obligations, commitments and unsettled claims that the Contractor has previously undertaken or incurred in good faith in connection with said Work. The Contractor shall, as a condition of receiving the payments referred to in this Article 15, execute and deliver all such papers and take all such steps, including the legal assignment of his contractual rights, as the Owner may require for the purpose of fully vesting in himself the rights and benefits of the Contractor under such obligations or commitments.

© 1978 • THE AMERICAN INSTITUTE OF ARCHITECTS, 1735 NEW YORK AVE., N.W., WASHINGTON, D.C. 20006

FIG. 8.4 (*continued*)

ARTICLE 16

MISCELLANEOUS PROVISIONS

16.1 Terms used in this Agreement which are defined in the Contract Documents shall have the meanings designated in those Contract Documents.

16.2 The Contract Documents, which constitute the entire agreement between the Owner and the Contractor, are listed in Article 1 and, except for Modifications issued after execution of this Agreement, are enumerated as follows:

(List below the Agreement, the Conditions of the Contract, [General, Supplementary, and other Conditions], the Drawings, the Specifications, and any Addenda and accepted alternates, showing page or sheet numbers in all cases and dates where applicable.)

This Agreement entered into as of the day and year first written above.

OWNER	CONTRACTOR
______________________	______________________
______________________	______________________
______________________	______________________

© 1978 • THE AMERICAN INSTITUTE OF ARCHITECTS, 1735 NEW YORK AVE., N.W., WASHINGTON, D.C. 20006

8.4.1
FORMS OF FEE

Under the cost plus form of contract the contractor is reimbursed for incurred direct costs and is also paid a fee to cover the firm's general overhead and profit. The various forms of this type of contract vary primarily in the way in which this fee is established. The following are included among the more widely used forms of this type of contract:

Cost plus a percentage of cost. Although this form of contract is still used on some private projects, laws have been enacted that largely prohibit its use on public works. Under this type of contract the contractor receives a percentage of direct construction cost as a fee. In other words, the greater the costs of the project, the greater the contractor's fee. Such an arrangement offers no incentive for the contractor to control costs. In fact it provides a fairly powerful incentive to increase the costs as much as possible. The use of this form of contract should be restricted to conditions where the owner has had prior dealings with the contractor and has developed a high level of confidence in the firm's integrity.

Cost plus fixed fee. Under this form of contract the contractor's general overhead and profit are covered by a fixed amount of dollars. Use of this form requires that the approximate size of the project must be established ahead of time since the size of the fee should be related to the size of the total project. It would not be reasonable to expect a contractor to construct a $10,000,000 project for the same fixed fee established for a $5,000,000 project. The fixed form of contract offers no incentive for the contractor to increase the total cost of the project since the size of the fee to be received would not be increased accordingly. However, it does provide some incentive to complete the project as quickly as possible since a delay in completion will increase the contractor's general overhead costs and therefore erode the anticipated profit figure.

Cost plus a sliding fee. There may be some situations where the owner is unsure regarding the total size of the project, perhaps because of financial questions, marketing forecasts, or several other factors. Under such conditions, a sliding fee may be used. This provides for a basic, minimum fee to be paid to the contractor with incremental increases if the size of the project exceeds the initial established amount. These increases in fee should be agreed upon ahead of time and listed in the original agreement.

Guaranteed maximum. One of the major arguments against the use of any type of cost plus contract is the fact that the owner does not know the total cost ahead of time. This concern can be modified by including within the contract a provision for a guaranteed maximum, sometimes referred to as an "upset" price. With the inclusion of such a provision, the contractor furnishes the owner with a price that the project must not exceed. If the cost does exceed this figure, the excess expense is paid by the contractor. It logically follows that the drawings and specifications under such a provision must be complete enough to enable the contractor to determine such a maximum figure.

Cost plus a fixed fee with upset price and profit sharing. This form of contract would appear to be the ultimate in cost plus contracts. It contains provisions for the contractor to be reimbursed for direct costs and to receive a fixed fee to cover general overhead and profit. In addition, the contractor gives a guaranteed maximum price beyond which the total project cannot go. The contract further contains agreement regarding the sharing of any "savings" realized with respect to the upset price. These savings may be shared in any manner to which the owner and the contractor agree. A small portion may go to the contractor and the major share to the owner, or the major portion may be awarded to the contractor and only a token percentage to the owner. Regardless of how the savings are divided, such an arrangement offers incentives to both the owner and the contractor to hold costs to a minimum. Both parties share a commonality of purpose to seek out economic alternatives in materials and methods that will be most beneficial.

8.4.2 REIMBURSABLE COSTS

Of extreme importance in any form of cost plus contract is the complete definition of what constitutes a reimbursable cost. Many potential disputes can be avoided if care is exercised in this regard. The contract should include statements describing and defining those items of direct costs that are considered reimbursable under the contract. Such action will diminish the danger of assumptions regarding more or less standard costs. Because of these provisions, most cost plus contract forms will be considerably longer than other forms of contract. Among those items that should be included under defined reimbursable costs are the following:

Project labor costs. In addition to payroll, this would include payroll taxes such as workmen's compensation and unemployment compensation, insurance, subsistence and travel allowances, and any fringe benefits that must be paid on labor.

Project equipment costs. This can be a particularly troublesome item and deserves extra attention. Large equipment costs can be based upon the contractor's own cost accounting records or on published rental rates for similar equipment. The contract should be specific as to the method that is to be in force for the project. Also included here are costs of transportation and setup of the equipment, as well as any special fees in connection with transportation.

Project material costs. In addition to the cost of the material itself, such expenses as sales tax, inspection, testing, transportation, and storage costs should be included.

Subcontracts. Each subcontract is a reimbursable expense as is the cost of any bonds required of the subcontractors. Although most cost plus contracts call for competitive bids to be taken on subcontract work, there are some instances where these are also negotiated. If such is the case, care should be taken to secure the owner's approval before subcontracts are awarded.

Field office and utilities. If a mobile job office is used, a proportionate charge should be agreed upon ahead of time. This may be based on either depreciation or rental charges. In addition, the cost of transporting the office to and from the job site, along with the costs of any auxiliary buildings such as storage or tool sheds, should be included. This item also includes the cost of temporary heat, power, and water for the job office as well as for the project itself, and any job telephones and long distance charges.

Consumables charges. This item will normally include such items as small tools and temporary protection tarpaulins.

Bonds and insurance. The premium cost for all required project bonds —such as performance, payment, and warranty bonds—is a cost of doing business and should be reimbursable. Also included should be the proportionate share of any insurance premiums that the contractor carries as well as premiums on any insurance taken out specifically for this project.

Overhead. Although general overhead expenses are covered under the fee portion of the contract and are thus not a reimbursable cost, any overhead items that can be identified with the particular project should be charged as a reimbursable cost.

Losses. Under a cost plus form of contract the major risk is to be borne by the owner. As a result, any losses that are not due to the fault or negligence of the contractor should be defined as a reimbursable cost. This would include losses caused by flood, fire, theft, or vandalism.

8.5 SINGLE VS. SEPARATE CONTRACTS

No discussion of construction contracts would be complete without consideration being given to the subject of single and separate contracts. This is a controversial issue that has led to much discussion and disagreement between general and specialty contractors. The subject has also been addressed by several state legislatures and the federal congress. Some states have enacted legislation requiring the award of single contracts for public projects while other states have mandated separate contracts.

8.5.1 SINGLE CONTRACTS

Under the single contract system, the owner awards only one prime contract. The prime contractor in turn awards a varying number of subcontracts. Under this approach the owner and the design professional deal directly with only one construction firm and hold that firm solely reponsible for the total project. Many owners express satisfaction with an arrangement that provides them with single responsibility. Some design professionals also express preference for the single contract method because of the ease of communication and coordination. A few design firms have created a difference in the fees charged under the two systems, the higher fee being charged for a project where separate contracts are awarded.

8.5.2 SEPARATE CONTRACTS

Under the separate contract approach, the owner awards a number of prime contracts. In states where laws require separate contracts for public projects, the most common example would be a building project. Prime contracts would normally be awarded for general, electrical, plumbing and mechanical work. Specialty contractors generally favor this approach because it removes them from bid shopping and bid peddling by the general contractor. Specialty contractors also try to support their position by claiming that most general contractors are not really qualified to supervise the work of the specialty contractors. On the other hand, general contractors claim that they are the ones who have traditionally provided the coordination for the total project and that they are still expected to do so even when separate contracts are used.

8.5.3 COST FACTORS

General contractors add a markup to the bids of specialty and subcontractors to cover the cost of coordination and administration of these contracts. Some owners may elect the separate contract system in an attempt to save the cost of this markup. However, coordination and administration of the project must still be performed. Under separate contracts, this responsibility may fall upon the shoulders of the owner or the design professional, or by default upon the general contractor. As previously stated, some design professionals charge a higher fee for a separate contract project to cover these additional duties. It is not expected that an early resolution of this controversy will be realized.

9 DOCUMENTS SUPPORTING THE AGREEMENT

FIGURES

9.1 LETTER OF INTENT

There may be occasions when an owner intends to award a contract and proceed with the construction of a project but for various reasons is delayed in doing so. The delay may be caused by the institution providing the financing, by hearings of planning and zoning commissions, or by further requirements of a governmental agency. Whatever the reason, the owner fully expects the project to proceed after the hindrance has been removed. Under these circumstances, the owner is precluded from signing a formal contract but does not wish to lose the bid prices that have been received. In such a situation, the owner may issue a letter of intent, which is actually an interim contract between the contractor and the owner defining what the contractor is authorized to do and how payment shall be made. This document is followed by a formal agreement at a later date.

9.1.1 INFORMATION TO BE INCLUDED

Although the exact content of the letter of intent will vary according to the specific project, it will generally contain the following information:

The name, title, or number of the project for which the letter of intent is being issued.

A statement explaining that this is but a temporary expedient and that it is the intention of both parties to enter into a formal agreement at a later date. The reason for the delay may or may not be given. It is often desirable that a time limit be established for the signing of a formal contract.

Authorization for the contractor to proceed with the work. Although on some occasions this may be an open-ended instruction, it will usually clearly define the amount of work that the contractor is authorized to perform. This may be limited to site preparation work, or it may include placing material orders. In any case, a definite statement needs to be made regarding the amount of work that may be done under this document.

Method and amount of payment to be rendered if a formal agreement is never executed. Although it is the intent of both parties to enter into a formal contract, if circumstances make this impossible, provisions should be made ahead of time for payment of the work that has been performed under the letter of intent. The contractor should be extremely cautious about doing any work or making commitments to any subcontracts beyond the limits stated in the letter of intent.

Signatures of both parties. Since the letter of intent is a legal contract, it should be signed by both the owner and the contractor. The practice of permitting the document to be issued by the design professional with signature acceptance by the contractor should be discouraged since this may lead to legal complications.

The date on which the letter of intent is signed. This may be especially important if the document contains instructions for the work to proceed within a given time limit.

9.2 NOTICE TO PROCEED

In some instances a contract is signed by the owner and the contractor when work cannot begin on the project immediately. This delay may be caused by any one of a number of formalities that must be met. Under such a situation, the contract would normally contain a provision that the owner will furnish notification to the contractor when the work may begin. Such a notice is generally referred to as a "Notice to Proceed."

9.2.1 PURPOSE

The purpose of the notice to proceed is to furnish authorization to the contractor to begin work. It can also provide authorization for the contractor to be on the site if the delay has been caused by lack of a clear title to the property. It may further serve as the offficial starting date for any time calculations if there are penalty or bonus clauses involved in the agreement.

9.2.2 FORM AND CONTENT

The notice to proceed is quite often in the form of a letter sent by the owner to the contractor, although it may be issued by the design professional under the owner's authorization. It should contain a statement that the contractor is to proceed with the work as well as a time limitation within which work is to begin, ten days being the normal time limit used within the construction industry. A sample form that may be used for the notice to proceed is shown in Fig. 9.1.

9.3 PERFORMANCE BOND

To insure the contract against default by the contractor, a performance bond may be required. While such protection is of interest to any owner, it is of particular concern to the public owner who is using tax monies for the construction of a project. Once a construction contract has been signed, the owner is entitled to receive the project as described in the contract documents for the amount of money stipulated. The purpose of the performance bond is to protect this reasonable expectation.

9.3.1 WHERE REQUIRED

A performance bond is required on all federally funded projects over a stated minimum amount, usually $2,000. Most states and municipalities have passed companion legislation regarding the protection of publicly funded contracts. Private contracts may be similarly protected depending upon the wishes of the individual owners. Information regarding performance bond requirements should be given in the bidding documents. A private owner may in some instances wish to place a qualification upon the bidders in an attempt to ensure some minimum financial standards. Under those circumstances, the bidding documents can state that the successful bidder may be asked to furnish a performance bond, rather than stating that one is absolutely required. This would permit the owner to require one if there is any doubt regarding the financial capacity of any of the bidders.

9.3.2 AMOUNT AND FORM

The amount of the performance bond is usually equal to 100 percent of the contract. There are instances, however, where the amount may be much less, perhaps as low as 10 percent of the contract. If any additive changes are made to the contract, the amount of the performance bond can be increased accordingly. Any additive changes to a contract covered by a performance bond, or any changes affecting payments or cash flow, should be approved by the surety. This approval is necessary since such changes may affect the contractor's bonding capacity, as will be discussed later in this chapter. Changes made without the surety's knowledge and approval may jeopardize chances of collecting under the bond.

FIG. 9.1 Notice to proceed

NOTICE TO PROCEED
(Example)

TO: ______________________ DATE: ______________________

____
Contractor

Attention: ______________________

Gentlemen:
Subject: Notice to Proceed with the Construction of ______________________
(Division or Section)
You are hereby directed to proceed with the construction of the subject improvement in accordance with the terms of the contract documents, plans and specifications entered into by ____________ and ____________
(Contractor) (Town or Association)
The signed contract is dated__________, 19____. The stipulated time for commencing work on __________
____
(Division or Section)
shall begin__________, 19____, in accordance with the contract terms.

Sincerely yours,

____
Owner

By______________________

Title______________________

There are many different forms used for a performance bond. Many associations publish and encourage the use of their standard forms, and most bonding companies and a variety of governmental agencies and private owners have developed their own forms. In some instances a combined form is used for both the performance and the payment bonds. An example of a performance bond form is shown in Fig. 9.2.

9.3.3 DATE AND SIGNATURE REQUIREMENTS

The performance bond should have the same date as the contract and should be delivered to the owner at the time the contract is signed. The bond must be signed by the surety and by the contractor, also known as the Principal. It is issued to the owner, also known as the Obligee. The charge made by the surety for furnishing the bond is called a premium. Although the premium is usually paid by the contractor, it is indirectly paid by the owner since it is a business cost for the construction company. The performance bond remains in effect for the life of the contract. After the contract conditions have been satisfied, the bond becomes void. Signatures on the bond should be acknowledged by a notary public, with the corporate power of attorney accompanying the signature of any officer other than the president of a corporation. Bond premiums should always be paid in advance.

9.3.4 REGULATIONS AND SELECTION

A bonding company operates under regulations similar to those that apply to insurance companies. They must obtain a charter from the state or states in which they do business and they must publish their premium rates, which are subject to review. Most contractors select one bonding company and have that

FIG. 9.2 Performance bond

THE AMERICAN INSTITUTE OF ARCHITECTS

AIA Document A311

Performance Bond

KNOW ALL MEN BY THESE PRESENTS: that

(Here insert full name and address or legal title of Contractor)

as Principal, hereinafter called Contractor, and,

(Here insert full name and address or legal title of Surety)

as Surety, hereinafter called Surety, are held and firmly bound unto

(Here insert full name and address or legal title of Owner)

as Obligee, hereinafter called Owner, in the amount of

Dollars ($),

for the payment whereof Contractor and Surety bind themselves, their heirs, executors, administrators, successors and assigns, jointly and severally, firmly by these presents.

WHEREAS,

Contractor has by written agreement dated 19 , entered into a contract with Owner for

(Here insert full name, address and description of project)

in accordance with Drawings and Specifications prepared by

(Here insert full name and address or legal title of Architect)

which contract is by reference made a part hereof, and is hereinafter referred to as the Contract.

This document has been reproduced with the permission of the American Institute of Architects under application number 79082. Further reproduction, in part or in whole, is not authorized. Because AIA documents are revised from time to time, users should ascertain from AIA the current edition of the document reproduced above.

FIG. 9.2 *(continued)*

PERFORMANCE BOND

NOW, THEREFORE, THE CONDITION OF THIS OBLIGATION is such that, if Contractor shall promptly and faithfully perform said Contract, then this obligation shall be null and void; otherwise it shall remain in full force and effect.

The Surety hereby waives notice of any alteration or extension of time made by the Owner.

Whenever Contractor shall be, and declared by Owner to be in default under the Contract, the Owner having performed Owner's obligations thereunder, the Surety may promptly remedy the default, or shall promptly

1) Complete the Contract in accordance with its terms and conditions, or

2) Obtain a bid or bids for completing the Contract in accordance with its terms and conditions, and upon determination by Surety of the lowest responsible bidder, or, if the Owner elects, upon determination by the Owner and the Surety jointly of the lowest responsible bidder, arrange for a contract between such bidder and Owner, and make available as Work progresses (even though there should be a default or a succession of defaults under the contract or contracts of completion arranged under this paragraph) sufficient funds to pay the cost of completion less the balance of the contract price; but not exceeding, including other costs and damages for which the Surety may be liable hereunder, the amount set forth in the first paragraph hereof. The term "balance of the contract price," as used in this paragraph, shall mean the total amount payable by Owner to Contractor under the Contract and any amendments thereto, less the amount properly paid by Owner to Contractor.

Any suit under this bond must be instituted before the expiration of two (2) years from the date on which final payment under the Contract falls due.

No right of action shall accrue on this bond to or for the use of any person or corporation other than the Owner named herein or the heirs, executors, administrators or successors of the Owner.

Signed and sealed this day of 19

(Principal) (Seal)

(Witness)

(Title)

(Surety) (Seal)

(Witness)

(Title)

firm furnish all their bonds whenever possible. This is a preferred arrangement since the bonding company must be intimately familiar with the contractor's financial situation. In some instances, an owner may specify that the bond is to be furnished by a particular bonding company. This may be done for a variety of reasons, one of which may be that the owner has greater confidence in the integrity of that particular surety. Such a situation may cause some potential bidders difficulty or may even cause them to decide not to bid, since they must subject their firm to financial scrutiny by an "outside" organization.

9.3.5 PURPOSE AND LIMITATIONS

Since the purpose of the performance bond is to ensure the performance of the contract, the prime beneficiary is the owner, not the contractor. It is issued by a surety company on behalf of the construction firm. Under the terms of the bond, the surety guarantees to the owner that the contract will be performed in accordance with the stated requirements. If the contractor should default, the surety is obligated to see that the contract is completed in compliance with the original conditions. This tends to increase the credibility of the construction company's bid, a situation that is similar to a bank requiring a cosigner on a loan. Court decisions reflect some disagreement as to the degree of liability that a bonding company has under such a bond. Some decisions have held the surety liable for total completion of the contract regardless of the costs, while other decisions have made the surety liable only up to the face amount of the bond. In other words, if the contract and bond are in the amount of $10,000,000 and the contractor defaults when the project is only half completed, the surety is obligated to complete the project by paying all additional expenditures up to the full $10,000,000.

9.3.6 COMPLETION BY THE SURETY

Bonding companies have the option of selecting their own method of completing the project in case of default by the contractor. They may make use of the labor force and equipment of the original contractor, they may take competitive bids for the balance of the work, or they may enter into a negotiated contract with another construction firm. The work must, of course, be performed and completed in accordance with the original contract documents. This would include any penalty or liquidated damage provisions. In return, the surety is entitled to collect from the owner all monies still due and payable under the terms of the original contract. Although court decisions conflict somewhat about who is entitled to collect the retained percentages from the work performed by the original contractor, most courts award these to the surety company. In addition, if the bonding company incurs any loss of monies in the completion of the contract, they are entitled, by means of litigation, to collect such amounts from the defaulting contractor.

9.3.7 INVESTIGATION BY THE SURETY

Prior to agreeing to furnish bonds on behalf of a construction company, a surety will usually conduct a two-stage investigation. The first stage determines whether the surety wishes to bond the construction firm as a regular client, and the second stage determines whether a bond should be issued for that firm on a particular project on which the firm wishes to bid. The second stage is conducted prior to the bid being submitted since the issuance of a bid bond for a firm carries with it the implied obligation to furnish the contract bonds.

9.3.8 FIRST-STAGE INVESTIGATION

During the first-stage examination the general condition of the company is analyzed. The following items would be among those investigated:

The professional ability of the principals of the firm as well as the professional ability of the key personnel and the firm as a whole. This may include evidence of professional training as reflected by formal education and professional registration. While most states do not register construction engineers as professional engineers (Iowa is one that does), several states do give examinations for registration as a contractor. Some states register individuals, while others register the firm.

The integrity and personal habits of the firm's principals. This would include interviews with persons well acquainted with the principals to determine whether they are honest as well as moral in their personal lives. Consideration may be given to an individual's drinking and gambling habits, marital status, and extramarital involvement.

The financial standing and line of bank credit for the firm. This would necessitate a study of the firm's balance sheet and income statement with particular attention to liquid assets and liquidity of assets. It would also involve contacting the firm's bank to confirm their line of credit.

The financial standing and credit rating of the principals of the firm. Even though the firm is the organization being considered for bonding, and even though the firm may be incorporated, the basis for the bonding decision still rests on the shoulders of the principals. In some borderline situations, the officers of a corporation may be asked to post their personal assets to secure bonding on a particular project.

The experience of the firm as well as the experience of the firm's principals and key personnel in the line of work being proposed for bonding. This would include a consideration of balanced experience with respect to both type and size of projects being considered.

If the surety company is satisfied with the results of their examination of these items, they will generally agree to furnish bonds for the construction firm. This is a general agreement, however, and is not an automatic assurance that the contractor will receive a bond on all projects being considered. From the time when the bonding company gives notification of a willingness to furnish bonds, the construction company should keep the surety continuously informed regarding their work load and financial situation. A good bonding agent can furnish valuable counsel and advice to the contractor on many financial aspects of the business.

9.3.9 SECOND-STAGE INVESTIGATION

When the contractor has an interest in bidding a particular project, the bonding agent should be contacted immediately. At that time the bonding company will conduct the second stage of the examination to determine whether a bond will be authorized for the project. The bonding company evaluates the following items during this stage:

The amount of work that the contractor presently has on hand. This will consist of work that is uncompleted under existing contracts. Work that is bonded as well as work for which a bond was not required will be included.

The contractor should calculate this figure as accurately as possible. While there may be a temptation to reduce this figure in order to secure a greater bonding company limit, such action can mislead the bonding company and it is not in the best interest of the contractor. The figure of work on hand should also include recent bids that the contractor has submitted and that may result in a contract being received.

The contractor's recent bidding record. If recent contracts have been the result of bids where a lot of money was "left on the table," it may raise the question of the accuracy of these bids. A spread of more than 5 or 6 percent between the contractor's low bid and the second bid may cause the bonding company to ask some very searching questions. Too many jobs taken at too low a price usually marks the first step toward financial disaster.

The amount of present working capital. This figure is equal to current assets less current liabilities. It is a prime indicator of whether or not the contractor has enough cash to handle the proposed additional project. In case of doubts, a cash flow study should be conducted.

The contractor's availability of credit. Most financial advisers agree that the worst time to borrow money is when you need it. For this reason, the contractor should always have a line of credit established against the time when a loan may be needed. The amount of the available loan will fluctuate according to various factors. The bonding company will be interested in what that current amount is. They will also be concerned with the proposed interest rate and any unusual time restrictions regarding repayment that may be imposed.

The terms of payment under the proposed contract. The form of the contract that will be used on the project should be included within the bidding documents. This will give information relative to how often the contractor will be paid and what the retained percentage will be. Such conditions have a marked effect upon the contractor's cash flow and will be studied carefully by the bonding company. For example, if the proposed retainage is 20 percent rather than the standard 5 to 10 percent, it may impose restrictions on the contractor's cash flow that will require either that a loan will have to be secured or that the contractor may not be able to handle the work.

The contractor's previous experience with work similar to that proposed in this project. This item will consider both the type of work and the size of project being proposed. Therefore if a firm is interested in increasing both the scope and volume of their work, they should undertake smaller projects when the work is different from previous experience and projects that are only a realistic percentage larger than similar projects done in the past.

The amount of the project that will be subcontracted. By subcontracting some of the work, the contractor's own risk is diminished. This in turn decreases the amount of exposure for the bonding company. Under certain circumstances, the bonding company may require the general contractor to require bonds of some of the subcontractors in order to reduce the overall risk.

9.3.10 BASIS FOR PREMIUMS

Surety bond premium rates are based upon the classification of the project being proposed. These classifications are A-1, A, B, and miscellaneous. Bonding companies may offer "deviated" or lower rates to construction firms that have built a good reputation for operations. Certain types of contracts may also qualify for these lower rates. Rates are generally quoted for a twelve-month period or for the life of the contract, whichever is shorter. If a longer bonding time is required, additional premiums are charged.

9.4 PAYMENT BOND

Although subcontractors, material suppliers, and laborers who have not received payment may file mechanics liens on private projects, this procedure is prohibited by law on public funded projects. To afford this much needed protection and thereby encourage individuals and firms within the construction industry to work on public projects, a secondary or back-up bond is required. This is known as a Payment Bond or a Labor and Material Bond and is required on federal projects under the terms of the Miller Act. Most states and many municipalities have passed "little Miller Acts" to provide similar protection on projects under their jurisdiction. In the unlikely absence of a payment bond on a public project, the only recourse for the unpaid parties is to file a "stop notice." This is a legal, formal notice to the owner that the general contractor has not paid the injured party. The owner would then bring pressure upon the contractor to make payment.

Although payment bonds are not as common on private projects as they are on public projects, they are sometimes called for when the owner has some doubt about the financial stability of the contractor. In this way, the risk of the owner having liens filed against the project is "passed on" to the surety who is obligated to pay in case of default by the contractor.

9.4.1 AMOUNT AND CLAIMS

The payment bond should be in an amount sufficient to cover the cost of labor and materials furnished by third parties. While it may be for any amount up to 100 percent of the contract, it is usually issued for 50 percent of the total contract amount. Claims against the payment bond must satisfy the following requirements:

Claims are to be filed in the proper court.

Claims must be filed within a period of one year after the project has been completed by the general contractor.

If the person or firm making the claim does not have a contract with the general contractor, a formal written notice must be given to the general contractor, the owner, and the surety.

Barring legal requirements to the contrary, claims must be made no later than 90 days after the claimant last performed work on the project. This is generally interpreted to mean "on site" work.

The payment bond is sometimes a separate document issued in conjunction with the performance bond. It may also be issued as a combined document with the performance bond, as mentioned earlier in this chapter. An example of the payment bond as a separate document is shown in Fig. 9.3

FIG. 9.3 Payment bond

THE AMERICAN INSTITUTE OF ARCHITECTS

AIA Document A311

Labor and Material Payment Bond

THIS BOND IS ISSUED SIMULTANEOUSLY WITH PERFORMANCE BOND IN FAVOR OF THE OWNER CONDITIONED ON THE FULL AND FAITHFUL PERFORMANCE OF THE CONTRACT

KNOW ALL MEN BY THESE PRESENTS: that
(Here insert full name and address or legal title of Contractor)

as Principal, hereinafter called Principal, and,
(Here insert full name and address or legal title of Surety)

as Surety, hereinafter called Surety, are held and firmly bound unto
(Here insert full name and address or legal title of Owner)

as Obligee, hereinafter called Owner, for the use and benefit of claimants as hereinbelow defined, in the amount of
(Here insert a sum equal to at least one-half of the contract price) Dollars ($),
for the payment whereof Principal and Surety bind themselves, their heirs, executors, administrators, successors and assigns, jointly and severally, firmly by these presents.

WHEREAS,

Principal has by written agreement dated 19 , entered into a contract with Owner for
(Here insert full name, address and description of project)

in accordance with Drawings and Specifications prepared by
(Here insert full name and address or legal title of Architect)

which contract is by reference made a part hereof, and is hereinafter referred to as the Contract.

This document has been reproduced with the permission of the American Institute of Architects under application number 79082. Further reproduction, in part or in whole, is not authorized. Because AIA documents are revised from time to time, users should ascertain from AIA the current edition of the document reproduced above.

FIG. 9.3 *(continued)*

LABOR AND MATERIAL PAYMENT BOND

NOW, THEREFORE, THE CONDITION OF THIS OBLIGATION is such that, if Principal shall promptly make payment to all claimants as hereinafter defined, for all labor and material used or reasonably required for use in the performance of the Contract, then this obligation shall be void; otherwise it shall remain in full force and effect, subject, however, to the following conditions:

1. A claimant is defined as one having a direct contract with the Principal or with a Subcontractor of the Principal for labor, material, or both, used or reasonably required for use in the performance of the Contract, labor and material being construed to include that part of water, gas, power, light, heat, oil, gasoline, telephone service or rental of equipment directly applicable to the Contract.

2. The above named Principal and Surety hereby jointly and severally agree with the Owner that every claimant as herein defined, who has not been paid in full before the expiration of a period of ninety (90) days after the date on which the last of such claimant's work or labor was done or performed, or materials were furnished by such claimant, may sue on this bond for the use of such claimant, prosecute the suit to final judgment for such sum or sums as may be justly due claimant, and have execution thereon. The Owner shall not be liable for the payment of any costs or expenses of any such suit.

3. No suit or action shall be commenced hereunder by any claimant:

a) Unless claimant, other than one having a direct contract with the Principal, shall have given written notice to any two of the following: the Principal, the Owner, or the Surety above named, within ninety (90) days after such claimant did or performed the last of the work or labor, or furnished the last of the materials for which said claim is made, stating with substantial accuracy the amount claimed and the name of the party to whom the materials were furnished, or for whom the work or labor was done or performed. Such notice shall be served by mailing the same by registered mail or certified mail, postage prepaid, in an envelope addressed to the Principal, Owner or Surety, at any place where an office is regularly maintained for the transaction of business, or served in any manner in which legal process may be served in the state in which the aforesaid project is located, save that such service need not be made by a public officer.

b) After the expiration of one (1) year following the date on which Principal ceased Work on said Contract, it being understood, however, that if any limitation embodied in this bond is prohibited by any law controlling the construction hereof such limitation shall be deemed to be amended so as to be equal to the minimum period of limitation permitted by such law.

c) Other than in a state court of competent jurisdiction in and for the county or other political subdivision of the state in which the Project, or any part thereof, is situated, or in the United States District Court for the district in which the Project, or any part thereof, is situated, and not elsewhere.

4. The amount of this bond shall be reduced by and to the extent of any payment or payments made in good faith hereunder, inclusive of the payment by Surety of mechanics' liens which may be filed of record against said improvement, whether or not claim for the amount of such lien be presented under and against this bond.

Signed and sealed this day of 19

	______________________ *(Principal)* (Seal)
______________________ *(Witness)*	
	______________________ *(Title)*
	______________________ *(Surety)* (Seal)
______________________ *(Witness)*	
	______________________ *(Title)*

AIA DOCUMENT A311 • PERFORMANCE BOND AND LABOR AND MATERIAL PAYMENT BOND • AIA ®
FEBRUARY 1970 ED. • THE AMERICAN INSTITUTE OF ARCHITECTS, 1735 N.Y. AVE., N.W., WASHINGTON, D. C. 20006

9.5 WARRANTY BOND

Although the performance bond technically runs for one year following completion of the project to cover any latent short-term deficiencies in labor and/or material, this is often too short a period for some of the items of work. To cover these circumstances, a warranty bond may be called for to cover these items, a separate one usually issued for each item. These bonds may be issued for different periods of time by different sureties.

9.5.1 ROOFING BOND

Built-up roofing is one item where a warranty bond is often required. Roofing bonds may be issued for varying periods—from 2 up to 25 years. They generally guarantee that the quality of material furnished will be in accordance with the specifications and that the installation will be done by a competent roofer who will comply with accepted practices. The material manufacturer may have an inspector present during construction to ensure compliance.

If the roofing should fail to meet the requirements because of a deficiency in either the labor or material, the surety furnishing the bond is obligated to take corrective action. There have been instances where this corrective action has cost more than the original installation. Failures caused by natural disasters, commonly referred to as "acts of God," are not covered under the bond. Experience has shown that most roof failures can be attributed to quality of workmanship rather than deficiencies in material. A separate premium is charged for a roofing bond and this is paid for by the owner, either directly or indirectly. Fig. 9.4 shows an example of a roofing bond.

9.5.2 OTHER WARRANTY BONDS

There are several other types of bonds that may be used on various projects. In fact, an owner can secure and require bonds for just about anything on the project if there is sufficient cause to do so. Examples of other types of bonds include subdivision bonds, equipment bonds, fidelity bonds, and license bonds.

9.6 CHANGE ORDERS

After the owner and contractor have signed the agreement, any changes in construction or in the contract conditions are put into effect by means of a change order. A change order should be used whenever there is *any* change in the current conditions, not just when it involves a change in price or a change in completion time. For example, if the room finish schedule calls for a particular wall to be painted purple and the owner states that this should be changed to green, a change order should be executed, even though the same type of paint will be used and the same amount of labor will be required. Otherwise, the contractor is without any protection if the owner wants the original color after seeing the substitute. Change orders and addenda serve a similar function, except that addenda are used to document changes during the bidding stage and change orders are used to document changes after the agreement has been signed.

9.6.1 WHO INITIATES THEM

The initial reason for a change order being issued may originate with the design professional, the owner, the contractor, or any one of the subcontractors or material suppliers. In some instances a change order may originate with an outside source, such as an inspector or the permit authority. The actual change order is usually prepared by the design professional. It serves as an amendment to the agreement in that it modifies or changes the original conditions. Because of this, the fairly common practice of the design professional issuing it to the contractor for an acceptance signature may cause problems.

FIG. 9.4 Roofing bond

ROOFING BOND

KNOW ALL MEN BY THESE PRESENTS: that

____
(Name of Surety)

____
(Address of Surety)

hereinafter called the Surety, is held and firmly bound to the Owner named below, or the successors and assigns of said Owner, in a sum not exceeding the principal amount set forth below, for the payment of which we, our successors and assigns hereby bind ourselves by this Bond No.______, said principal amount being $________________.
The conditions of this obligation are as follows:

Owner______________________________

Building title and location______________________________

No. of squares______________________Date of completion______________________
Term of bond from date of completion: Roof______years
Flashing______years

Roofing contractor:______________________________
Address:______________________________

Whereas, subject to the conditions set forth below in this Bond,______________________agrees that
(Name of Surety)

during the term of this Bond, it will at its own expense, in the aggregate not exceeding the principal amount thereof, make or cause to be made any repairs that may become necessary solely by reason of ordinary wear and tear by the elements in order to maintain said roof including composition flashing if included above but excluding metalwork and any materials supplied by others, in a watertight condition:

(1) ______________________________shall not be liable for any damage
(Surety)
to the building or its contents or damage to the roof membrane caused by
(2) lightning, hail, windstorm or other casualty;
(3) settlement, distortion, failure, or cracking of the roof deck or insulation, or building foundations or walls;
(4) infiltration, expansion or condensation of moisture in, through or around the walls, coping, or building structure;
(5) traffic of any nature over the roof or resulting from the roof being used as a storage area;
(6) ______________________________.
(Other conditions)

In Witness Thereof, ______________________________has caused
(Surety)

this instrument to be executed by duly authorized representatives this ______ day of ____________________, 19______.

____
Surety
By______________________
Attorney-in-Fact
(As per attached Power of Attorney)

____
(Witness as to Surety)

____
(Title)

The change order should be signed by both the owner and the contractor. In instances where the contract documents give the design professional the proper authority, the design professional could sign the change order as the owner's agent. Since a change order constitutes a change to the original terms of the agreement, it should be remembered the terms of the change order should be negotiated between the owner and the contractor. Technically, the contractor has the legal right to refuse a change order, barring original contract provisions to the contrary. In actual practice, however, this is seldom done in the interest of good working relations.

9.6.2 INFORMATION TO BE INCLUDED

When prepared in proper form, a change order should contain the following information:

The name, title, or number of the project.

The date of the change order.

The number of the change order. Change orders should be numbered consecutively according to the contract on which they are written.

Changes required under this order. A complete description is indicated, giving information regarding previous as well as new requirements. Reference should be made to specific drawings, sections of specifications, or documents that are affected. The same procedure applies to changes involving technical requirements as well as general contract conditions.

Change in contract price. This should indicate whether the change in price is additive or deductive. Extra care may be needed if the change is to a unit price contract.

Any change in time of completion. Many a contractor has learned the hard lesson that time extension is not automatic with a change order. If any change in completion time is required, it should always be specifically covered in the change order.

Required signatures. As previously mentioned, the owner and the contractor must sign, unless arrangements have been made for the design professional to sign on behalf of the owner. Provisions should be made for any formalities that should accompany the signatures, such as a corporate seal or power of attorney.

Compare these requirements with the sample form shown in Fig. 9.5

9.6.3 REASONS FOR ISSUING

There are several reasons for issuing a change order. If doubt exists as to whether one is needed, it is usually safer to use—rather than not to use—a change order. Among the reasons for using a change order would be the following:

The owner has secured additional financing. Quite often the owner takes bids on additive alternates but is not able to include them in the contract because of monetary restrictions. If more funds become available during the construction phase, they may then be added to the contract.

The emergence of unforeseen conditions during the construction of the project. These may very often be subsurface conditions that test borings did not disclose. In such a case, previously quoted unit prices from the original bid may be used as the basis for a price change for the extra work.

Inability to secure material in conformance with original specifications. This may be caused by strikes, embargoes, market conditions, etc. Accordingly, substitutions will normally be approved, with documentation of the approval being the change order.

Changes needed to correct errors and/or omissions in the original documents. Although these are quite often discovered during the bidding phase, it is not uncommon for some to surface during construction.

FIG. 9.5 Change order

CONTRACT CHANGE ORDER
(Example)

	ORDER NO.
	DATE
	STATE
Contract for	COUNTY

Owner

To______________________________
Contractor

You are hereby requested to comply with the following changes from the contract plans and specifications:

Description of Changes (Supplemental Plans & Specifications Attached)	DECREASE in Contract Price	INCREASE in Contract Price
	$	$
TOTALS	$____	
NET CHANGE IN CONTRACT PRICE	$____	

JUSTIFICATION:

The amount of the Contract will be (decreased) (increased) by the sum of:

______________________ dollars ($______________________).

The contract total including this and previous Change Orders will be:

______________________ dollars ($______________________).

The contract period provided for completion will be (increased) (decreased) (unchanged):______________days.

This document will become a supplement to the contract and all provisions will apply hereto.

Requested______________________ (Owner) ________ (Date)

Recommended______________________ (Owner's Architect/Engineer) ________ (Date)

Accepted______________________ (Contractor) ________ (Date)

Approved by FHA______________________ (Name and Title)

Changes requested or suggested by the owner, contractor, or design professional. One of the three parties may think of a change in material and/or methods during the construction phase. If the change is approved by the other two, it may be adopted through a change order.

9.7 PROGRESS SCHEDULES

Two elements are vital to good management practices—for the design professional as well as the manager of construction. The first is advanced planning; the second is a good communication system. Many of the scheduling techniques used in the construction industry have the potential to include both of these elements within one system. Analyzing and planning a construction project in advance is necessary if the project is to proceed smoothly and efficiently. This process can be compared to the practice that is essential before a new surgeon can perform operations or before a pilot is granted a license.

Progress schedules are an accepted means of communicating planning decisions to all parties involved with the project. Some of the scheduling techniques contribute to the planning process and permit a more thorough analysis of the job to be performed. Progress schedules are in widespread use by owners, contractors, and design professionals throughout the construction industry. Many bidding documents mandate that a construction schedule be prepared, and they may even specify the scheduling method to be used.

9.7.1 BAR CHARTS

One of the oldest methods of construction scheduling that is still in use today is the Gantt chart, more commonly referred to as a "bar chart." An example is shown in Fig. 9.6. This is one of the more simple scheduling systems and is easily understood even by those with limited education. This factor should be kept in mind since some of the field personnel who will be required to use the schedule may have a poor educational background.

A bar chart schedule consists of a series of horizontal bars, or elements, with each bar representing a separate branch of the work. These are presented on a time grid. The scale of the time grid will vary, depending upon the size of the project and the degree of control desired. In medium-sized projects the time unit would be in working days. A degree of completion curve is quite often superimposed over the bar chart to show the overall percentage of completion for the project. A representative bar chart would show the following information:

A list of the branches of work or parts of the project into which the contract has been divided. These should be listed as nearly as possible in chronological order.

A scale of time measurement appropriate for the project. While this will normally be in working days, under special circumstances the units could be hours or weeks. The time scale should be related to calendar dates for ease in checking the project for satisfactory progress.

A set of horizontal bars indicating starting and completion times for each branch of the work. Not all of these will be continuous since some will necessarily be interrupted for other parts of the project. Many firms will leave a space under each bar for the placing of another bar that is used to indicate actual progress of the work. This allows easy and immediate comparison of proposed and actual progress.

The name, title, or number of the project along with the date when the schedule was issued. Space may also be provided for revision dates as necessary. In most cases the name of the issuing firm is also given.

Some schedulers use a column immediately following the branches of the work to show the percentage of the total project within each branch. This may be computed on either a working time or dollar basis.

As previously mentioned, a degree of completion curve may be superimposed over the bar chart. This will show the planned percentage of completion at each stage of the project. Here again, both proposed and actual curves may be used.

Some written instructions normally accompany the schedule. It is difficult, if not impossible, to show such information as crew sizes, equipment selection, and similar items directly on the schedule, but this information is needed to make the communication more complete.

9.7.2 CRITICAL PATH METHOD

Beginning in the late 1950s, several so-called advanced scheduling techniques made their appearance. Perhaps the first of these was the Program Evaluation and Review Technique (PERT), developed by a special team for use in the design and construction of a missile system for the U.S. Navy. This was followed closely by a system developed by the DuPont Corporation, which became known as Critical Path Method (CPM). This in turn spawned a virtual family of alphabetically designated scheduling systems, each one a variation and possible improvement over its predecessor. In actuality, research has shown that none of them were completely original. During World War I, the military forces on both sides used a scheduling technique, which controlled the movement of men and material, based upon the same principles.

9.7.2.1 CPM STRUCTURE The Critical Path Method is perhaps the most representative of these new scheduling techniques and has experienced the most widespread use within the construction industry. CPM utilizes a network diagraming system that consists of arrow lines and nodes. Each arrow line represents an "activity," which is a defined block of work, separate and distinct from all other blocks of work. The nodes represent "events," which serve as checkpoints in time. Dotted arrow lines, or "dummy arrows," are used to show sequence and dependency between the various activities. A duration time is calculated for each activity, based upon crew sizes determined during the planning stage of the project. The activity arrows are then arranged graphically in a network diagram so that the blocks of work are performed in an orderly sequence.

9.7.2.2 CPM GUIDELINES A few basic rules must be observed in the construction of the network diagram if satisfactory results are to be obtained. They include the following guidelines:

Time usually flows from left to right. Although an activity arrow may go up or down vertically, it should not go from right to left. This convention is not absolute but represents good drafting practice.

FIG. 9.6 Bar chart

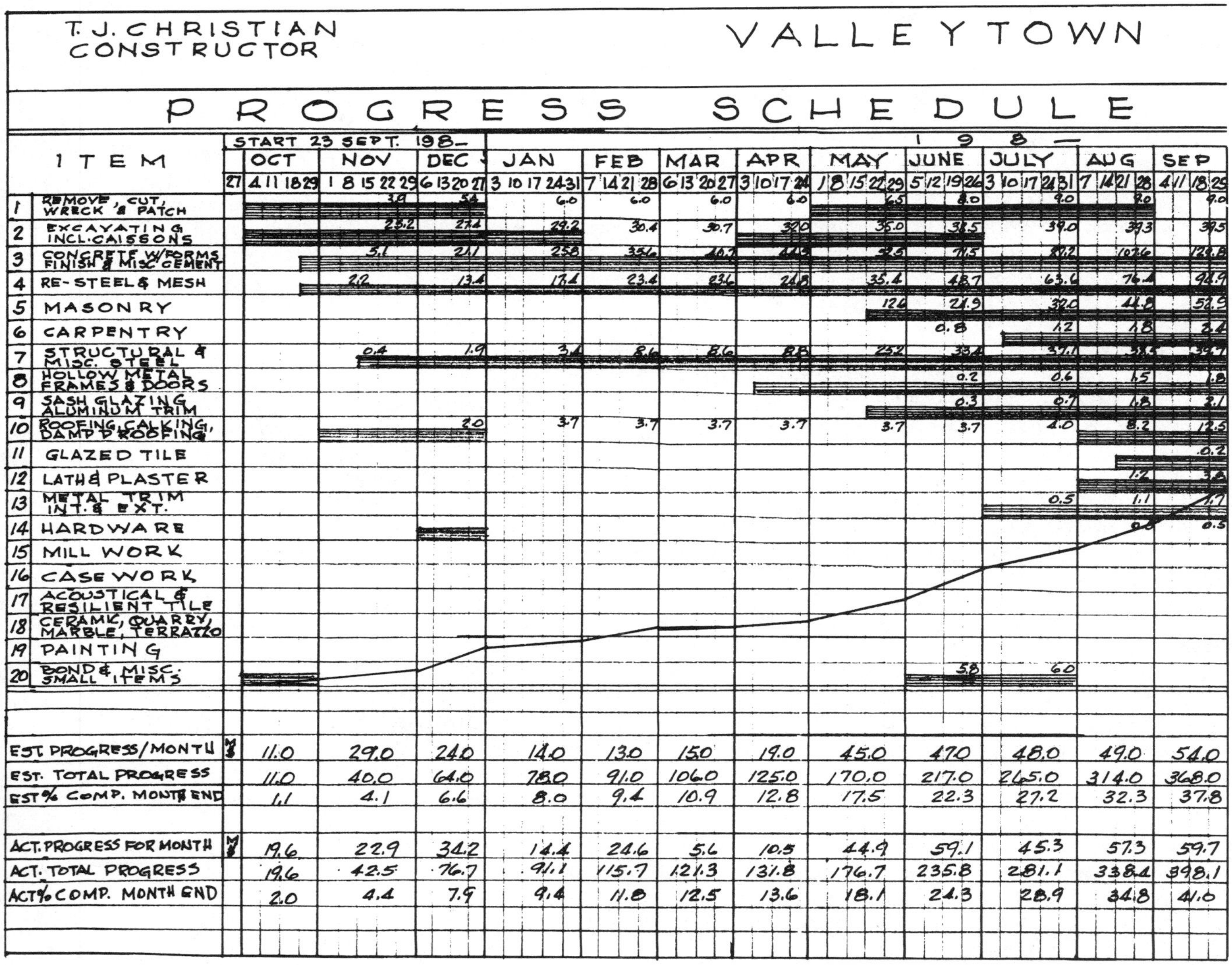

T. J. CHRISTIAN
CONSTRUCTOR

VALLEYTOWN

PROGRESS SCHEDULE

START 23 SEPT. 198–

	ITEM
1	REMOVE, CUT, WRECK & PATCH
2	EXCAVATING INCL. CAISSONS
3	CONCRETE W/FORMS FINISH & MISC. CEMENT
4	RE-STEEL & MESH
5	MASONRY
6	CARPENTRY
7	STRUCTURAL & MISC. STEEL
8	HOLLOW METAL FRAMES & DOORS
9	SASH GLAZING ALUMINUM TRIM
10	ROOFING, CALKING, DAMPPROOFING
11	GLAZED TILE
12	LATH & PLASTER
13	METAL TRIM INT. & EXT.
14	HARDWARE
15	MILL WORK
16	CASE WORK
17	ACOUSTICAL & RESILIENT TILE
18	CERAMIC, QUARRY, MARBLE, TERRAZZO
19	PAINTING
20	BOND & MISC. SMALL ITEMS

	OCT	NOV	DEC	JAN	FEB	MAR	APR	MAY	JUNE	JULY	AUG	SEP
EST. PROGRESS/MONTH M$	11.0	29.0	24.0	14.0	13.0	15.0	19.0	45.0	47.0	48.0	49.0	54.0
EST. TOTAL PROGRESS	11.0	40.0	64.0	78.0	91.0	106.0	125.0	170.0	217.0	265.0	314.0	368.0
EST % COMP. MONTH END	1.1	4.1	6.6	8.0	9.4	10.9	12.8	17.5	22.3	27.2	32.3	37.8
ACT. PROGRESS FOR MONTH M$	19.6	22.9	34.2	14.4	24.6	5.6	10.5	44.9	59.1	45.3	57.3	59.7
ACT. TOTAL PROGRESS	19.6	42.5	76.7	91.1	115.7	121.3	131.8	176.7	235.8	281.1	338.4	398.1
ACT % COMP. MONTH END	2.0	4.4	7.9	9.4	11.8	12.5	13.6	18.1	24.3	28.9	34.8	41.0

FIG. 9.6 *(continued)*

HOSPITAL

TEEJAY ASSOCIATES
ARCHITECTS

PLANNED TIME
ACTUAL TIME

1 9 8 — COMP. 12 SEPT. 198–

OCT	NOV	DEC	JAN	FEB	MAR	APR	MAY	JUNE	JULY	AUG	SEP	VALUE	PROGRESS CURVE
2 9 16 23 30	6 13 20 27	4 11 18 25	1 8 15 22 29	5 12 19 26	5 12 19 26	2 9 16 23 30	7 14 21 28	4 11 18 25	2 9 16 23 30	6 13 20 27	6 13 20 27		%
9.0	9.0	9.0	9.0	9.5	9.5	9.8	10.5	10.6	10.8	11.0		11200	100
39.7	40.0	40.0	40.0	40.0	40.0	40.0	41.7	42.0	42.0	42.0		42000	95
151.3	163.2	164.8	165.7	166.0	166.0	166.0	169.6	170.0	170.0	170.0		170000	90
104.6	109.6	110.8	110.8	110.8	110.8	110.8	111.8	112.0	112.0	112.0		112000	85
91.3	115.7	121.4	127.6	129.4	136.3	139.9	141.2	141.2	142.0	142.0		142000	80
2.7	2.9	3.1	4.6	5.2	5.4	5.6	5.8	6.3	6.6	6.7		6800	75
40.6	41.3	41.9	42.3	42.9	44.9	44.9	45.7	46.4	46.2	46.2		46200	70
5.2	8.2	9.4	10.2	10.4	10.6	10.8	11.3	11.3	11.3	11.7		11700	65
22.7	49.5	56.3	56.6	56.6	57.0	57.4	58.8	58.8	58.8	59.0		59500	60
16.2	17.3	21.8	23.6	23.9	23.9	24.5	24.5	24.5	24.5	24.7		25000	55
5.0	10.7	12.5	19.7	23.2	28.7	31.0	31.9	31.9	33.0	33.0		33000	50
5.4	5.4	5.4	6.8	13.4	31.3	43.8	47.1	49.5	50.7	52.3		52800	45
9.3	14.4	19.9	29.6	31.3	32.9	35.1	36.7	38.8	39.9	44.1		48500	40
0.5	0.5	0.5	0.5	0.5	1.2	3.4	11.3	13.2	17.5	19.9		22000	35
			0.7	0.7	0.7	3.2	8.7	19.1	26.1	28.8		30400	30
						3.8	12.8	12.8	33.2	38.9		40400	25
	0.8	0.8	0.8	0.8	0.8	1.7	16.4	27.2	34.3	41.7		47200	20
				16.6	31.2	41.7	45.8	45.8	45.8	47.2		47500	15
				1.0	3.1	6.4	10.9	11.9	12.9	15.3		17200	10
						7.3	7.4	7.6	7.8	8.0		8037	5
												973437	
62.0	60.0	55.0	44.0	53.0	60.0	72.0	78.0	60.0	41.0	20.437			
430.0	490.0	545.0	589.0	642.0	702.0	774.0	852.0	912.0	953.0	973.437			
44.2	50.4	56.0	60.5	66.0	72.1	79.5	87.5	93.5	98.0	100.0			
110.9	79.2	38.9	29.2	34.2	51.4	45.2	62.8	28.2	57.2	19.6			
509.0	588.2	627.1	656.3	690.5	741.9	787.1	849.9	878.1	935.4	955.0			
52.3	60.5	64.5	67.5	71.0	76.3	80.9	87.4	90.2	96.2	98.7			

The designation in the node at the tail of the activity arrow will normally be less than the designation in the node at the head of the activity arrow. Designation is usually by numbers but may be by letters in a smaller network. The designations do not have to be consecutive, but it is suggested that they be of increasing magnitude. There are some programs that permit the designation at the head of the arrow to be smaller than that at the tail of the arrow. However, most programs follow the practice of increasing magnitude. It is a common practice to omit some numbers in the scheduling process so that additional activities may be added later if found to be necessary.

No two activities can share the same set of node designations unless there is some other device used to differentiate between them. While several activities may share one node designation, the set should be unique for each activity. One of the major functions of the dummy arrow is to facilitate each activity having its own set of node designations.

No activity emanating from an event can begin until all activities coming into that event have been completed.

If all the above guidelines are observed, the longest possible path through the network diagram will be the earliest the project can be completed. This is called the "critical path." It is possible to have more than one critical path in complex networks.

9.7.2.3 TIME CALCULATIONS Calculations may now be performed giving the earliest and latest occurrence times for each event. These in turn may be used to generate the earliest and latest starting and finishing times for each activity. Earliest time calculations are determined by proceeding forward through the network, and the latest times are determined by going backwards through the network. These are referred to as a "right hand pass" and "left hand pass" respectively. It should be noted that in most cases the time available in which to perform the activity is greater than the estimated amount of time required for the activity. This excess is called "float time." A good manager may benefit from float time by reducing crew sizes or shifting personnel to more crucial activities.

Activities having no float time are called "critical" activities and lie on the critical path. A delay in the completion of any one of the critical activities will result in a delay in the completion of the total project. An earlier completion of one of the critical activities *may* result in earlier completion for the project. This is not automatic since there may be more than one critical path, or a new critical path may be created by the change.

9.7.2.4 CPM ADVANTAGES Along with similar scheduling techniques, CPM offers several advantages to management. It is a more definitive system than the bar chart and therefore allows for more complete communication of information. It is a system in which mistakes, obstructions, and "bottlenecks" become readily apparent and as a result may be corrected before construction begins. CPM necessitates a more thorough analysis of the project before the schedule can be prepared, and this very analysis will result in a better understanding of the work to be done. This should lead to a more efficient operation, with savings in time and money.

The system also makes it possible to incorporate the activities of the owner and the design professional as well as the contractor, subcontractor, and material supplier into the schedule. This makes it easier to determine responsibility and to take appropriate corrective action in the case of delays. This would include such items as delivery of materials and checking of shop drawings. A fairly small, non-complex sample of a critical path network schedule is shown in Fig. 9.7. Some schedules are extremely extensive and can contain several hundred activities.

9.7.3 PRECEDENCE DIAGRAMING METHOD

Early development of CPM presented two basic methods of diagraming the network. The first method was to use the lines (or arrows) to represent the activities and the nodes to represent connection points and checkpoints in time (events). This method is in widespread use as CPM. In the second method, the nodes represent the activities and the lines show sequence and relationship. This method of diagraming has evolved into the Precedence Diagraming Method (PDM). Although CPM is the method in more widespread use today, an increasing number of companies are switching to PDM.

9.7.3.1 PDM STRUCTURE PDM uses the network node to represent the activities. While the nodes may be of any shape, they are usually squares or rectangles. The node will contain the name of the activity and its time or activity duration. Activities may also be identified by a number or letter designation. This is somewhat simpler than the method used in CPM since a single rather than a double identification is required. Under CPM no activity can be started until all preceding activities have been completed. This is only true for some activities under PDM. This represents the "Finish to Start" relationship. In other words, an activity cannot start until the preceding activity is finished. Under PDM, however, "Start to Start" and "Finish to Finish" relationships are also possible. Expressed another way, some activities can be started before the activity preceding them has been finished. An example of this is shown by activities M and L in Fig. 9.8. In this diagram, activity M (slab on grade) may be started before activity L (capillary fill) is completed. However, M cannot start until one-half of L has been accomplished. Using CPM, the activities would have to be divided to show this relationship, thus increasing the number of activities.

Furthermore, some activities, which have been started before the preceding activity is finished, cannot be completed until the preceding activity is finished. Using the same two activities (L and M) to illustrate this point, the slab on grade (M) cannot be finished until the capillary fill (L) is in place.

9.7.3.2 TIME CALCULATIONS The procedure for determining the critical activities in a PDM network is similar to the method used for CPM. Earliest start and finish, latest start and finish, and the amount of float time available can all be determined. The second part of Fig. 9.8 shows the results of time calculations for the sample PDM network.

9.7.3.3 PDM ADVANTAGES Some firms and individuals believe that PDM is easier to learn and is more readily understood than CPM. They point out that a smaller number of activities is required for the same project with a corresponding reduction in the number of calculations. They also claim that the network is easier to construct and to read, and that it offers a higher level of flexibility.

FIG. 9.7 CPM schedule

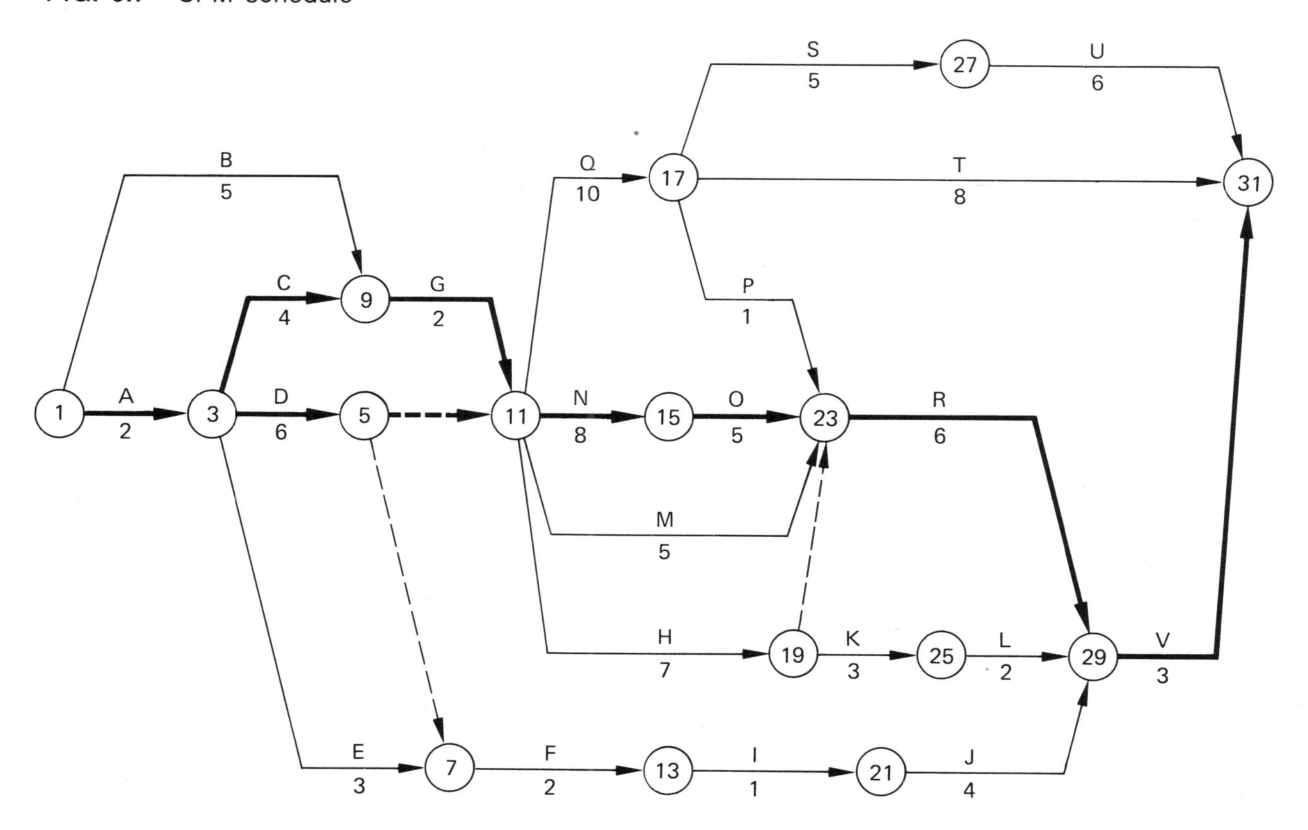

Darker lines indicate critical paths

i	j	ACTIVITY	DUR	ES	EF	LS	LF	TF
1	3	A	2	0	2	0	2	0
1	9	B	5	0	5	1	6	1
3	5	D	6	2	8	2	8	0
3	7	E	3	2	5	17	20	15
3	9	C	4	2	6	2	6	0
5	7	–––	0	–––	–––	–––	–––	–––
5	11	–––	0	–––	–––	–––	–––	–––
7	13	F	2	8	10	20	22	12
9	11	G	2	6	8	6	8	0
11	15	N	8	8	16	8	16	0
11	17	Q	10	8	18	9	19	1
11	19	H	7	8	15	14	21	6
11	23	M	5	8	13	16	21	8
13	21	I	1	10	11	22	23	12
15	23	O	5	16	21	16	21	0
17	23	P	1	18	19	20	21	2
17	27	S	5	18	23	19	24	1
17	31	T	8	18	26	22	30	4
19	23	–––	0	–––	–––	–––	–––	–––
19	25	K	3	15	18	22	25	7
21	29	J	4	11	15	23	27	12
23	29	R	6	21	27	21	27	0
25	29	L	2	18	20	25	27	7
27	31	U	6	23	29	24	30	1
29	31	V	3	27	30	27	30	0

FIG. 9.8 PDM network

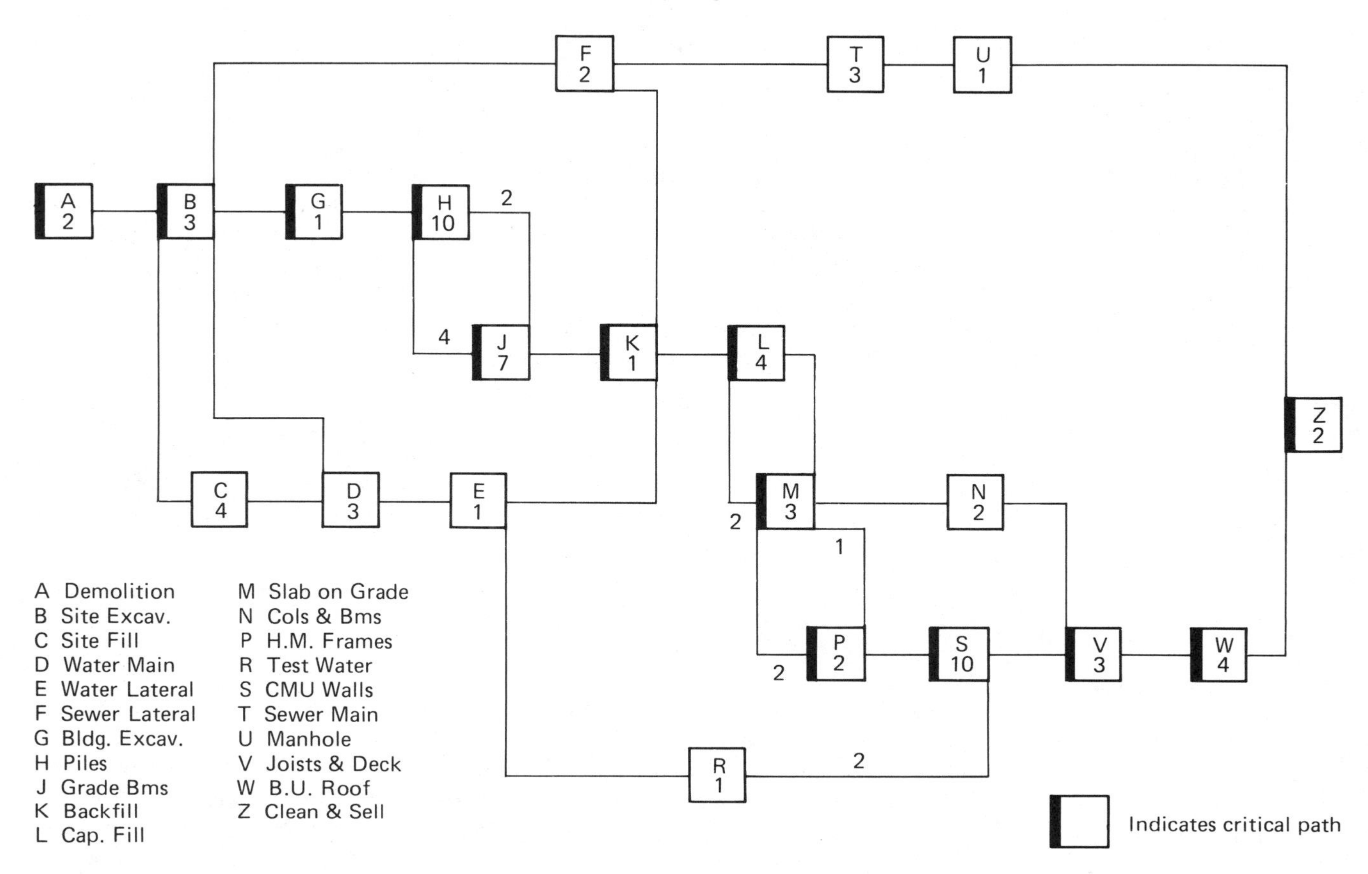

PDM TIME COMPUTATIONS

DES.	ACTIVITY	DUR	ES	EF	LS	LF	TF	CRI
A	Demolition	2	0	2	0	2	0	*
B	Site Exc.	3	2	5	2	5	0	*
C	Site Fill	4	2	6	11	15	9	
G	Bldg. Exc.	1	5	6	5	6	0	*
D	Water Main	3	6	9	15	18	9	
E	Water Lateral	1	9	10	18	19	9	
R	Test Water	1	10	11	32	33	22	
H	Piles	10	6	16	6	16	0	*
J	Grade Beams	7	10	18	11	18	1/0	*
K	Backfill	1	18	19	18	19	0	*
F	Sewer Lateral	2	5	7	17	19	12	
T	Sewer Main	3	7	10	38	41	31	
U	Manhole	1	10	11	41	42	31	
L	Cap. Fill	4	19	23	19	24	0/1	*
M	SOG	3	21	24	21	24	0	*
P	H.M. Frames	2	23	25	23	24	0	*
N	Columns & Beams	2	24	26	33	35	9	
S	CMU Walls	10	25	35	25	35	0	*
V	Joists & Deck	3	35	38	35	38	0	*
W	Roofing	4	38	42	38	42	0	*
Z	Clean-up & Sell	2	42	44	42	44	0	*

Reproduced by permission of Michael L. Whelan.

9.7.4 COMPUTER APPLICATIONS

Most of the advanced scheduling techniques adapt quite easily to manipulation by computer. Although the only basic talent that a computer has is the ability to add, subtract, multiply, and divide, it does so at an almost astronomical rate and through arrangement of commands can be made to perform many different functions. The amount of information generated from a construction schedule can be increased many times per unit of time by using a computer. In addition, it allows for much faster feedback than a manual system, with a decrease in errors. The computer also facilitates the tying together of the schedule with the cost control system. Computer driven graphic devices can be used to generate charts and graphs of various aspects of the schedule that are often more easily understood by some personnel than numerical printouts.

9.8 PAYMENT SCHEDULES

Most construction contracts require the contractor to submit a proposed schedule of payments prior to the beginning of construction. While this is most commonly done after the signing of the contract, there may be instances when submission must accompany the proposal. The payment schedule is subject to approval by the owner and/or the design professional. It should be of concern to all three parties that the payment schedule be as closely as possible an accurate projection of realistic percentages of payment as the job progresses.

On occasion an unbalanced schedule may be submitted. Under an unbalanced schedule, the contractor is proposing to "front load" the payment plan by recovering a greater amount of monies at the beginning of the job. If such a schedule is approved, the owner may be faced with an insufficient amount of unpaid funds under the contract to secure its completion in case of default by the contractor. In cases where it can be shown that the design professional was negligent in approving such a schedule, the courts have found the design professional liable for the resulting loss to the owner or the surety company.

The contractor who knowingly submits such a schedule is most often courting financial problems since companies as well as individuals tend to spend the funds on hand rather than hewing to a realistic budget. It is logical for payment to be unbalanced only to the extent that the contractor should be able to recover the project set-up and start-up expenses at the beginning, rather than distributing them over the life of the job.

9.8.1 NEED FOR ACCURACY

The payment schedule is of importance to both the owner and the contractor. The owner needs the information for financing and budget purposes. Banks will often arrange financing for an owner based upon a payment schedule, with one rate of interest being applied up to a predetermined expenditure, and another rate applying to the balance. For such financing plans to work properly, the payment schedule must be reliable.

In the case of a public funded project, the rate at which the work is to be performed is sometimes tied directly to the rate at which funds from a particular revenue bond issue become available. Under this situation, it is impossible for the owner to pay at a faster rate than planned on. If the contractor's payment schedule contains a serious error in timing, the contractor may be faced with the necessity of financing parts of the project until funds become available. Payments under the payment schedule are disbursed by means of a form such as that shown in Fig. 9.9.

FIG. 9.9 Application and certificate for payment

APPLICATION AND CERTIFICATE FOR PAYMENT

AIA DOCUMENT G702

PAGE ONE OF PAGES

TO (Owner):

PROJECT:

APPLICATION NO:

PERIOD FROM:
TO:

Distribution to:
☐ OWNER
☐ ARCHITECT
☐ CONTRACTOR
☐
☐

ATTENTION:

CONTRACT FOR:

ARCHITECT'S PROJECT NO:

CONTRACT DATE:

CONTRACTOR'S APPLICATION FOR PAYMENT

CHANGE ORDER SUMMARY			
Change Orders approved in previous months by Owner TOTAL		ADDITIONS	DEDUCTIONS
Approved this Month			
Number	Date Approved		
TOTALS			
Net change by Change Orders			

The undersigned Contractor certifies that to the best of his knowledge, information and belief the Work covered by this Application for Payment has been completed in accordance with the Contract Documents, that all amounts have been paid by him for Work for which previous Certificates for Payment were issued and payments received from the Owner, and that current payment shown herein is now due.

CONTRACTOR:

By: ______________ Date: ______

Application is made for Payment, as shown below, in connection with the Contract. **Continuation Sheet, AIA Document G703,** is attached.

The present status of the account for this Contract is as follows:

ORIGINAL CONTRACT SUM .... $______

Net change by Change Orders .... $______

CONTRACT SUM TO DATE .... $______

TOTAL COMPLETED & STORED TO DATE .... $______
(Column G on G703)

RETAINAGE ______% .... $______
or total in Column I on G703

TOTAL EARNED LESS RETAINAGE .... $______

LESS PREVIOUS CERTIFICATES FOR PAYMENT .... $______

CURRENT PAYMENT DUE .... $______

State of: County of:
Subscribed and sworn to before me this day of , 19
Notary Public:
My Commission expires:

ARCHITECT'S CERTIFICATE FOR PAYMENT

In accordance with the Contract Documents, based on on-site observations and the data comprising the above application, the Architect certifies to the Owner that the Work has progressed to the point indicated; that to the best of his knowledge, information and belief, the quality of the Work is in accordance with the Contract Documents; and that the Contractor is entitled to payment of the AMOUNT CERTIFIED.

AMOUNT CERTIFIED .... $______

(Attach explanation if amount certified differs from the amount applied for.)

ARCHITECT:

By: ______________ Date: ______

This Certificate is not negotiable. The AMOUNT CERTIFIED is payable only to the Contractor named herein. Issuance, payment and acceptance of payment are without prejudice to any rights of the Owner or Contractor under this Contract.

AIA DOCUMENT G702 • APPLICATION AND CERTIFICATE FOR PAYMENT • APRIL 1978 EDITION • AIA® • © 1978
THE AMERICAN INSTITUTE OF ARCHITECTS, 1735 NEW YORK AVENUE, N.W., WASHINGTON, D.C. 20006

G702 — 1978

This document has been reproduced with the permission of the American Institute of Architects under application number 79082. Further reproduction, in part or in whole, is not authorized. Because AIA documents are revised from time to time, users should ascertain from AIA the current edition of the document reproduced above.

FIG. 9.9 *(continued)*

CONTINUATION SHEET *AIA DOCUMENT G703* PAGE OF PAGES

AIA Document G702, APPLICATION AND CERTIFICATE FOR PAYMENT, containing
Contractor's signed Certification is attached.
In tabulations below, amounts are stated to the nearest dollar.
Use Column I on Contracts where variable retainage for line items may apply.

APPLICATION NUMBER:
APPLICATION DATE:
PERIOD FROM:
TO:
ARCHITECT's PROJECT NO:

A	B	C	D	E	F	G		H	I
ITEM No.	DESCRIPTION OF WORK	SCHEDULED VALUE	WORK COMPLETED			TOTAL COMPLETED AND STORED TO DATE (D+E+F)	% (G ÷ C)	BALANCE TO FINISH (C–G)	RETAINAGE
				This Application					
			Previous Applications	Work in Place	Stored Materials (not in D or E)				

AIA DOCUMENT G703 • CONTINUATION SHEET • APRIL 1978 EDITION • AIA® • © 1978
THE AMERICAN INSTITUTE OF ARCHITECTS, 1735 NEW YORK AVE., N.W., WASHINGTON, D.C. 20006

G703 — 1978

The contractor's cash flow projections are based largely upon payment schedules for the projects under contract. These are the major sources of information regarding anticipated income and expenses, at least those that are directly related to a project. Without the input from the payment schedules, the contractor would find it difficult to determine amounts of working capital at various times in the future, which is vital to bidding decisions. Such information is also required by the surety company that is furnishing bonds for the contractor.

9.9 CONSTRUCTION INSURANCE

A contractor may carry a number of different types of insurance for a variety of reasons. Some types are required by law, primarily because the contractor is an employer. Examples would include unemployment compensation and workmen's compensation. Although some may include social security within this category, it is better classified as a tax rather than as a type of insurance.

The contractor may also elect to carry several types of insurance to reduce the risk of doing business. This category may include fire insurance, life insurance on key personnel, insurance covering the loss of valuable papers and documents, vehicle insurance, equipment insurance, and several other types. Most of the insurance in this group is optional and is carried as a result of the contractor's decision to pay someone else to underwrite a given risk. Such assumption of risk is accomplished by a conditional contract called a policy. Under the terms of the policy, the insurer promises to reimburse the policyholder for a specified loss or liability in exchange for the payment of a premium, or fee.

Another classification of insurance carried by a contractor is that which is required by the terms of the contract with the owner. This group would normally include property insurance on the project and various types of liability insurance to protect the owner. The type of insurance and the amounts of each type will be given in the bidding documents and in the form of the agreement.

9.9.1 CERTIFICATE OF INSURANCE

So that the owner's interests may be properly protected, the design professional requires the contractor to submit or have a surety submit a certificate of insurance. This certificate is prepared and signed by the insurance company and sent to the owner. By this means the insurer attests that the contractor is carrying the stated types and amounts of insurance listed therein. The design professional should check the certificate to determine that the coverage listed meets the requirements given in the contract documents. The certificate must often be submitted prior to the signing of the contract. At the very latest, its submission should be required before the field operations begin.

An example of a certificate of insurance is shown in Fig. 9.10. This sample contains a cancellation clause that is necessary to protect the owner's interests. Under the terms of this clause, the owner must be notified prior to the cancellation or change of any of the listed coverage. Such a provision will allow an owner time to make any necessary arrangements so that the project and the owner's interest are not unprotected.

FIG. 9.10 Certificate of insurance

CERTIFICATE OF INSURANCE

AIA DOCUMENT G705

This certifies to the Addressee shown below that the following described policies, subject to their terms, conditions and exclusions, have been issued to:

NAME & ADDRESS OF INSURED

COVERING (SHOW PROJECT NAME AND/OR NUMBER AND LOCATION)

Addressee:
(Owner)

Date

KIND OF INSURANCE	POLICY NUMBER	Inception/Expiration Date	LIMITS OF LIABILITY	
1. (a) Workmen's Comp.			$//////////////	Statutory Workmen's Compensation
(b) Employers' Liability			$	One Accident and Aggregate Disease
2. Comprehensive General Liability			$	Each Occurrence—Premises and Operations
			$	Each Occurrence—Independent Contractors
			$	Each Occurrence—COMPLETED OPERATIONS AND PRODUCTS
(a) Bodily Injury			$	Each Occurrence—Contractual
			$	Aggregate—COMPLETED OPERATIONS AND PRODUCTS
(b) Personal Injury			$	Each Person Aggregate
			$	General Aggregate
(c) Property Damage			$	Each Occurrence—Premises—Operations
			$	Each Occurrence—INDEPENDENT CONTRACTOR
			$	Each Occurrence—COMPLETED OPERATIONS AND PRODUCTS
			$	Each Occurrence—Contractual
			$	Aggregate—
			$	Aggregate—OPERATIONS, INDEPENDENT CONTRACTOR, PRODUCTS AND CONTRACTUAL
3. Comprehensive Automobile Liability				
			$	Each Person—
(a) Bodily Injury			$	Each Occurrence—
(b) Property Damage			$	Each Occurrence—
4. (Other)				

UNDER GENERAL LIABILITY POLICY OR POLICIES

Yes No

1. Does Property Damage Liability Insurance shown include coverage for XC and U hazards? . . . . ________ ________
2. Is Occurrence Basis Coverage provided under Property Damage Liability? . . . . ________ ________
3. Is Broad Form Property Damage Coverage provided for this Project? . . . . ________ ________
4. Does Personal Injury Liability Insurance include coverage for personal injury sustained by any person as a result of an offense directly or indirectly related to the employment of such person by the Insured? . . . . ________ ________
5. Is coverage provided for Contractual Liability (including indemnification provision) assumed by Insured? . . . . ________ ________

UNDER AUTOMOBILE LIABILITY POLICY OR POLICIES

1. Does coverage above apply to non-owned and hired automobiles? . . . . ________ ________
2. Is Occurrence Basis Coverage provided under Property Damage Liability? . . . . ________ ________

CANCELLATION OR NON-RENEWAL
In the event of cancellation or non-renewal of any of the foregoing, fifteen (15) days written notice shall be given to the party to whom this certificate is addressed.

EXTENT OF CERTIFICATION
This certificate is issued as a matter of information only and confers no rights upon the holder. By its issuance the company does not alter, change, modify or extend any of the provisions of the above policies.

NAME OF INSURANCE COMPANY

ADDRESS

SIGNATURE OF AUTHORIZED REPRESENTATIVE

AIA DOCUMENT G705 • CERTIFICATE OF INSURANCE • FEBRUARY 1973 EDITION • AIA® • ©1973
THE AMERICAN INSTITUTE OF ARCHITECTS, 1735 NEW YORK AVE., NW, WASHINGTON, D.C. 20006 ONE PAGE

This document has been reproduced with the permission of the American Institute of Architects under application number 79082. Further reproduction, in part or in whole, is not authorized. Because AIA documents are revised from time to time, users should ascertain from AIA the current edition of the document reproduced above.

It is standard procedure (some regard it as mandatory) for the general contractor to extend to the subcontractors insurance requirements that are similar to those required of the general contractor. Such coverage is checked by requiring a certificate of insurance, which should be sent to the general contractor. This procedure will ensure that each subcontractor carries and maintains the insurance required by the individual subcontracts.

10 SUBCONTRACTS AND PURCHASE ORDERS

FIGURES

10.1 THE SUBCONTRACTING PROCESS

The construction firm that has a direct contractual relationship with the owner is called a prime contractor. A firm having a contractual relationship with a prime contractor to perform on-site work is called a subcontractor. There is yet another tier known as the sub-subcontractor. This is a firm having a contractual relationship with the subcontractor. Although in past years it was common for a construction firm to perform most of the work on a project with its own forces, the practice today is to make widespread use of the subcontracting system.

The use of subcontractors is partly due to the wish of the general contractor to reduce the risk of exposure from a financial standpoint. This is accomplished through the delegation of responsibility by means of the subcontract terms. Increased use of the subcontract system is also the result of the increased specialization that has taken place in the construction industry, as it has in almost all other industries in the country. This increased specialization has tended to improve productivity and efficiency. By utilizing the subcontract system, the general contractor reduces the risk of exposure by delegating parts of the responsibility for the total project to other firms. While the ultimate responsibility remains that of the general contractor, the sharing of this risk by the subcontractors is real and effective. The majority of the work for a civil project is still performed by the general contractor, but on building projects it is not unusual for 80 percent or more of the project to be subcontracted.

10.1.1 RELATIONSHIP BETWEEN PRIME AND SUBCONTRACTORS

There is no contractual relationship between the owner and the subcontractors. As a result, the owner still looks to the general contractor for guarantee that the work will be done promptly and in accordance with the contract documents. The prime contractor has the responsibility to see that the subcontractors meet their obligations under the subcontracts. This responsibility must be met so that the terms of the contract with the owner can be fulfilled. At the same time, the prime contractor has certain responsibilities to the subcontractors:

To award the subcontract to the lowest responsible bidder in each branch of the work. Just as the prime contractor expects to be treated fairly in this respect, so should similar consideration be extended to the subcontractors. The prime contractor should have no difficulty recognizing the cost of bidding incurred by the subcontractors and the attendant expectation of a contract award to the low bid. Failure to observe this practice will generally disadvantage the prime contractor more in the long run than it will the individual subcontractor.

To provide competent and responsible supervision to coordinate the work of the subcontractors with that of the prime contractor and with other subcontractors. The award of a subcontract is not an abdication of responsibility by the prime contractor for that branch of the work. Rather, it is a partial delegation of responsibility, with the ultimate responsibility remaining with the prime contractor. In order for this dual responsibility to be met, proper supervision must be furnished.

To pay legitimate statements under the subcontract in a prompt manner. Each construction firm is in business to realize a fair and reasonable profit. This is true whether the firm is operating as a prime or as a sub. Since the subcontractor is dependent upon the prime contractor for payment, it should be forthcoming in a fair and equitable manner. Although some prime contractors partly finance their projects by late payments to the subcontractors, this unfortunate practice should be discouraged.

To serve as an advocate for the subcontractor in disputes with the owner or design professional when the subcontractor has a valid claim. Since the subcontractor has no contractual relationship with owner, the subcontractor may be at a disadvantage in pressing for settlement of a legitimate claim. It should be the responsibility of the prime contractor to protect the subcontractor and ensure fair dealings.

To establish a communication system for the project that will inform the subcontractor sufficiently in advance when work is expected to be performed under the subcontract. This is closely related to the prime contractor's responsibility to furnish proper supervision, but it is ignored in a sufficient number of cases to warrant special mention. Along with a project schedule furnished to the subcontractor, the prime contractor should provide update information as it becomes necessary.

10.1.2 BONDING OF SUBCONTRACTORS

It is sometimes necessary to require a subcontractor to furnish a performance bond. In some instances this is required because the prime contractor is not familiar with the subcontractor and wishes additional assurance that the subcontract will be executed properly. In others it may be required by the prime contractor's bonding company. If the prime contractor is approaching the bonding limit established by the surety, the surety may request that the risk of exposure be reduced by requiring performance bonds from the major subcontractors. Any decision requiring bonds of the subcontractors should be made known to them before bids are taken. It should also be determined in advance whether the premiums for the bonds are to be paid for by the prime contractor or by the subcontractor.

10.1.3 FORMS OF SUBCONTRACT

There are many forms of subcontracts that can be used. Some prime contractors prepare their own form in consultation with an attorney, while others prefer to use one of the standard forms that have been prepared by an association. The Associated General Contractors of America, the American Subcontractors Association, the American Institute of Architects, and the National Society of Professional Engineers are among those organizations that publish and promulgate standard subcontract forms. One such form of subcontract is shown in Fig. 10.1. These standard forms are generally acceptable to the subcontractor and have the advantage of utilizing standard clauses and phrases whose meanings and intentions have often already been tested in the courts. Regardless of which form of subcontract is used, it should be reviewed periodically to determine that it meets the needs and desires of the firm. Many problems, and even litigation, can be avoided by a well-prepared form of subcontract.

(text continues on page 203)

FIG. 10.1 Subcontract

STANDARD SUBCONTRACT AGREEMENT

THIS AGREEMENT made this day of in the year Nineteen

Hundred and by and between

hereinafter

called the Subcontractor and

hereinafter called the Contractor.

WITNESSETH, That the Subcontractor and Contractor for the consideration hereinafter named agree as follows:

(Developed as a guide by The Associated General Contractors of America, The National Electrical Contractors Association, The Mechanical Contractors Association of America. The Sheet Metal and Air Conditioning Contractors National Association and the National Association of Plumbing - Heating - Cooling Contractors ©1966 by the Associated General Contractors of America and the Associated Specialty Contractors, Inc.)

Reproduced by permission of The Associated General Contractors of America and The Associated Specially Contractors, Inc.

FIG. 10.1 *(continued)*

ARTICLE I

The Subcontractor agrees to furnish all material and perform all work as described in Article II hereof for

(Here name the project.)

for

(Here name the Contractor.)

at

(Here insert the location of the work and name of Owner.)

in accordance with this Agreement, the Agreement between the Owner and Contractor, and in accordance with the General Conditions of the Contract, Supplementary General Conditions, the Drawings and Specifications and addenda prepared by

hereinafter called the Architect or Owner's authorized agent, all of which documents, signed by the parties thereto or identified by the Architect or Owner's authorized agent, form a part of a Contract

between the Contractor and the Owner dated , 19 , and hereby become a part of this contract, and herein referred to as the Contract Documents, and shall be made available to the Subcontractor upon his request prior to and at anytime subsequent to signing this Subcontract.

ARTICLE II

The Subcontractor and the Contractor agree that the materials and equipment to be furnished and work to be done by the Subcontractor are:

(Here insert a precise description of the work, preferably by reference to the numbers of the drawings and the pages of the specifications including addenda and accepted alternates.)

FIG. 10.1 *(continued)*

ARTICLE III

Time is of the essence and the Subcontractor agrees to commence and to complete the work as described in Article II as follows:

(Here insert any information pertaining to the method of notification for commencement of work, starting and completion dates, or duration, and any liquidated damage requirements.)

(a) No extension of time of this contract will be recognized without the written consent of the Contractor which consent shall not be withheld unreasonably consistent with Article X-4 of this Contract, subject to the arbitration provisions herein provided.

ARTICLE IV

The Contractor agrees to pay the Subcontractor for the performance of this work the

sum of ($)
in current funds, subject to additions and deductions for changes as may be agreed upon in writing, and to make monthly payments on account thereof in accordance with Article X, Sections 20-23 inclusive.

(Here insert additional details—unit prices, etc., payment procedure including date of monthly applications for payment, payment procedure if other than on a monthly basis, consideration of materials safely and suitably stored at the site or at some other location agreed upon in writing by the parties—and any provisions made for limiting or reducing the amount retained after the work reaches a certain stage of completion which should be consistent with the Contract Documents.)

FIG. 10.1 *(continued)*

ARTICLE V

Final payment shall be due when the work described in this contract is fully completed and performed in accordance with the Contract Documents, and payment to be consistent with Article IV and Article X, Sections 18, 20-23 inclusive of this contract.

Before issuance of the final payment the Subcontractor if required shall submit evidence satisfactory to the Contractor that all payrolls, material bills, and all known indebtedness connected with the Subcontractor's work have been satisfied.

ARTICLE VI
Performance and Payment Bonds

(Here insert any requirement for the furnishing of performance and payment bonds.)

FIG. 10.1 *(continued)*

ARTICLE VII
Temporary Site Facilities

(Here insert any requirements and terms concerning temporary site facilities, i.e., storage, sheds, water, heat, light, power, toilets, hoists, elevators, scaffolding, cold weather protection, ventilating, pumps, watchman service, etc.)

ARTICLE VIII
Insurance

Unless otherwise provided herein, the Subcontractor shall have a direct liability for the acts of his employees and agents for which he is legally responsible, and the Subcontractor shall not be required to assume the liability for the acts of any others.

Prior to starting work the insurance required to be furnished shall be obtained from a responsible company or companies to provide proper and adequate coverage and satisfactory evidence will be furnished to the Contractor that the Subcontractor has complied with the requirements as stated in this Section.

(Here insert any insurance requirements and Subcontractor's responsibility for obtaining, maintaining and paying for necessary insurance, not less than limits as may be specified in the Contract Documents or required by laws. This to include fire insurance and extended coverage, consideration of public liability, property damage, employer's liability, and workmen's compensation insurance for the Subcontractor and his employees. The insertion should provide the agreement of the Contractor and the Subcontractor on subrogation waivers, provision for notice of cancellation, allocation of insurance proceeds, and other aspects of insurance.)

(It is recommended that the AGC Insurance and Bonds Checklist (AGC Form No. 29) be referred to as a guide for other insurance coverages.)

FIG. 10.1 *(continued)*

ARTICLE IX
Job Conditions

(Here insert any applicable arrangements and necessary cooperation concerning labor matters for the project.)

ARTICLE X

In addition to the foregoing provisions the parties also agree:
That the Subcontractor shall:

(1) Be bound to the Contractor by the terms of the Contractor Documents and this Agreement, and assume toward the Contractor all the obligations and responsibilities that the Contractor, by those documents, assumes toward the Owner, as applicable to this Subcontract. (a) Not discriminate against any employee or applicant for employment because of race, creed, color, or national origin.

(2) Submit to the Contractor applications for payment at such times as stipulated in Article IV so as to enable the Contractor to apply for payment.

If payments are made on valuations of work done, the Subcontractor shall, before the first application, submit to the Contractor a schedule of values of the various parts of the work, aggregating the total sum of the Contract, made out in such detail as the Subcontractor and Contractor may agree upon, or as required by the Owner, and, if required, supported by such evidence as to its correctness as the Contractor may direct. This schedule, when approved by the Contractor, shall be used as a basis for Certificates for Payment, unless it be found to be in error. In applying for payment, the Subcontractor shall submit a statement based upon this schedule.

If payments are made on account of materials not incorporated in the work but delivered and suitably stored at the site, or at some other location agreed upon in writing, such payments shall be in accordance with the terms and conditions of the Contract Documents.

(3) Pay for all materials and labor used in, or in connection with, the performance of this contract, through the period covered by previous payments received from the Contractor, and furnish satisfactory evidence when requested by the Contractor, to verify compliance with the above requirements.

(4) Make all claims for extras, for extensions of time and for damage for delays or otherwise, promptly to the Contractor consistent with the Contract Documents.

(5) Take necessary precaution to properly protect the finished work of other trades.

(6) Keep the building and premises clean at all times of debris arising out of the operation of this subcontract. The Subcontractor shall not be held responsible for unclean conditions caused by other contractors or subcontractors, unless otherwise provided for.

(7) Comply with all statutory and/or contractual safety requirements applying to his work and/or initiated by the Contractor, and shall report within 3 days to the Contractor any injury to the Subcontractor's employees at the site of the project.

FIG. 10.1 *(continued)*

(8) (a) Not assign this subcontract or any amounts due or to become due thereunder without the written consent of the contractor. (b) Nor subcontract the whole of this subcontract without the written consent of the contractor. (c) Nor further subcontract portions of this subcontract without written notification to the contractor when such notification is requested by the contractor.

(9) Guarantee his work against all defects of materials and/or workmanship as called for in the plans, specifications and addenda, or if no guarantee is called for, then for a period of one year from the dates of partial or total acceptance of the Subcontractor's work by the Owner.

(10) And does hereby agree that if the Subcontractor should neglect to prosecute the work diligently and properly or fail to perform any provision of this contract, the Contractor, after three days written notice to the Subcontractor, may, without prejudice to any other remedy he may have, make good such deficiencies and may deduct the cost thereof from the payment then or thereafter due the Subcontractor, provided, however, that if such action is based upon faulty workmanship the Architect or Owner's authorized agent, shall first have determined that the workmanship and/or materials is defective.

(11) And does hereby agree that the Contractor's equipment will be available to the Subcontractor only at the Contractor's discretion and on mutually satisfactory terms.

(12) Furnish periodic progress reports of the work as mutually agreed including the progress of materials or equipment under this Agreement that may be in the course of preparation or manufacture.

(13) Make any and all changes or deviations from the original plans and specifications without nullifying the original contract when specifically ordered to do so in writing by the Contractor. The Subcontractor prior to the commencement of this revised work, shall submit promptly to the Contractor written copies of the cost or credit proposal for such revised work in a manner consistent with the Contract Documents.

(14) Cooperate with the Contractor and other Subcontractors whose work might interfere with the Subcontractor's work and to participate in the preparation of coordinated drawings in areas of congestion as required by the Contract Documents, specifically noting and advising the Contractor of any such interference.

(15) Cooperate with the Contractor in scheduling his work so as not conflict or interfere with the work of others. To promptly submit shop drawings, drawings, and samples, as required in order to carry on said work efficiently and at speed that will not cause delay in the progress of the Contractor's work or other branches of the work carried on by other Subcontractors.

(16) Comply with all Federal, State and local laws and ordinances applying to the building or structure and to comply and give adequate notices relating to the work to proper authorities and to secure and pay for all necessary licenses or permits to carry on the work as described in the Contract Documents as applicable to this Subcontract.

(17) Comply with Federal, State and local tax laws, Social Security laws and Unemployment Compensation laws and Workmen's Compensation Laws insofar as applicable to the performance of this subcontract.

(18) And does hereby agree that all work shall be done subject to the final approval of the Architect or Owner's authorized agent, and his decision in matters relating to artistic effect shall be final, if within the terms of the Contract Documents.

That the Contractor shall—

(19) Be bound to the Subcontractor by all the obligations that the Owner assumes to the Contractor under the Contract Documents and by all the provisions thereof affording remedies and redress to the Contractor from the Owner insofar as applicable to this Subcontract.

(20) Pay the Subcontractor within seven days, unless otherwise provided in the Contract Documents, upon the payment of certificates issued under the Contractor's schedule of values, or as described in Article IV herein. The amount of the payment shall be equal to the percentage of completion certified by the Owner or his authorized agent for the work of this Subcontractor applied to the amount set forth under Article IV and allowed to the Contractor on account of the Subcontractor's work to the extent of the Subcontractor's interest therein.

(21) Permit the Subcontractor to obtain direct from the Architect or Owner's authorized agent, evidence of percentages of completion certified on his account.

(22) Pay the Subcontractor on demand for his work and/or materials as far as executed and fixed in place, less the retained percentage, at the time the payment should be made to the Subcontractor if the Architect or Owner's authorized agent fails to issue the certificate for any fault of the Contractor and not the fault of the Subcontractor or as otherwise provided herein.

(23) And does hereby agree that the failure to make payments to the Subcontractor as herein provided for any cause not the fault of the Subcontractor, within 7 days from the Contractor's receipt of payment or from time payment should be made as

FIG. 10.1 *(continued)*

provided in Article X, Section 22, or maturity, then the Subcontractor may upon 7 days written notice to the Contractor stop work without prejudice to any other remedy he may have.

(24) Not issue or give any instructions, order or directions directly to employees or workmen of the Subcontractor other than to the persons designated as the authorized representative(s) of the Subcontractor.

(25) Make no demand for liquidated damages in any sum in excess of such amount as may be specifically named in the subcontract, provided, however, no liquidated damages shall be assessed for delays or causes attributable to other Subcontractors or arising outside the scope of this Subcontract.

(26) And does hereby agree that no claim for services rendered or materials furnished by the Contractor to the Subcontractor shall be valid unless written notice thereof is given by the Contractor to the Subcontractor during the first ten days of the calendar month following that in which the claim originated.

(27) Give the Subcontractor an opportunity to be present and to submit evidence in any arbitration involving his rights.

(28) Name as arbitor under arbitration proceedings as provided in the General Conditions the person nominated by the Subcontractor, if the sole cause of dispute is the work, materials, rights or responsibilities of the Subcontractor; or if, of the Subcontractor and any other Subcontractor jointly, to name as such arbitrator the person upon whom they agree.

That the Contractor and the Subcontractor agree—

(29) That in the matter of arbitration, their rights and obligations and all procedure shall be analogous to those set forth in the Contract Documents provided, however, that a decision by the Architect or Owner's authorized agent, shall not be a condition precedent to arbitration.

(30) This subcontract is solely for the benefit of the signatories hereto.

ARTICLE XI

IN WITNESS WHEREOF the parties hereto have executed this Agreement under seal, the day and year first above written.

Attest: ________________ Subcontractor

________________ (Seal) ________________ By (Title)

Attest: ________________ Contractor

________________ (Seal) ________________ By (Title)

10.1.4 RELATING SUBCONTRACT TO PRIME CONTRACT

It is desirable that the subcontract include many of the same provisions contained in the prime contract. These may be specifically mentioned or they may be included by reference. Among those provisions usually included by reference are the General Conditions of the Agreement. These provisions ensure that the subcontractor will operate under the same general guidelines as those that apply to the prime contractor. Following this practice will also aid in furnishing the project to the owner in compliance with the prime contract. It is also desirable that the subcontract make reference to certain sections or paragraphs of the specifications when defining the work that is to be performed under the subcontract. As an example, instead of stating that the subcontractor is to perform all structural steel work, it would be better to state in the subcontract that all work as described and required under Section 5A of the specifications is to be included. This will ensure that any conditions and provisions peculiar to the project are not omitted.

10.1.5 METHOD OF PAYMENT

A particularly troublesome area in connection with subcontracts is the method of payment that is to be used. The standard practice for prime contractors within the construction industry is for bills of the preceding month to be submitted by the first of the following month, with payment being made by the twentieth of the month. A similar practice is probably the best procedure to follow with respect to subcontracts. Subcontractors initiate the payment process by submitting an application for payment. A typical example of the many forms in use is shown in Fig. 10.2.

Some subcontracts contain the provision that the prime contractor is not obligated to pay the subcontractor until the prime has received payment from the owner. Many subcontractors feel that this is unfair since they have no control over the situation and do not have a contractual relationship with the owner. On the other side of the argument is the fact that the prime contractor may not be financially able to pay the subcontractor until payment has been received from the owner. This aspect of subcontracts has been the subject of much heated discussion and is still unresolved. Even court decisions have failed to furnish a clear-cut guideline since conflicting rulings have been handed down by different courts. Prime contractors will generally discover that if they deal fairly with the subcontractors in all other aspects of the work, this particular item will seldom become a major obstacle.

10.1.6 RETAINAGE

The method of paying subcontractors is related to the subject of retainage under the subcontracts, a matter of vital concern. The general contract will usually allow the owner to retain a portion of the monthly billing until the end of the project, although many owners are now reducing the percentage of the retainage after one-half of the project is completed. The owner uses this retained percentage as a lever to ensure completion and/or correction of uncompleted or improperly done portions of the work. The prime contractor has in the past followed a similar practice for similar reasons with respect to the subcontractors. This generally causes no problems with subcontractors whose work is performed during the latter stages of the project.

FIG. 10.2 Subcontractor's application for payment

SUBCONTRACTOR'S APPLICATION FOR PAYMENT

(Developed as a guide by The Associated General Contractors of America, The National Electrical Contractors Association, The Mechanical Contractors Association of America. The Sheet Metal and Air Conditioning Contractors National Association and The National Association of Plumbing-Heating-Cooling Contractors.)

TO: ______________________________

FROM: ______________________________

____

PROJECT:

PAYMENT REQUEST NO. ______________________________

PERIOD ____________, 19____, to ____________, 19____.

STATEMENT OF CONTRACT ACCOUNT:

1.	Original Contract Amount	$________
2.	Approved Change Order Nos. ________ (As per attached breakdown) (Net)	$________
3.	Adjusted Contract Amount	$________
4.	Value of Work Completed to Date: (As per attached breakdown)	$________
5.	Value of Approved Change Orders Completed: (As per attached breakdown)	$________
6.	Materials Stored on Site: (As per attached breakdown)	$________
7.	Total to Date	$________
8.	Less Amount Retained (________%)	($________)
9.	Total Less Retainage	$________
10.	Total Previously Certified (Deduct)	$________
11.	AMOUNT DUE THIS REQUEST	$________

CERTIFICATE OF THE SUBCONTRACTOR:

I hereby certify that the work performed and the materials supplied to date, as shown on the above represent the actual value of accomplishment under the terms of the Contract (and all authorized changes thereto) between the undersigned and ______________________ relating to the above referenced project.

I also certify that payments, less applicable retention, have been made through the period covered by previous payments received from the contractor, to (1) all my subcontractors (sub-subcontractors) and (2) for all materials and labor used in or in connection with the performance of this Contract. I further certify I have complied with Federal, State and local tax laws, including Social Security laws and Unemployment Compensation laws and Workmen's Compensation laws insofar as applicable to the performance of this Contract.

Date ______________________________

Subscribed and sworn before me this ________ day of

____________________, 19____

Notary Public: ______________________________

My Commission Expires:

____
SUBCONTRACTOR

BY: ______________________________
(authorized signature)

TITLE: ______________________________

Reproduced by permission of The Association General Contractors of America.

However, for those whose work is performed and completed during the early stages of the project, it may work a financial hardship to have 10 or 20 percent of the subcontract retained until the entire project is completed. This is especially true if the project is one with a duration time of three to four years and the subcontractor's work has been completed within the first few months. Under these conditions the subcontractor may argue that any defects in the subcontractor's performance should be able to be discovered a considerable time before final completion. In an attempt to alleviate this situation, some subcontracts put a time limit within which final payment to the subcontractor shall be made. In other instances, the total amount of the retained percentage is reduced at varying stages of the project in order to effect more complete payment to the subcontractors.

10.1.7 OTHER INCLUSIONS

There are a myriad of other provisions from the prime contract that may have to be included in the subcontracts: a minority hiring provision, wage restrictions such as Davis-Bacon, state preference laws, and many others. The prime contractor may wish to include provisions for arbitration in case of contract disputes or for termination of the subcontract in the event of nonperformance by the subcontractor.

10.2 PURCHASE ORDERS

A purchase order is the method used by the contractor to purchase material. The order is limited to materials and does not contain provisions for any on-site labor. If this is to be included, a subcontract must be used. When the purchase order is signed by both the contractor and the material vendor, it becomes a material contract. As such, it is binding on both signatory parties and can be enforced in the same way as any other contract.

Purchase orders provide a system for the contractor to secure project materials in an orderly and efficient manner. They can be incorporated into the contractor's accounting system so that control can be exercised over them with regard to the delivery of material and payment for that material.

10.2.1 PURCHASING POLICY

Most of a contractor's material is purchased on a project basis. Each construction project contains unique requirements with respect to materials, and the contractor cannot realistically procure the material in question until the contract has been received. For this reason, it is impractical for the contractor to stockpile any but a small group of standard materials that are required on most of the firm's projects. In addition, stockpiling of materials may result in a loss to the contractor through obsolescence and deterioration. The cost of warehousing and additional handling and transportation costs also work against the advisability of stockpiling. An additional factor is that the contractor will have funds tied up in such an inventory. This can have a detrimental effect upon cash flow and bonding capacity.

Purchasing should be a centralized activity for the contractor. It should be handled by an individual in a smaller firm, or by a separate department in a larger firm. A strict system of control should be established to ensure that collusion and kickbacks are not practiced between the purchasing department and the material suppliers. Those persons responsible for purchasing should also maintain a comprehensive list of potential suppliers for the different materials required by the firm. This should be kept current and readily available to the estimating department.

10.2.2 PURCHASE ORDER FORMS

Most contractors prefer to prepare and use their own purchase order forms. In this way they are able to control the conditions of purchase and ensure that conditions that are favorable to them are included. As with all other forms prepared by the contractor, it is advisable to have them checked by an attorney before using them. They should contain provisions that will assist the contractor in maintaining control of the project but will not at the same time place unfair conditions upon the material suppliers. An example of a purchase order form is shown in Fig. 10.3. The following information should be included in such a form:

Complete description of the material ordered. This will include reference to any applicable brand names or other indicators of quality, and the amount of material being ordered. It is desirable that the description of the material refer to the appropriate sections of the specifications as a guarantee of accuracy.

Price to be paid. Information should be given regarding any unit prices or total price as well as the inclusion or deletion of such items as sales tax and delivery charges.

Delivery conditions. Both the time and the location regarding delivery of the material should be specified. Delivery time may be related to the project schedule and provisions should be made for minor changes in this time for the benefit of the contractor. However, reasonable limitations should be imposed upon this right to change. Delivery location should be specific. If nearest rail siding is acceptable, or if job site delivery is required, this should be clearly stated.

Inspection provisions. If inspection provisions are different from or have been added to those mentioned in the referenced specification sections, they should be clearly delineated in the purchase order. This would include possible plant inspections as well as job site delivery inspection.

Signatures. Provisions should be made for the necessary signatures. This may include not only the authorizing signature for the contractor but also the acceptance signature for the vendor. Provisions should also be made for dating each signature.

10.2.3 CONRACTOR FORM VS. VENDOR FORM

In some situations the vendor may refuse to sign the contractor's purchase order form. If this refusal comes about because an error is suspected in price or in some other condition, then the original quotation must be checked to determine the proper information and to make the necessary changes. The refusal may occur, however, because the vendor refuses to do business unless the vendor's form of purchase order is used. In some areas an entire group of vendors covering a particular material—for example, redi-mix concrete suppliers—may adopt this attitude. Under these circumstances there is usually little that the contractors can do to change the situation. When faced with this problem, the contractor must be especially careful in reviewing the conditions that are contained in the seller's form. Many such forms will include price escalation clauses that can work a hardship upon the contractor when the project in-

FIG. 10.3 Purchase order

PURCHASE ORDER

T. J. CHRISTIAN CONSTRUCTION

785 Willow Street

______, ______

Purchase Order No.________________ (This number must be placed on all invoices, packages and correspondence.)
To ____

____

____

Enter our order as follows:

DATE	REQUISITION NO.	SHIP VIA	TERMS	REQUIRED BY

Quantity	Description and Conditions	Price	Unit	Amount

Upon receipt, acknowledge order and advise regard shipping date	Authorized Signature

volves a long period of time. If there are particularly onerous conditions contained in the form, negotiations carried on at the association level will often be successful in getting them changed.

10.2.4 CASH DISCOUNTS

Most material suppliers offer a discount to the contractor if payment is made prior to the standard 30-day period. This is referred to as a cash discount and serves as an incentive to encourage early payment by the contractor, which will have the effect of improving the cash flow for the material supplier. A fairly common cash discount is 2% 10 da. 30 net, which translates into a 2 percent discount of the total bill if payment is made within 10 days after delivery; otherwise the total, or net, amount is due within 30 days after delivery. Many bonding companies and lending or financial institutions use the handling of cash discounts by the contractor as one measure of financial condition. It is also an excellent way for the contractor to improve the profit picture on a project.

DIVISION V

CONTRACT CONDITIONS

GENERAL CONDITIONS 11

FIGURES

11.1 DEFINITION

Over the years it has been determined that there are several provisions that can apply to most construction contracts. While a given construction project may contain unique requirements, the manner in which the contract is administered tends to conform closely with the manner in which most other construction contracts are administered. Those provisions that have been found to be of wide application have been collected into documents labeled as "General Conditions of the Contract" or some similar title.

11.2 PURPOSE

The general conditions spell out the contract conditions by defining the duties and responsibilities of those who are signatory to the contract, as well as the duties and responsibilities of those affected by the contract. These general conditions also define the relationship of the various parties involved in the construction project. They have as their purpose the establishment of the rules that will govern the administration of the construction contract, subcontracts, and all other activities in connection with the construction project.

The general conditions are usually bound within the same cover as the specifications, and should be included by reference as a part of each section of the trade specifications. They are also made a part of the agreement between the owner and the contractor, again by reference. The same procedure should be followed by the general contractor with respect to the subcontracts so that the subcontractors are legally bound to contract conditions that apply to the general contractor.

11.3 STANDARD FORMS

The content and subject matter of the general conditions do not fall within the field of expertise of many design professionals. For this reason, they should be prepared by an attorney, or at least in consultation with one. Many professional and industry associations have prepared and published standard versions of the general conditions. These standard forms possess several advantages:

They have been reviewed by persons having many years of experience within the construction industry. This experience has given them a good idea about where the trouble spots within the general conditions are and how to avoid them.

They have been prepared in consultation with legal personnel experienced in the use of construction contract documents.

They use, as much as possible, a language that has already been tested in the courts. This gives users the advantage of being able to predict, with some degree of assurance, how the courts will rule in case of a dispute.

They permit contractors and other users of the general conditions to work with a document that has become familiar to them. It therefore will save time for the bidders and tends to produce a more favorable price for the owner.

Standard forms of the general conditions have been prepared not only by groups such as the Associated General Contractors of America, the American Institute of Architects, the Consulting Engineers Council, and the American Society of Civil Engineers, but also by several governmental agencies. The list of governmental agencies includes the General Services Administration, the U.S. Corps of Engineers, and the departments of transportation in many states.

11.3.1 USE OF STANDARD FORMS

Whenever a specification writer elects to use one of the standardized versions of the general conditions, they should be used in their entirety. Each version presents a complete coverage of contract conditions. If the specification writer wishes to vary any of these requirements, changes should be effected through the special or supplemental general conditions. The practice of using only a part of a standardized version should be discouraged, since it defeats the purpose of its use. When a standardized version is used, a copy should also be included within the cover with each set of the specifications. Including the general conditions by reference may well cause the bidders to question which version is being used, since some of the standardized versions have undergone a number of revisions.

It is also not advisable for the design office to retype the standardized versions of the general conditions. Copies printed and sold by the association that has prepared them should be secured instead and included with the specifications. Retyping may raise a question as to whether the general conditions are in the original form, or whether they have been modified to include some unusual provisions. (See the section on escape clauses in Chapter 17.) This point may be reinforced by referring to the two examples of general conditions included in the text. Figure 11.1 illustrates the general conditions published by the American Institute of Architects, and Fig. 11.2, the general conditions for design-build agreements furnished by the Associated General Contractors of America. Note that Section 4 of each document refers to the contractor. However, the content of the two sections is considerably different because of the change in types of contract.

11.4 CONTENT

Although the specific language and to a lesser degree the content of different versions of the general conditions will vary, for the most part the general conditions should contain the following:

- Definition of the contract documents
- Duties and responsibilities of the parties to the contract
- Duties and responsibilities of the design professional
- Subcontractors
- Separate contracts
- Time aspects of the contract
- Payments and completion
- Insurance and bonds
- Changes in the work
- Arbitration and termination of contract
- Miscellaneous provisions

11.4.1 DEFINITION OF CONTRACT DOCUMENTS

That part of the general conditions dealing with the definition of the contract documents will usually include a listing of such documents. In addition, it is common practice to include a statement relative to the relationship between the documents. Of particular importance is information specifying which document has precedence in the case of conflicts. Statements regarding intent and interpretation of the documents are also included in this portion of the general condi-

(text continues on page 255)

FIG. 11.1 General conditions

THE AMERICAN INSTITUTE OF ARCHITECTS

AIA Document A201

General Conditions of the Contract for Construction

THIS DOCUMENT HAS IMPORTANT LEGAL CONSEQUENCES; CONSULTATION WITH AN ATTORNEY IS ENCOURAGED WITH RESPECT TO ITS MODIFICATION

1976 EDITION

TABLE OF ARTICLES

This document has been approved and endorsed by The Associated General Contractors of America.

Copyright 1911, 1915, 1918, 1925, 1937, 1951, 1958, 1961, 1963, 1966, 1967, 1970, © 1976 by The American Institute of Architects, 1735 New York Avenue, N.W., Washington, D. C. 20006. Reproduction of the material herein or substantial quotation of its provisions without permission of the AIA violates the copyright laws of the United States and will be subject to legal prosecution.

This document has been reproduced with the permission of the American Institute of Architects under application number 79082. Further reproduction, in part or in whole, is not authorized. Because AIA documents are revised from time to time, useres should ascertain from AIA the current edition of the document reproduced above.

FIG. 11.1 *(continued)*

INDEX

AIA DOCUMENT A201 • GENERAL CONDITIONS OF THE CONTRACT FOR CONSTRUCTION • THIRTEENTH EDITION • AUGUST 1976
AIA® • © 1976 • THE AMERICAN INSTITUTE OF ARCHITECTS, 1735 NEW YORK AVENUE, N.W., WASHINGTON, D.C. 20006

FIG. 11.1 *(continued)*

FIG. 11.1 *(continued)*

4 A201-1976 AIA DOCUMENT A201 • GENERAL CONDITIONS OF THE CONTRACT FOR CONSTRUCTION • THIRTEENTH EDITION • AUGUST 1976
AIA® • © 1976 • THE AMERICAN INSTITUTE OF ARCHITECTS, 1735 NEW YORK AVENUE, N.W., WASHINGTON, D.C. 20006

FIG. 11.1 *(continued)*

GENERAL CONDITIONS OF THE CONTRACT FOR CONSTRUCTION

ARTICLE 1

CONTRACT DOCUMENTS

1.1 DEFINITIONS

1.1.1 THE CONTRACT DOCUMENTS

The Contract Documents consist of the Owner-Contractor Agreement, the Conditions of the Contract (General, Supplementary and other Conditions), the Drawings, the Specifications, and all Addenda issued prior to and all Modifications issued after execution of the Contract. A Modification is (1) a written amendment to the Contract signed by both parties, (2) a Change Order, (3) a written interpretation issued by the Architect pursuant to Subparagraph 2.2.8, or (4) a written order for a minor change in the Work issued by the Architect pursuant to Paragraph 12.4. The Contract Documents do not include Bidding Documents such as the Advertisement or Invitation to Bid, the Instructions to Bidders, sample forms, the Contractor's Bid or portions of Addenda relating to any of these, or any other documents, unless specifically enumerated in the Owner-Contractor Agreement.

1.1.2 THE CONTRACT

The Contract Documents form the Contract for Construction. This Contract represents the entire and integrated agreement between the parties hereto and supersedes all prior negotiations, representations, or agreements, either written or oral. The Contract may be amended or modified only by a Modification as defined in Subparagraph 1.1.1. The Contract Documents shall not be construed to create any contractual relationship of any kind between the Architect and the Contractor, but the Architect shall be entitled to performance of obligations intended for his benefit, and to enforcement thereof. Nothing contained in the Contract Documents shall create any contractual relationship between the Owner or the Architect and any Subcontractor or Sub-subcontractor.

1.1.3 THE WORK

The Work comprises the completed construction required by the Contract Documents and includes all labor necessary to produce such construction, and all materials and equipment incorporated or to be incorporated in such construction.

1.1.4 THE PROJECT

The Project is the total construction of which the Work performed under the Contract Documents may be the whole or a part.

1.2 EXECUTION, CORRELATION AND INTENT

1.2.1 The Contract Documents shall be signed in not less than triplicate by the Owner and Contractor. If either the Owner or the Contractor or both do not sign the Conditions of the Contract, Drawings, Specifications, or any of the other Contract Documents, the Architect shall identify such Documents.

1.2.2 By executing the Contract, the Contractor represents that he has visited the site, familiarized himself with the local conditions under which the Work is to be performed, and correlated his observations with the requirements of the Contract Documents.

1.2.3 The intent of the Contract Documents is to include all items necessary for the proper execution and completion of the Work. The Contract Documents are complementary, and what is required by any one shall be as binding as if required by all. Work not covered in the Contract Documents will not be required unless it is consistent therewith and is reasonably inferable therefrom as being necessary to produce the intended results. Words and abbreviations which have well-known technical or trade meanings are used in the Contract Documents in accordance with such recognized meanings.

1.2.4 The organization of the Specifications into divisions, sections and articles, and the arrangement of Drawings shall not control the Contractor in dividing the Work among Subcontractors or in establishing the extent of Work to be performed by any trade.

1.3 OWNERSHIP AND USE OF DOCUMENTS

1.3.1 All Drawings, Specifications and copies thereof furnished by the Architect are and shall remain his property. They are to be used only with respect to this Project and are not to be used on any other project. With the exception of one contract set for each party to the Contract, such documents are to be returned or suitably accounted for to the Architect on request at the completion of the Work. Submission or distribution to meet official regulatory requirements or for other purposes in connection with the Project is not to be construed as publication in derogation of the Architect's common law copyright or other reserved rights.

ARTICLE 2

ARCHITECT

2.1 DEFINITION

2.1.1 The Architect is the person lawfully licensed to practice architecture, or an entity lawfully practicing architecture identified as such in the Owner-Contractor Agreement, and is referred to throughout the Contract Documents as if singular in number and masculine in gender. The term Architect means the Architect or his authorized representative.

2.2 ADMINISTRATION OF THE CONTRACT

2.2.1 The Architect will provide administration of the Contract as hereinafter described.

2.2.2 The Architect will be the Owner's representative during construction and until final payment is due. The Architect will advise and consult with the Owner. The Owner's instructions to the Contractor shall be forwarded

FIG. 11.1 *(continued)*

through the Architect. The Architect will have authority to act on behalf of the Owner only to the extent provided in the Contract Documents, unless otherwise modified by written instrument in accordance with Subparagraph 2.2.18.

2.2.3 The Architect will visit the site at intervals appropriate to the stage of construction to familiarize himself generally with the progress and quality of the Work and to determine in general if the Work is proceeding in accordance with the Contract Documents. However, the Architect will not be required to make exhaustive or continuous on-site inspections to check the quality or quantity of the Work. On the basis of his on-site observations as an architect, he will keep the Owner informed of the progress of the Work, and will endeavor to guard the Owner against defects and deficiencies in the Work of the Contractor.

2.2.4 The Architect will not be responsible for and will not have control or charge of construction means, methods, techniques, sequences or procedures, or for safety precautions and programs in connection with the Work, and he will not be responsible for the Contractor's failure to carry out the Work in accordance with the Contract Documents. The Architect will not be responsible for or have control or charge over the acts or omissions of the Contractor, Subcontractors, or any of their agents or employees, or any other persons performing any of the Work.

2.2.5 The Architect shall at all times have access to the Work wherever it is in preparation and progress. The Contractor shall provide facilities for such access so the Architect may perform his functions under the Contract Documents.

2.2.6 Based on the Architect's observations and an evaluation of the Contractor's Applications for Payment, the Architect will determine the amounts owing to the Contractor and will issue Certificates for Payment in such amounts, as provided in Paragraph 9.4.

2.2.7 The Architect will be the interpreter of the requirements of the Contract Documents and the judge of the performance thereunder by both the Owner and Contractor.

2.2.8 The Architect will render interpretations necessary for the proper execution or progress of the Work, with reasonable promptness and in accordance with any time limit agreed upon. Either party to the Contract may make written request to the Architect for such interpretations.

2.2.9 Claims, disputes and other matters in question between the Contractor and the Owner relating to the execution or progress of the Work or the interpretation of the Contract Documents shall be referred initially to the Architect for decision which he will render in writing within a reasonable time.

2.2.10 All interpretations and decisions of the Architect shall be consistent with the intent of and reasonably inferable from the Contract Documents and will be in writing or in the form of drawings. In his capacity as interpreter and judge, he will endeavor to secure faithful performance by both the Owner and the Contractor, will not show partiality to either, and will not be liable for the result of any interpretation or decision rendered in good faith in such capacity.

2.2.11 The Architect's decisions in matters relating to artistic effect will be final if consistent with the intent of the Contract Documents.

2.2.12 Any claim, dispute or other matter in question between the Contractor and the Owner referred to the Architect, except those relating to artistic effect as provided in Subparagraph 2.2.11 and except those which have been waived by the making or acceptance of final payment as provided in Subparagraphs 9.9.4 and 9.9.5, shall be subject to arbitration upon the written demand of either party. However, no demand for arbitration of any such claim, dispute or other matter may be made until the earlier of (1) the date on which the Architect has rendered a written decision, or (2) the tenth day after the parties have presented their evidence to the Architect or have been given a reasonable opportunity to do so, if the Architect has not rendered his written decision by that date. When such a written decision of the Architect states (1) that the decision is final but subject to appeal, and (2) that any demand for arbitration of a claim, dispute or other matter covered by such decision must be made within thirty days after the date on which the party making the demand receives the written decision, failure to demand arbitration within said thirty days' period will result in the Architect's decision becoming final and binding upon the Owner and the Contractor. If the Architect renders a decision after arbitration proceedings have been initiated, such decision may be entered as evidence but will not supersede any arbitration proceedings unless the decision is acceptable to all parties concerned.

2.2.13 The Architect will have authority to reject Work which does not conform to the Contract Documents. Whenever, in his opinion, he considers it necessary or advisable for the implementation of the intent of the Contract Documents, he will have authority to require special inspection or testing of the Work in accordance with Subparagraph 7.7.2 whether or not such Work be then fabricated, installed or completed. However, neither the Architect's authority to act under this Subparagraph 2.2.13, nor any decision made by him in good faith either to exercise or not to exercise such authority, shall give rise to any duty or responsibility of the Architect to the Contractor, any Subcontractor, any of their agents or employees, or any other person performing any of the Work.

2.2.14 The Architect will review and approve or take other appropriate action upon Contractor's submittals such as Shop Drawings, Product Data and Samples, but only for conformance with the design concept of the Work and with the information given in the Contract Documents. Such action shall be taken with reasonable promptness so as to cause no delay. The Architect's approval of a specific item shall not indicate approval of an assembly of which the item is a component.

2.2.15 The Architect will prepare Change Orders in accordance with Article 12, and will have authority to order minor changes in the Work as provided in Subparagraph 12.4.1.

AIA® • © 1976 • THE AMERICAN INSTITUTE OF ARCHITECTS, 1735 NEW YORK AVENUE, N.W., WASHINGTON, D.C. 20006

FIG. 11.1 *(continued)*

2.2.16 The Architect will conduct inspections to determine the dates of Substantial Completion and final completion, will receive and forward to the Owner for the Owner's review written warranties and related documents required by the Contract and assembled by the Contractor, and will issue a final Certificate for Payment upon compliance with the requirements of Paragraph 9.9.

2.2.17 If the Owner and Architect agree, the Architect will provide one or more Project Representatives to assist the Architect in carrying out his responsibilities at the site. The duties, responsibilities and limitations of authority of any such Project Representative shall be as set forth in an exhibit to be incorporated in the Contract Documents.

2.2.18 The duties, responsibilities and limitations of authority of the Architect as the Owner's representative during construction as set forth in the Contract Documents will not be modified or extended without written consent of the Owner, the Contractor and the Architect.

2.2.19 In case of the termination of the employment of the Architect, the Owner shall appoint an architect against whom the Contractor makes no reasonable objection whose status under the Contract Documents shall be that of the former architect. Any dispute in connection with such appointment shall be subject to arbitration.

ARTICLE 3

OWNER

3.1 DEFINITION

3.1.1 The Owner is the person or entity identified as such in the Owner-Contractor Agreement and is referred to throughout the Contract Documents as if singular in number and masculine in gender. The term Owner means the Owner or his authorized representative.

3.2 INFORMATION AND SERVICES REQUIRED OF THE OWNER

3.2.1 The Owner shall, at the request of the Contractor, at the time of execution of the Owner-Contractor Agreement, furnish to the Contractor reasonable evidence that he has made financial arrangements to fulfill his obligations under the Contract. Unless such reasonable evidence is furnished, the Contractor is not required to execute the Owner-Contractor Agreement or to commence the Work.

3.2.2 The Owner shall furnish all surveys describing the physical characteristics, legal limitations and utility locations for the site of the Project, and a legal description of the site.

3.2.3 Except as provided in Subparagraph 4.7.1, the Owner shall secure and pay for necessary approvals, easements, assessments and charges required for the construction, use or occupancy of permanent structures or for permanent changes in existing facilities.

3.2.4 Information or services under the Owner's control shall be furnished by the Owner with reasonable promptness to avoid delay in the orderly progress of the Work.

3.2.5 Unless otherwise provided in the Contract Documents, the Contractor will be furnished, free of charge, all copies of Drawings and Specifications reasonably necessary for the execution of the Work.

3.2.6 The Owner shall forward all instructions to the Contractor through the Architect.

3.2.7 The foregoing are in addition to other duties and responsibilities of the Owner enumerated herein and especially those in respect to Work by Owner or by Separate Contractors, Payments and Completion, and Insurance in Articles 6, 9 and 11 respectively.

3.3 OWNER'S RIGHT TO STOP THE WORK

3.3.1 If the Contractor fails to correct defective Work as required by Paragraph 13.2 or persistently fails to carry out the Work in accordance with the Contract Documents, the Owner, by a written order signed personally or by an agent specifically so empowered by the Owner in writing, may order the Contractor to stop the Work, or any portion thereof, until the cause for such order has been eliminated; however, this right of the Owner to stop the Work shall not give rise to any duty on the part of the Owner to exercise this right for the benefit of the Contractor or any other person or entity, except to the extent required by Subparagraph 6.1.3.

3.4 OWNER'S RIGHT TO CARRY OUT THE WORK

3.4.1 If the Contractor defaults or neglects to carry out the Work in accordance with the Contract Documents and fails within seven days after receipt of written notice from the Owner to commence and continue correction of such default or neglect with diligence and promptness, the Owner may, after seven days following receipt by the Contractor of an additional written notice and without prejudice to any other remedy he may have, make good such deficiencies. In such case an appropriate Change Order shall be issued deducting from the payments then or thereafter due the Contractor the cost of correcting such deficiencies, including compensation for the Architect's additional services made necessary by such default, neglect or failure. Such action by the Owner and the amount charged to the Contractor are both subject to the prior approval of the Architect. If the payments then or thereafter due the Contractor are not sufficient to cover such amount, the Contractor shall pay the difference to the Owner.

ARTICLE 4

CONTRACTOR

4.1 DEFINITION

4.1.1 The Contractor is the person or entity identified as such in the Owner-Contractor Agreement and is referred to throughout the Contract Documents as if singular in number and masculine in gender. The term Contractor means the Contractor or his authorized representative.

4.2 REVIEW OF CONTRACT DOCUMENTS

4.2.1 The Contractor shall carefully study and compare the Contract Documents and shall at once report to the Architect any error, inconsistency or omission he may discover. The Contractor shall not be liable to the Owner or

FIG. 11.1 *(continued)*

the Architect for any damage resulting from any such errors, inconsistencies or omissions in the Contract Documents. The Contractor shall perform no portion of the Work at any time without Contract Documents or, where required, approved Shop Drawings, Product Data or Samples for such portion of the Work.

4.3 SUPERVISION AND CONSTRUCTION PROCEDURES

4.3.1 The Contractor shall supervise and direct the Work, using his best skill and attention. He shall be solely responsible for all construction means, methods, techniques, sequences and procedures and for coordinating all portions of the Work under the Contract.

4.3.2 The Contractor shall be responsible to the Owner for the acts and omissions of his employees, Subcontractors and their agents and employees, and other persons performing any of the Work under a contract with the Contractor.

4.3.3 The Contractor shall not be relieved from his obligations to perform the Work in accordance with the Contract Documents either by the activities or duties of the Architect in his administration of the Contract, or by inspections, tests or approvals required or performed under Paragraph 7.7 by persons other than the Contractor.

4.4 LABOR AND MATERIALS

4.4.1 Unless otherwise provided in the Contract Documents, the Contractor shall provide and pay for all labor, materials, equipment, tools, construction equipment and machinery, water, heat, utilities, transportation, and other facilities and services necessary for the proper execution and completion of the Work, whether temporary or permanent and whether or not incorporated or to be incorporated in the Work.

4.4.2 The Contractor shall at all times enforce strict discipline and good order among his employees and shall not employ on the Work any unfit person or anyone not skilled in the task assigned to him.

4.5 WARRANTY

4.5.1 The Contractor warrants to the Owner and the Architect that all materials and equipment furnished under this Contract will be new unless otherwise specified, and that all Work will be of good quality, free from faults and defects and in conformance with the Contract Documents. All Work not conforming to these requirements, including substitutions not properly approved and authorized, may be considered defective. If required by the Architect, the Contractor shall furnish satisfactory evidence as to the kind and quality of materials and equipment. This warranty is not limited by the provisions of Paragraph 13.2.

4.6 TAXES

4.6.1 The Contractor shall pay all sales, consumer, use and other similar taxes for the Work or portions thereof provided by the Contractor which are legally enacted at the time bids are received, whether or not yet effective.

4.7 PERMITS, FEES AND NOTICES

4.7.1 Unless otherwise provided in the Contract Documents, the Contractor shall secure and pay for the building permit and for all other permits and governmental fees, licenses and inspections necessary for the proper execution and completion of the Work which are customarily secured after execution of the Contract and which are legally required at the time the bids are received.

4.7.2 The Contractor shall give all notices and comply with all laws, ordinances, rules, regulations and lawful orders of any public authority bearing on the performance of the Work.

4.7.3 It is not the responsibility of the Contractor to make certain that the Contract Documents are in accordance with applicable laws, statutes, building codes and regulations. If the Contractor observes that any of the Contract Documents are at variance therewith in any respect, he shall promptly notify the Architect in writing, and any necessary changes shall be accomplished by appropriate Modification.

4.7.4 If the Contractor performs any Work knowing it to be contrary to such laws, ordinances, rules and regulations, and without such notice to the Architect, he shall assume full responsibility therefor and shall bear all costs attributable thereto.

4.8 ALLOWANCES

4.8.1 The Contractor shall include in the Contract Sum all allowances stated in the Contract Documents. Items covered by these allowances shall be supplied for such amounts and by such persons as the Owner may direct, but the Contractor will not be required to employ persons against whom he makes a reasonable objection.

4.8.2 Unless otherwise provided in the Contract Documents:

- **.1** these allowances shall cover the cost to the Contractor, less any applicable trade discount, of the materials and equipment required by the allowance delivered at the site, and all applicable taxes;
- **.2** the Contractor's costs for unloading and handling on the site, labor, installation costs, overhead, profit and other expenses contemplated for the original allowance shall be included in the Contract Sum and not in the allowance;
- **.3** whenever the cost is more than or less than the allowance, the Contract Sum shall be adjusted accordingly by Change Order, the amount of which will recognize changes, if any, in handling costs on the site, labor, installation costs, overhead, profit and other expenses.

4.9 SUPERINTENDENT

4.9.1 The Contractor shall employ a competent superintendent and necessary assistants who shall be in attendance at the Project site during the progress of the Work. The superintendent shall represent the Contractor and all communications given to the superintendent shall be as binding as if given to the Contractor. Important communications shall be confirmed in writing. Other communications shall be so confirmed on written request in each case.

4.10 PROGRESS SCHEDULE

4.10.1 The Contractor, immediately after being awarded the Contract, shall prepare and submit for the Owner's and Architect's information an estimated progress sched-

AIA® • © 1976 • THE AMERICAN INSTITUTE OF ARCHITECTS, 1735 NEW YORK AVENUE, N.W., WASHINGTON, D.C. 20006

FIG. 11.1 *(continued)*

ule for the Work. The progress schedule shall be related to the entire Project to the extent required by the Contract Documents, and shall provide for expeditious and practicable execution of the Work.

4.11 DOCUMENTS AND SAMPLES AT THE SITE

4.11.1 The Contractor shall maintain at the site for the Owner one record copy of all Drawings, Specifications, Addenda, Change Orders and other Modifications, in good order and marked currently to record all changes made during construction, and approved Shop Drawings, Product Data and Samples. These shall be available to the Architect and shall be delivered to him for the Owner upon completion of the Work.

4.12 SHOP DRAWINGS, PRODUCT DATA AND SAMPLES

4.12.1 Shop Drawings are drawings, diagrams, schedules and other data specially prepared for the Work by the Contractor or any Subcontractor, manufacturer, supplier or distributor to illustrate some portion of the Work.

4.12.2 Product Data are illustrations, standard schedules, performance charts, instructions, brochures, diagrams and other information furnished by the Contractor to illustrate a material, product or system for some portion of the Work.

4.12.3 Samples are physical examples which illustrate materials, equipment or workmanship and establish standards by which the Work will be judged.

4.12.4 The Contractor shall review, approve and submit, with reasonable promptness and in such sequence as to cause no delay in the Work or in the work of the Owner or any separate contractor, all Shop Drawings, Product Data and Samples required by the Contract Documents.

4.12.5 By approving and submitting Shop Drawings, Product Data and Samples, the Contractor represents that he has determined and verified all materials, field measurements, and field construction criteria related thereto, or will do so, and that he has checked and coordinated the information contained within such submittals with the requirements of the Work and of the Contract Documents.

4.12.6 The Contractor shall not be relieved of responsibility for any deviation from the requirements of the Contract Documents by the Architect's approval of Shop Drawings, Product Data or Samples under Subparagraph 2.2.14 unless the Contractor has specifically informed the Architect in writing of such deviation at the time of submission and the Architect has given written approval to the specific deviation. The Contractor shall not be relieved from responsibility for errors or omissions in the Shop Drawings, Product Data or Samples by the Architect's approval thereof.

4.12.7 The Contractor shall direct specific attention, in writing or on resubmitted Shop Drawings, Product Data or Samples, to revisions other than those requested by the Architect on previous submittals.

4.12.8 No portion of the Work requiring submission of a Shop Drawing, Product Data or Sample shall be commenced until the submittal has been approved by the Architect as provided in Subparagraph 2.2.14. All such portions of the Work shall be in accordance with approved submittals.

4.13 USE OF SITE

4.13.1 The Contractor shall confine operations at the site to areas permitted by law, ordinances, permits and the Contract Documents and shall not unreasonably encumber the site with any materials or equipment.

4.14 CUTTING AND PATCHING OF WORK

4.14.1 The Contractor shall be responsible for all cutting, fitting or patching that may be required to complete the Work or to make its several parts fit together properly.

4.14.2 The Contractor shall not damage or endanger any portion of the Work or the work of the Owner or any separate contractors by cutting, patching or otherwise altering any work, or by excavation. The Contractor shall not cut or otherwise alter the work of the Owner or any separate contractor except with the written consent of the Owner and of such separate contractor. The Contractor shall not unreasonably withhold from the Owner or any separate contractor his consent to cutting or otherwise altering the Work.

4.15 CLEANING UP

4.15.1 The Contractor at all times shall keep the premises free from accumulation of waste materials or rubbish caused by his operations. At the completion of the Work he shall remove all his waste materials and rubbish from and about the Project as well as all his tools, construction equipment, machinery and surplus materials.

4.15.2 If the Contractor fails to clean up at the completion of the Work, the Owner may do so as provided in Paragraph 3.4 and the cost thereof shall be charged to the Contractor.

4.16 COMMUNICATIONS

4.16.1 The Contractor shall forward all communications to the Owner through the Architect.

4.17 ROYALTIES AND PATENTS

4.17.1 The Contractor shall pay all royalties and license fees. He shall defend all suits or claims for infringement of any patent rights and shall save the Owner harmless from loss on account thereof, except that the Owner shall be responsible for all such loss when a particular design, process or the product of a particular manufacturer or manufacturers is specified, but if the Contractor has reason to believe that the design, process or product specified is an infringement of a patent, he shall be responsible for such loss unless he promptly gives such information to the Architect.

4.18 INDEMNIFICATION

4.18.1 To the fullest extent permitted by law, the Contractor shall indemnify and hold harmless the Owner and the Architect and their agents and employees from and against all claims, damages, losses and expenses, including but not limited to attorneys' fees, arising out of or resulting from the performance of the Work, provided that any such claim, damage, loss or expense (1) is attributable to bodily injury, sickness, disease or death, or to injury to or destruction of tangible property (other than the Work itself) including the loss of use resulting therefrom,

AIA® • © 1976 • THE AMERICAN INSTITUTE OF ARCHITECTS, 1735 NEW YORK AVENUE, N.W., WASHINGTON, D.C. 20006

FIG. 11.1 *(continued)*

and (2) is caused in whole or in part by any negligent act or omission of the Contractor, any Subcontractor, anyone directly or indirectly employed by any of them or anyone for whose acts any of them may be liable, regardless of whether or not it is caused in part by a party indemnified hereunder. Such obligation shall not be construed to negate, abridge, or otherwise reduce any other right or obligation of indemnity which would otherwise exist as to any party or person described in this Paragraph 4.18.

4.18.2 In any and all claims against the Owner or the Architect or any of their agents or employees by any employee of the Contractor, any Subcontractor, anyone directly or indirectly employed by any of them or anyone for whose acts any of them may be liable, the indemnification obligation under this Paragraph 4.18 shall not be limited in any way by any limitation on the amount or type of damages, compensation or benefits payable by or for the Contractor or any Subcontractor under workers' or workmen's compensation acts, disability benefit acts or other employee benefit acts.

4.18.3 The obligations of the Contractor under this Paragraph 4.18 shall not extend to the liability of the Architect, his agents or employees, arising out of (1) the preparation or approval of maps, drawings, opinions, reports, surveys, change orders, designs or specifications, or (2) the giving of or the failure to give directions or instructions by the Architect, his agents or employees provided such giving or failure to give is the primary cause of the injury or damage.

ARTICLE 5

SUBCONTRACTORS

5.1 DEFINITION

5.1.1 A Subcontractor is a person or entity who has a direct contract with the Contractor to perform any of the Work at the site. The term Subcontractor is referred to throughout the Contract Documents as if singular in number and masculine in gender and means a Subcontractor or his authorized representative. The term Subcontractor does not include any separate contractor or his subcontractors.

5.1.2 A Sub-subcontractor is a person or entity who has a direct or indirect contract with a Subcontractor to perform any of the Work at the site. The term Sub-subcontractor is referred to throughout the Contract Documents as if singular in number and masculine in gender and means a Sub-subcontractor or an authorized representative thereof.

5.2 AWARD OF SUBCONTRACTS AND OTHER CONTRACTS FOR PORTIONS OF THE WORK

5.2.1 Unless otherwise required by the Contract Documents or the Bidding Documents, the Contractor, as soon as practicable after the award of the Contract, shall furnish to the Owner and the Architect in writing the names of the persons or entities (including those who are to furnish materials or equipment fabricated to a special design) proposed for each of the principal portions of the Work. The Architect will promptly reply to the Contractor in writing stating whether or not the Owner or the Architect, after due investigation, has reasonable objection to any such proposed person or entity. Failure of the Owner or Architect to reply promptly shall constitute notice of no reasonable objection.

5.2.2 The Contractor shall not contract with any such proposed person or entity to whom the Owner or the Architect has made reasonable objection under the provisions of Subparagraph 5.2.1. The Contractor shall not be required to contract with anyone to whom he has a reasonable objection.

5.2.3 If the Owner or the Architect has reasonable objection to any such proposed person or entity, the Contractor shall submit a substitute to whom the Owner or the Architect has no reasonable objection, and the Contract Sum shall be increased or decreased by the difference in cost occasioned by such substitution and an appropriate Change Order shall be issued; however, no increase in the Contract Sum shall be allowed for any such substitution unless the Contractor has acted promptly and responsively in submitting names as required by Subparagraph 5.2.1.

5.2.4 The Contractor shall make no substitution for any Subcontractor, person or entity previously selected if the Owner or Architect makes reasonable objection to such substitution.

5.3 SUBCONTRACTUAL RELATIONS

5.3.1 By an appropriate agreement, written where legally required for validity, the Contractor shall require each Subcontractor, to the extent of the Work to be performed by the Subcontractor, to be bound to the Contractor by the terms of the Contract Documents, and to assume toward the Contractor all the obligations and responsibilities which the Contractor, by these Documents, assumes toward the Owner and the Architect. Said agreement shall preserve and protect the rights of the Owner and the Architect under the Contract Documents with respect to the Work to be performed by the Subcontractor so that the subcontracting thereof will not prejudice such rights, and shall allow to the Subcontractor, unless specifically provided otherwise in the Contractor-Subcontractor agreement, the benefit of all rights, remedies and redress against the Contractor that the Contractor, by these Documents, has against the Owner. Where appropriate, the Contractor shall require each Subcontractor to enter into similar agreements with his Sub-subcontractors. The Contractor shall make available to each proposed Subcontractor, prior to the execution of the Subcontract, copies of the Contract Documents to which the Subcontractor will be bound by this Paragraph 5.3, and identify to the Subcontractor any terms and conditions of the proposed Subcontract which may be at variance with the Contract Documents. Each Subcontractor shall similarly make copies of such Documents available to his Sub-subcontractors.

ARTICLE 6

WORK BY OWNER OR BY SEPARATE CONTRACTORS

6.1 OWNER'S RIGHT TO PERFORM WORK AND TO AWARD SEPARATE CONTRACTS

6.1.1 The Owner reserves the right to perform work related to the Project with his own forces, and to award

AIA® • © 1976 • THE AMERICAN INSTITUTE OF ARCHITECTS, 1735 NEW YORK AVENUE, N.W., WASHINGTON, D.C. 20006

FIG. 11.1 (*continued*)

separate contracts in connection with other portions of the Project or other work on the site under these or similar Conditions of the Contract. If the Contractor claims that delay or additional cost is involved because of such action by the Owner, he shall make such claim as provided elsewhere in the Contract Documents.

6.1.2 When separate contracts are awarded for different portions of the Project or other work on the site, the term Contractor in the Contract Documents in each case shall mean the Contractor who executes each separate Owner-Contractor Agreement.

6.1.3 The Owner will provide for the coordination of the work of his own forces and of each separate contractor with the Work of the Contractor, who shall cooperate therewith as provided in Paragraph 6.2.

6.2 MUTUAL RESPONSIBILITY

6.2.1 The Contractor shall afford the Owner and separate contractors reasonable opportunity for the introduction and storage of their materials and equipment and the execution of their work, and shall connect and coordinate his Work with theirs as required by the Contract Documents.

6.2.2 If any part of the Contractor's Work depends for proper execution or results upon the work of the Owner or any separate contractor, the Contractor shall, prior to proceeding with the Work, promptly report to the Architect any apparent discrepancies or defects in such other work that render it unsuitable for such proper execution and results. Failure of the Contractor so to report shall constitute an acceptance of the Owner's or separate contractors' work as fit and proper to receive his Work, except as to defects which may subsequently become apparent in such work by others.

6.2.3 Any costs caused by defective or ill-timed work shall be borne by the party responsible therefor.

6.2.4 Should the Contractor wrongfully cause damage to the work or property of the Owner, or to other work on the site, the Contractor shall promptly remedy such damage as provided in Subparagraph 10.2.5.

6.2.5 Should the Contractor wrongfully cause damage to the work or property of any separate contractor, the Contractor shall upon due notice promptly attempt to settle with such other contractor by agreement, or otherwise to resolve the dispute. If such separate contractor sues or initiates an arbitration proceeding against the Owner on account of any damage alleged to have been caused by the Contractor, the Owner shall notify the Contractor who shall defend such proceedings at the Owner's expense, and if any judgment or award against the Owner arises therefrom the Contractor shall pay or satisfy it and shall reimburse the Owner for all attorneys' fees and court or arbitration costs which the Owner has incurred.

6.3 OWNER'S RIGHT TO CLEAN UP

6.3.1 If a dispute arises between the Contractor and separate contractors as to their responsibility for cleaning up as required by Paragraph 4.15, the Owner may clean up and charge the cost thereof to the contractors responsible therefor as the Architect shall determine to be just.

ARTICLE 7

MISCELLANEOUS PROVISIONS

7.1 GOVERNING LAW

7.1.1 The Contract shall be governed by the law of the place where the Project is located.

7.2 SUCCESSORS AND ASSIGNS

7.2.1 The Owner and the Contractor each binds himself, his partners, successors, assigns and legal representatives to the other party hereto and to the partners, successors, assigns and legal representatives of such other party in respect to all covenants, agreements and obligations contained in the Contract Documents. Neither party to the Contract shall assign the Contract or sublet it as a whole without the written consent of the other, nor shall the Contractor assign any moneys due or to become due to him hereunder, without the previous written consent of the Owner.

7.3 WRITTEN NOTICE

7.3.1 Written notice shall be deemed to have been duly served if delivered in person to the individual or member of the firm or entity or to an officer of the corporation for whom it was intended, or if delivered at or sent by registered or certified mail to the last business address known to him who gives the notice.

7.4 CLAIMS FOR DAMAGES

7.4.1 Should either party to the Contract suffer injury or damage to person or property because of any act or omission of the other party or of any of his employees, agents or others for whose acts he is legally liable, claim shall be made in writing to such other party within a reasonable time after the first observance of such injury or damage.

7.5 PERFORMANCE BOND AND LABOR AND MATERIAL PAYMENT BOND

7.5.1 The Owner shall have the right to require the Contractor to furnish bonds covering the faithful performance of the Contract and the payment of all obligations arising thereunder if and as required in the Bidding Documents or in the Contract Documents.

7.6 RIGHTS AND REMEDIES

7.6.1 The duties and obligations imposed by the Contract Documents and the rights and remedies available thereunder shall be in addition to and not a limitation of any duties, obligations, rights and remedies otherwise imposed or available by law.

7.6.2 No action or failure to act by the Owner, Architect or Contractor shall constitute a waiver of any right or duty afforded any of them under the Contract, nor shall any such action or failure to act constitute an approval of or acquiescence in any breach thereunder, except as may be specifically agreed in writing.

FIG. 11.1 *(continued)*

7.7 TESTS

7.7.1 If the Contract Documents, laws, ordinances, rules, regulations or orders of any public authority having jurisdiction require any portion of the Work to be inspected, tested or approved, the Contractor shall give the Architect timely notice of its readiness so the Architect may observe such inspection, testing or approval. The Contractor shall bear all costs of such inspections, tests or approvals conducted by public authorities. Unless otherwise provided, the Owner shall bear all costs of other inspections, tests or approvals.

7.7.2 If the Architect determines that any Work requires special inspection, testing, or approval which Subparagraph 7.7.1 does not include, he will, upon written authorization from the Owner, instruct the Contractor to order such special inspection, testing or approval, and the Contractor shall give notice as provided in Subparagraph 7.7.1. If such special inspection or testing reveals a failure of the Work to comply with the requirements of the Contract Documents, the Contractor shall bear all costs thereof, including compensation for the Architect's additional services made necessary by such failure; otherwise the Owner shall bear such costs, and an appropriate Change Order shall be issued.

7.7.3 Required certificates of inspection, testing or approval shall be secured by the Contractor and promptly delivered by him to the Architect.

7.7.4 If the Architect is to observe the inspections, tests or approvals required by the Contract Documents, he will do so promptly and, where practicable, at the source of supply.

7.8 INTEREST

7.8.1 Payments due and unpaid under the Contract Documents shall bear interest from the date payment is due at such rate as the parties may agree upon in writing or, in the absence thereof, at the legal rate prevailing at the place of the Project.

7.9 ARBITRATION

7.9.1 All claims, disputes and other matters in question between the Contractor and the Owner arising out of, or relating to, the Contract Documents or the breach thereof, except as provided in Subparagraph 2.2.11 with respect to the Architect's decisions on matters relating to artistic effect, and except for claims which have been waived by the making or acceptance of final payment as provided by Subparagraphs 9.9.4 and 9.9.5, shall be decided by arbitration in accordance with the Construction Industry Arbitration Rules of the American Arbitration Association then obtaining unless the parties mutually agree otherwise. No arbitration arising out of or relating to the Contract Documents shall include, by consolidation, joinder or in any other manner, the Architect, his employees or consultants except by written consent containing a specific reference to the Owner-Contractor Agreement and signed by the Architect, the Owner, the Contractor and any other person sought to be joined. No arbitration shall include by consolidation, joinder or in any other manner, parties other than the Owner, the Contractor and any other persons substantially involved in a common question of fact or law, whose presence is required if complete relief is to be accorded in the arbitration. No person other than the Owner or Contractor shall be included as an original third party or additional third party to an arbitration whose interest or responsibility is insubstantial. Any consent to arbitration involving an additional person or persons shall not constitute consent to arbitration of any dispute not described therein or with any person not named or described therein. The foregoing agreement to arbitrate and any other agreement to arbitrate with an additional person or persons duly consented to by the parties to the Owner-Contractor Agreement shall be specifically enforceable under the prevailing arbitration law. The award rendered by the arbitrators shall be final, and judgment may be entered upon it in accordance with applicable law in any court having jurisdiction thereof.

7.9.2 Notice of the demand for arbitration shall be filed in writing with the other party to the Owner-Contractor Agreement and with the American Arbitration Association, and a copy shall be filed with the Architect. The demand for arbitration shall be made within the time limits specified in Subparagraph 2.2.12 where applicable, and in all other cases within a reasonable time after the claim, dispute or other matter in question has arisen, and in no event shall it be made after the date when institution of legal or equitable proceedings based on such claim, dispute or other matter in question would be barred by the applicable statute of limitations.

7.9.3 Unless otherwise agreed in writing, the Contractor shall carry on the Work and maintain its progress during any arbitration proceedings, and the Owner shall continue to make payments to the Contractor in accordance with the Contract Documents.

ARTICLE 8

TIME

8.1 DEFINITIONS

8.1.1 Unless otherwise provided, the Contract Time is the period of time allotted in the Contract Documents for Substantial Completion of the Work as defined in Subparagraph 8.1.3, including authorized adjustments thereto.

8.1.2 The date of commencement of the Work is the date established in a notice to proceed. If there is no notice to proceed, it shall be the date of the Owner-Contractor Agreement or such other date as may be established therein.

8.1.3 The Date of Substantial Completion of the Work or designated portion thereof is the Date certified by the Architect when construction is sufficiently complete, in accordance with the Contract Documents, so the Owner can occupy or utilize the Work or designated portion thereof for the use for which it is intended.

8.1.4 The term day as used in the Contract Documents shall mean calendar day unless otherwise specifically designated.

8.2 PROGRESS AND COMPLETION

8.2.1 All time limits stated in the Contract Documents are of the essence of the Contract.

AIA® • © 1976 • THE AMERICAN INSTITUTE OF ARCHITECTS, 1735 NEW YORK AVENUE, N.W., WASHINGTON, D.C. 20006

FIG. 11.1 *(continued)*

8.2.2 The Contractor shall begin the Work on the date of commencement as defined in Subparagraph 8.1.2. He shall carry the Work forward expeditiously with adequate forces and shall achieve Substantial Completion within the Contract Time.

8.3 DELAYS AND EXTENSIONS OF TIME

8.3.1 If the Contractor is delayed at any time in the progress of the Work by any act or neglect of the Owner or the Architect, or by any employee of either, or by any separate contractor employed by the Owner, or by changes ordered in the Work, or by labor disputes, fire, unusual delay in transportation, adverse weather conditions not reasonably anticipatable, unavoidable casualties, or any causes beyond the Contractor's control, or by delay authorized by the Owner pending arbitration, or by any other cause which the Architect determines may justify the delay, then the Contract Time shall be extended by Change Order for such reasonable time as the Architect may determine.

8.3.2 Any claim for extension of time shall be made in writing to the Architect not more than twenty days after the commencement of the delay; otherwise it shall be waived. In the case of a continuing delay only one claim is necessary. The Contractor shall provide an estimate of the probable effect of such delay on the progress of the Work.

8.3.3 If no agreement is made stating the dates upon which interpretations as provided in Subparagraph 2.2.8 shall be furnished, then no claim for delay shall be allowed on account of failure to furnish such interpretations until fifteen days after written request is made for them, and not then unless such claim is reasonable.

8.3.4 This Paragraph 8.3 does not exclude the recovery of damages for delay by either party under other provisions of the Contract Documents.

ARTICLE 9

PAYMENTS AND COMPLETION

9.1 CONTRACT SUM

9.1.1 The Contract Sum is stated in the Owner-Contractor Agreement and, including authorized adjustments thereto, is the total amount payable by the Owner to the Contractor for the performance of the Work under the Contract Documents.

9.2 SCHEDULE OF VALUES

9.2.1 Before the first Application for Payment, the Contractor shall submit to the Architect a schedule of values allocated to the various portions of the Work, prepared in such form and supported by such data to substantiate its accuracy as the Architect may require. This schedule, unless objected to by the Architect, shall be used only as a basis for the Contractor's Applications for Payment.

9.3 APPLICATIONS FOR PAYMENT

9.3.1 At least ten days before the date for each progress payment established in the Owner-Contractor Agreement, the Contractor shall submit to the Architect an itemized Application for Payment, notarized if required, supported by such data substantiating the Contractor's right to payment as the Owner or the Architect may require, and reflecting retainage, if any, as provided elsewhere in the Contract Documents.

9.3.2 Unless otherwise provided in the Contract Documents, payments will be made on account of materials or equipment not incorporated in the Work but delivered and suitably stored at the site and, if approved in advance by the Owner, payments may similarly be made for materials or equipment suitably stored at some other location agreed upon in writing. Payments for materials or equipment stored on or off the site shall be conditioned upon submission by the Contractor of bills of sale or such other procedures satisfactory to the Owner to establish the Owner's title to such materials or equipment or otherwise protect the Owner's interest, including applicable insurance and transportation to the site for those materials and equipment stored off the site.

9.3.3 The Contractor warrants that title to all Work, materials and equipment covered by an Application for Payment will pass to the Owner either by incorporation in the construction or upon the receipt of payment by the Contractor, whichever occurs first, free and clear of all liens, claims, security interests or encumbrances, hereinafter referred to in this Article 9 as "liens"; and that no Work, materials or equipment covered by an Application for Payment will have been acquired by the Contractor, or by any other person performing Work at the site or furnishing materials and equipment for the Project, subject to an agreement under which an interest therein or an encumbrance thereon is retained by the seller or otherwise imposed by the Contractor or such other person.

9.4 CERTIFICATES FOR PAYMENT

9.4.1 The Architect will, within seven days after the receipt of the Contractor's Application for Payment, either issue a Certificate for Payment to the Owner, with a copy to the Contractor, for such amount as the Architect determines is properly due, or notify the Contractor in writing his reasons for withholding a Certificate as provided in Subparagraph 9.6.1.

9.4.2 The issuance of a Certificate for Payment will constitute a representation by the Architect to the Owner, based on his observations at the site as provided in Subparagraph 2.2.3 and the data comprising the Application for Payment, that the Work has progressed to the point indicated; that, to the best of his knowledge, information and belief, the quality of the Work is in accordance with the Contract Documents (subject to an evaluation of the Work for conformance with the Contract Documents upon Substantial Completion, to the results of any subsequent tests required by or performed under the Contract Documents, to minor deviations from the Contract Documents correctable prior to completion, and to any specific qualifications stated in his Certificate); and that the Contractor is entitled to payment in the amount certified. However, by issuing a Certificate for Payment, the Architect shall not thereby be deemed to represent that he has made exhaustive or continuous on-site inspections to check the quality or quantity of the Work or that he has reviewed the construction means, methods, techniques,

AIA® • © 1976 • THE AMERICAN INSTITUTE OF ARCHITECTS, 1735 NEW YORK AVENUE, N.W., WASHINGTON, D.C. 20006

FIG. 11.1 *(continued)*

sequences or procedures, or that he has made any examination to ascertain how or for what purpose the Contractor has used the moneys previously paid on account of the Contract Sum.

9.5 PROGRESS PAYMENTS

9.5.1 After the Architect has issued a Certificate for Payment, the Owner shall make payment in the manner and within the time provided in the Contract Documents.

9.5.2 The Contractor shall promptly pay each Subcontractor, upon receipt of payment from the Owner, out of the amount paid to the Contractor on account of such Subcontractor's Work, the amount to which said Subcontractor is entitled, reflecting the percentage actually retained, if any, from payments to the Contractor on account of such Subcontractor's Work. The Contractor shall, by an appropriate agreement with each Subcontractor, require each Subcontractor to make payments to his Subsubcontractors in similar manner.

9.5.3 The Architect may, on request and at his discretion, furnish to any Subcontractor, if practicable, information regarding the percentages of completion or the amounts applied for by the Contractor and the action taken thereon by the Architect on account of Work done by such Subcontractor.

9.5.4 Neither the Owner nor the Architect shall have any obligation to pay or to see to the payment of any moneys to any Subcontractor except as may otherwise be required by law.

9.5.5 No Certificate for a progress payment, nor any progress payment, nor any partial or entire use or occupancy of the Project by the Owner, shall constitute an acceptance of any Work not in accordance with the Contract Documents.

9.6 PAYMENTS WITHHELD

9.6.1 The Architect may decline to certify payment and may withhold his Certificate in whole or in part, to the extent necessary reasonably to protect the Owner, if in his opinion he is unable to make representations to the Owner as provided in Subparagraph 9.4.2. If the Architect is unable to make representations to the Owner as provided in Subparagraph 9.4.2 and to certify payment in the amount of the Application, he will notify the Contractor as provided in Subparagraph 9.4.1. If the Contractor and the Architect cannot agree on a revised amount, the Architect will promptly issue a Certificate for Payment for the amount for which he is able to make such representations to the Owner. The Architect may also decline to certify payment or, because of subsequently discovered evidence or subsequent observations, he may nullify the whole or any part of any Certificate for Payment previously issued, to such extent as may be necessary in his opinion to protect the Owner from loss because of:

.1 defective Work not remedied,

.2 third party claims filed or reasonable evidence indicating probable filing of such claims,

.3 failure of the Contractor to make payments properly to Subcontractors or for labor, materials or equipment,

.4 reasonable evidence that the Work cannot be completed for the unpaid balance of the Contract Sum,

.5 damage to the Owner or another contractor,

.6 reasonable evidence that the Work will not be completed within the Contract Time, or

.7 persistent failure to carry out the Work in accordance with the Contract Documents.

9.6.2 When the above grounds in Subparagraph 9.6.1 are removed, payment shall be made for amounts withheld because of them.

9.7 FAILURE OF PAYMENT

9.7.1 If the Architect does not issue a Certificate for Payment, through no fault of the Contractor, within seven days after receipt of the Contractor's Application for Payment, or if the Owner does not pay the Contractor within seven days after the date established in the Contract Documents any amount certified by the Architect or awarded by arbitration, then the Contractor may, upon seven additional days' written notice to the Owner and the Architect, stop the Work until payment of the amount owing has been received. The Contract Sum shall be increased by the amount of the Contractor's reasonable costs of shut-down, delay and start-up, which shall be effected by appropriate Change Order in accordance with Paragraph 12.3.

9.8 SUBSTANTIAL COMPLETION

9.8.1 When the Contractor considers that the Work, or a designated portion thereof which is acceptable to the Owner, is substantially complete as defined in Subparagraph 8.1.3, the Contractor shall prepare for submission to the Architect a list of items to be completed or corrected. The failure to include any items on such list does not alter the responsibility of the Contractor to complete all Work in accordance with the Contract Documents. When the Architect on the basis of an inspection determines that the Work or designated portion thereof is substantially complete, he will then prepare a Certificate of Substantial Completion which shall establish the Date of Substantial Completion, shall state the responsibilities of the Owner and the Contractor for security, maintenance, heat, utilities, damage to the Work, and insurance, and shall fix the time within which the Contractor shall complete the items listed therein. Warranties required by the Contract Documents shall commence on the Date of Substantial Completion of the Work or designated portion thereof unless otherwise provided in the Certificate of Substantial Completion. The Certificate of Substantial Completion shall be submitted to the Owner and the Contractor for their written acceptance of the responsibilities assigned to them in such Certificate.

9.8.2 Upon Substantial Completion of the Work or designated portion thereof and upon application by the Contractor and certification by the Architect, the Owner shall make payment, reflecting adjustment in retainage, if any, for such Work or portion thereof, as provided in the Contract Documents.

9.9 FINAL COMPLETION AND FINAL PAYMENT

9.9.1 Upon receipt of written notice that the Work is ready for final inspection and acceptance and upon receipt of a final Application for Payment, the Architect will

AIA® • © 1976 • THE AMERICAN INSTITUTE OF ARCHITECTS, 1735 NEW YORK AVENUE, N.W., WASHINGTON, D.C. 20006

FIG. 11.1 *(continued)*

promptly make such inspection and, when he finds the Work acceptable under the Contract Documents and the Contract fully performed, he will promptly issue a final Certificate for Payment stating that to the best of his knowledge, information and belief, and on the basis of his observations and inspections, the Work has been completed in accordance with the terms and conditions of the Contract Documents and that the entire balance found to be due the Contractor, and noted in said final Certificate, is due and payable. The Architect's final Certificate for Payment will constitute a further representation that the conditions precedent to the Contractor's being entitled to final payment as set forth in Subparagraph 9.9.2 have been fulfilled.

9.9.2 Neither the final payment nor the remaining retained percentage shall become due until the Contractor submits to the Architect (1) an affidavit that all payrolls, bills for materials and equipment, and other indebtedness connected with the Work for which the Owner or his property might in any way be responsible, have been paid or otherwise satisfied, (2) consent of surety, if any, to final payment and (3), if required by the Owner, other data establishing payment or satisfaction of all such obligations, such as receipts, releases and waivers of liens arising out of the Contract, to the extent and in such form as may be designated by the Owner. If any Subcontractor refuses to furnish a release or waiver required by the Owner, the Contractor may furnish a bond satisfactory to the Owner to indemnify him against any such lien. If any such lien remains unsatisfied after all payments are made, the Contractor shall refund to the Owner all moneys that the latter may be compelled to pay in discharging such lien, including all costs and reasonable attorneys' fees.

9.9.3 If, after Substantial Completion of the Work, final completion thereof is materially delayed through no fault of the Contractor or by the issuance of Change Orders affecting final completion, and the Architect so confirms, the Owner shall, upon application by the Contractor and certification by the Architect, and without terminating the Contract, make payment of the balance due for that portion of the Work fully completed and accepted. If the remaining balance for Work not fully completed or corrected is less than the retainage stipulated in the Contract Documents, and if bonds have been furnished as provided in Paragraph 7.5, the written consent of the surety to the payment of the balance due for that portion of the Work fully completed and accepted shall be submitted by the Contractor to the Architect prior to certification of such payment. Such payment shall be made under the terms and conditions governing final payment, except that it shall not constitute a waiver of claims.

9.9.4 The making of final payment shall constitute a waiver of all claims by the Owner except those arising from:

.1 unsettled liens,
.2 faulty or defective Work appearing after Substantial Completion,
.3 failure of the Work to comply with the requirements of the Contract Documents, or
.4 terms of any special warranties required by the Contract Documents.

9.9.5 The acceptance of final payment shall constitute a waiver of all claims by the Contractor except those previously made in writing and identified by the Contractor as unsettled at the time of the final Application for Payment.

ARTICLE 10

PROTECTION OF PERSONS AND PROPERTY

10.1 SAFETY PRECAUTIONS AND PROGRAMS

10.1.1 The Contractor shall be responsible for initiating, maintaining and supervising all safety precautions and programs in connection with the Work.

10.2 SAFETY OF PERSONS AND PROPERTY

10.2.1 The Contractor shall take all reasonable precautions for the safety of, and shall provide all reasonable protection to prevent damage, injury or loss to:

.1 all employees on the Work and all other persons who may be affected thereby;
.2 all the Work and all materials and equipment to be incorporated therein, whether in storage on or off the site, under the care, custody or control of the Contractor or any of his Subcontractors or Sub-subcontractors; and
.3 other property at the site or adjacent thereto, including trees, shrubs, lawns, walks, pavements, roadways, structures and utilities not designated for removal, relocation or replacement in the course of construction.

10.2.2 The Contractor shall give all notices and comply with all applicable laws, ordinances, rules, regulations and lawful orders of any public authority bearing on the safety of persons or property or their protection from damage, injury or loss.

10.2.3 The Contractor shall erect and maintain, as required by existing conditions and progress of the Work, all reasonable safeguards for safety and protection, including posting danger signs and other warnings against hazards, promulgating safety regulations and notifying owners and users of adjacent utilities.

10.2.4 When the use or storage of explosives or other hazardous materials or equipment is necessary for the execution of the Work, the Contractor shall exercise the utmost care and shall carry on such activities under the supervision of properly qualified personnel.

10.2.5 The Contractor shall promptly remedy all damage or loss (other than damage or loss insured under Paragraph 11.3) to any property referred to in Clauses 10.2.1.2 and 10.2.1.3 caused in whole or in part by the Contractor, any Subcontractor, any Sub-subcontractor, or anyone directly or indirectly employed by any of them, or by anyone for whose acts any of them may be liable and for which the Contractor is responsible under Clauses 10.2.1.2 and 10.2.1.3, except damage or loss attributable to the acts or omissions of the Owner or Architect or anyone directly or indirectly employed by either of them, or by anyone for whose acts either of them may be liable, and not attributable to the fault or negligence of the Contractor. The foregoing obligations of the Contractor are in addition to his obligations under Paragraph 4.18.

AIA® • © 1976 • THE AMERICAN INSTITUTE OF ARCHITECTS, 1735 NEW YORK AVENUE, N.W., WASHINGTON, D.C. 20006

FIG. 11.1 *(continued)*

10.2.6 The Contractor shall designate a responsible member of his organization at the site whose duty shall be the prevention of accidents. This person shall be the Contractor's superintendent unless otherwise designated by the Contractor in writing to the Owner and the Architect.

10.2.7 The Contractor shall not load or permit any part of the Work to be loaded so as to endanger its safety.

10.3 EMERGENCIES

10.3.1 In any emergency affecting the safety of persons or property, the Contractor shall act, at his discretion, to prevent threatened damage, injury or loss. Any additional compensation or extension of time claimed by the Contractor on account of emergency work shall be determined as provided in Article 12 for Changes in the Work.

ARTICLE 11

INSURANCE

11.1 CONTRACTOR'S LIABILITY INSURANCE

11.1.1 The Contractor shall purchase and maintain such insurance as will protect him from claims set forth below which may arise out of or result from the Contractor's operations under the Contract, whether such operations be by himself or by any Subcontractor or by anyone directly or indirectly employed by any of them, or by anyone for whose acts any of them may be liable:

.1 claims under workers' or workmen's compensation, disability benefit and other similar employee benefit acts;

.2 claims for damages because of bodily injury, occupational sickness or disease, or death of his employees;

.3 claims for damages because of bodily injury, sickness or disease, or death of any person other than his employees;

.4 claims for damages insured by usual personal injury liability coverage which are sustained (1) by any person as a result of an offense directly or indirectly related to the employment of such person by the Contractor, or (2) by any other person;

.5 claims for damages, other than to the Work itself, because of injury to or destruction of tangible property, including loss of use resulting therefrom; and

.6 claims for damages because of bodily injury or death of any person or property damage arising out of the ownership, maintenance or use of any motor vehicle.

11.1.2 The insurance required by Subparagraph 11.1.1 shall be written for not less than any limits of liability specified in the Contract Documents, or required by law, whichever is greater.

11.1.3 The insurance required by Subparagraph 11.1.1 shall include contractual liability insurance applicable to the Contractor's obligations under Paragraph 4.18.

11.1.4 Certificates of Insurance acceptable to the Owner shall be filed with the Owner prior to commencement of the Work. These Certificates shall contain a provision that coverages afforded under the policies will not be cancelled until at least thirty days' prior written notice has been given to the Owner.

11.2 OWNER'S LIABILITY INSURANCE

11.2.1 The Owner shall be responsible for purchasing and maintaining his own liability insurance and, at his option, may purchase and maintain such insurance as will protect him against claims which may arise from operations under the Contract.

11.3 PROPERTY INSURANCE

11.3.1 Unless otherwise provided, the Owner shall purchase and maintain property insurance upon the entire Work at the site to the full insurable value thereof. This insurance shall include the interests of the Owner, the Contractor, Subcontractors and Sub-subcontractors in the Work and shall insure against the perils of fire and extended coverage and shall include "all risk" insurance for physical loss or damage including, without duplication of coverage, theft, vandalism and malicious mischief. If the Owner does not intend to purchase such insurance for the full insurable value of the entire Work, he shall inform the Contractor in writing prior to commencement of the Work. The Contractor may then effect insurance which will protect the interests of himself, his Subcontractors and the Sub-subcontractors in the Work, and by appropriate Change Order the cost thereof shall be charged to the Owner. If the Contractor is damaged by failure of the Owner to purchase or maintain such insurance and to so notify the Contractor, then the Owner shall bear all reasonable costs properly attributable thereto. If not covered under the all risk insurance or otherwise provided in the Contract Documents, the Contractor shall effect and maintain similar property insurance on portions of the Work stored off the site or in transit when such portions of the Work are to be included in an Application for Payment under Subparagraph 9.3.2.

11.3.2 The Owner shall purchase and maintain such boiler and machinery insurance as may be required by the Contract Documents or by law. This insurance shall include the interests of the Owner, the Contractor, Subcontractors and Sub-subcontractors in the Work.

11.3.3 Any loss insured under Subparagraph 11.3.1 is to be adjusted with the Owner and made payable to the Owner as trustee for the insureds, as their interests may appear, subject to the requirements of any applicable mortgagee clause and of Subparagraph 11.3.8. The Contractor shall pay each Subcontractor a just share of any insurance moneys received by the Contractor, and by appropriate agreement, written where legally required for validity, shall require each Subcontractor to make payments to his Sub-subcontractors in similar manner.

11.3.4 The Owner shall file a copy of all policies with the Contractor before an exposure to loss may occur.

11.3.5 If the Contractor requests in writing that insurance for risks other than those described in Subparagraphs 11.3.1 and 11.3.2 or other special hazards be included in the property insurance policy, the Owner shall, if possible, include such insurance, and the cost thereof shall be charged to the Contractor by appropriate Change Order.

AIA® • © 1976 • THE AMERICAN INSTITUTE OF ARCHITECTS, 1735 NEW YORK AVENUE, N.W., WASHINGTON, D.C. 20006

FIG. 11.1 *(continued)*

11.3.6 The Owner and Contractor waive all rights against (1) each other and the Subcontractors, Sub-subcontractors, agents and employees each of the other, and (2) the Architect and separate contractors, if any, and their subcontractors, sub-subcontractors, agents and employees, for damages caused by fire or other perils to the extent covered by insurance obtained pursuant to this Paragraph 11.3 or any other property insurance applicable to the Work, except such rights as they may have to the proceeds of such insurance held by the Owner as trustee. The foregoing waiver afforded the Architect, his agents and employees shall not extend to the liability imposed by Subparagraph 4.18.3. The Owner or the Contractor, as appropriate, shall require of the Architect, separate contractors, Subcontractors and Sub-subcontractors by appropriate agreements, written where legally required for validity, similar waivers each in favor of all other parties enumerated in this Subparagraph 11.3.6.

11.3.7 If required in writing by any party in interest, the Owner as trustee shall, upon the occurrence of an insured loss, give bond for the proper performance of his duties. He shall deposit in a separate account any money so received, and he shall distribute it in accordance with such agreement as the parties in interest may reach, or in accordance with an award by arbitration in which case the procedure shall be as provided in Paragraph 7.9. If after such loss no other special agreement is made, replacement of damaged work shall be covered by an appropriate Change Order.

11.3.8 The Owner as trustee shall have power to adjust and settle any loss with the insurers unless one of the parties in interest shall object in writing within five days after the occurrence of loss to the Owner's exercise of this power, and if such objection be made, arbitrators shall be chosen as provided in Paragraph 7.9. The Owner as trustee shall, in that case, make settlement with the insurers in accordance with the directions of such arbitrators. If distribution of the insurance proceeds by arbitration is required, the arbitrators will direct such distribution.

11.3.9 If the Owner finds it necessary to occupy or use a portion or portions of the Work prior to Substantial Completion thereof, such occupancy or use shall not commence prior to a time mutually agreed to by the Owner and Contractor and to which the insurance company or companies providing the property insurance have consented by endorsement to the policy or policies. This insurance shall not be cancelled or lapsed on account of such partial occupancy or use. Consent of the Contractor and of the insurance company or companies to such occupancy or use shall not be unreasonably withheld.

11.4 LOSS OF USE INSURANCE

11.4.1 The Owner, at his option, may purchase and maintain such insurance as will insure him against loss of use of his property due to fire or other hazards, however caused. The Owner waives all rights of action against the Contractor for loss of use of his property, including consequential losses due to fire or other hazards however caused, to the extent covered by insurance under this Paragraph 11.4.

ARTICLE 12

CHANGES IN THE WORK

12.1 CHANGE ORDERS

12.1.1 A Change Order is a written order to the Contractor signed by the Owner and the Architect, issued after execution of the Contract, authorizing a change in the Work or an adjustment in the Contract Sum or the Contract Time. The Contract Sum and the Contract Time may be changed only by Change Order. A Change Order signed by the Contractor indicates his agreement therewith, including the adjustment in the Contract Sum or the Contract Time.

12.1.2 The Owner, without invalidating the Contract, may order changes in the Work within the general scope of the Contract consisting of additions, deletions or other revisions, the Contract Sum and the Contract Time being adjusted accordingly. All such changes in the Work shall be authorized by Change Order, and shall be performed under the applicable conditions of the Contract Documents.

12.1.3 The cost or credit to the Owner resulting from a change in the Work shall be determined in one or more of the following ways:

.1 by mutual acceptance of a lump sum properly itemized and supported by sufficient substantiating data to permit evaluation;
.2 by unit prices stated in the Contract Documents or subsequently agreed upon;
.3 by cost to be determined in a manner agreed upon by the parties and a mutually acceptable fixed or percentage fee; or
.4 by the method provided in Subparagraph 12.1.4.

12.1.4 If none of the methods set forth in Clauses 12.1.3.1, 12.1.3.2 or 12.1.3.3 is agreed upon, the Contractor, provided he receives a written order signed by the Owner, shall promptly proceed with the Work involved. The cost of such Work shall then be determined by the Architect on the basis of the reasonable expenditures and savings of those performing the Work attributable to the change, including, in the case of an increase in the Contract Sum, a reasonable allowance for overhead and profit. In such case, and also under Clauses 12.1.3.3 and 12.1.3.4 above, the Contractor shall keep and present, in such form as the Architect may prescribe, an itemized accounting together with appropriate supporting data for inclusion in a Change Order. Unless otherwise provided in the Contract Documents, cost shall be limited to the following: cost of materials, including sales tax and cost of delivery; cost of labor, including social security, old age and unemployment insurance, and fringe benefits required by agreement or custom; workers' or workmen's compensation insurance; bond premiums; rental value of equipment and machinery; and the additional costs of supervision and field office personnel directly attributable to the change. Pending final determination of cost to the Owner, payments on account shall be made on the Architect's Certificate for Payment. The amount of credit to be allowed by the Contractor to the Owner for any deletion

FIG. 11.1 *(continued)*

or change which results in a net decrease in the Contract Sum will be the amount of the actual net cost as confirmed by the Architect. When both additions and credits covering related Work or substitutions are involved in any one change, the allowance for overhead and profit shall be figured on the basis of the net increase, if any, with respect to that change.

12.1.5 If unit prices are stated in the Contract Documents or subsequently agreed upon, and if the quantities originally contemplated are so changed in a proposed Change Order that application of the agreed unit prices to the quantities of Work proposed will cause substantial inequity to the Owner or the Contractor, the applicable unit prices shall be equitably adjusted.

12.2 CONCEALED CONDITIONS

12.2.1 Should concealed conditions encountered in the performance of the Work below the surface of the ground or should concealed or unknown conditions in an existing structure be at variance with the conditions indicated by the Contract Documents, or should unknown physical conditions below the surface of the ground or should concealed or unknown conditions in an existing structure of an unusual nature, differing materially from those ordinarily encountered and generally recognized as inherent in work of the character provided for in this Contract, be encountered, the Contract Sum shall be equitably adjusted by Change Order upon claim by either party made within twenty days after the first observance of the conditions.

12.3 CLAIMS FOR ADDITIONAL COST

12.3.1 If the Contractor wishes to make a claim for an increase in the Contract Sum, he shall give the Architect written notice thereof within twenty days after the occurrence of the event giving rise to such claim. This notice shall be given by the Contractor before proceeding to execute the Work, except in an emergency endangering life or property in which case the Contractor shall proceed in accordance with Paragraph 10.3. No such claim shall be valid unless so made. If the Owner and the Contractor cannot agree on the amount of the adjustment in the Contract Sum, it shall be determined by the Architect. Any change in the Contract Sum resulting from such claim shall be authorized by Change Order.

12.3.2 If the Contractor claims that additional cost is involved because of, but not limited to, (1) any written interpretation pursuant to Subparagraph 2.2.8, (2) any order by the Owner to stop the Work pursuant to Paragraph 3.3 where the Contractor was not at fault, (3) any written order for a minor change in the Work issued pursuant to Paragraph 12.4, or (4) failure of payment by the Owner pursuant to Paragraph 9.7, the Contractor shall make such claim as provided in Subparagraph 12.3.1.

12.4 MINOR CHANGES IN THE WORK

12.4.1 The Architect will have authority to order minor changes in the Work not involving an adjustment in the Contract Sum or an extension of the Contract Time and not inconsistent with the intent of the Contract Documents. Such changes shall be effected by written order, and shall be binding on the Owner and the Contractor. The Contractor shall carry out such written orders promptly.

ARTICLE 13

UNCOVERING AND CORRECTION OF WORK

13.1 UNCOVERING OF WORK

13.1.1 If any portion of the Work should be covered contrary to the request of the Architect or to requirements specifically expressed in the Contract Documents, it must, if required in writing by the Architect, be uncovered for his observation and shall be replaced at the Contractor's expense.

13.1.2 If any other portion of the Work has been covered which the Architect has not specifically requested to observe prior to being covered, the Architect may request to see such Work and it shall be uncovered by the Contractor. If such Work be found in accordance with the Contract Documents, the cost of uncovering and replacement shall, by appropriate Change Order, be charged to the Owner. If such Work be found not in accordance with the Contract Documents, the Contractor shall pay such costs unless it be found that this condition was caused by the Owner or a separate contractor as provided in Article 6, in which event the Owner shall be responsible for the payment of such costs.

13.2 CORRECTION OF WORK

13.2.1 The Contractor shall promptly correct all Work rejected by the Architect as defective or as failing to conform to the Contract Documents whether observed before or after Substantial Completion and whether or not fabricated, installed or completed. The Contractor shall bear all costs of correcting such rejected Work, including compensation for the Architect's additional services made necessary thereby.

13.2.2 If, within one year after the Date of Substantial Completion of the Work or designated portion thereof or within one year after acceptance by the Owner of designated equipment or within such longer period of time as may be prescribed by law or by the terms of any applicable special warranty required by the Contract Documents, any of the Work is found to be defective or not in accordance with the Contract Documents, the Contractor shall correct it promptly after receipt of a written notice from the Owner to do so unless the Owner has previously given the Contractor a written acceptance of such condition. This obligation shall survive termination of the Contract. The Owner shall give such notice promptly after discovery of the condition.

13.2.3 The Contractor shall remove from the site all portions of the Work which are defective or non-conforming and which have not been corrected under Subparagraphs 4.5.1, 13.2.1 and 13.2.2, unless removal is waived by the Owner.

13.2.4 If the Contractor fails to correct defective or non-conforming Work as provided in Subparagraphs 4.5.1, 13.2.1 and 13.2.2, the Owner may correct it in accordance with Paragraph 3.4.

AIA® • © 1976 • THE AMERICAN INSTITUTE OF ARCHITECTS, 1735 NEW YORK AVENUE, N.W., WASHINGTON, D.C. 20006

FIG. 11.1 *(continued)*

13.2.5 If the Contractor does not proceed with the correction of such defective or non-conforming Work within a reasonable time fixed by written notice from the Architect, the Owner may remove it and may store the materials or equipment at the expense of the Contractor. If the Contractor does not pay the cost of such removal and storage within ten days thereafter, the Owner may upon ten additional days' written notice sell such Work at auction or at private sale and shall account for the net proceeds thereof, after deducting all the costs that should have been borne by the Contractor, including compensation for the Architect's additional services made necessary thereby. If such proceeds of sale do not cover all costs which the Contractor should have borne, the difference shall be charged to the Contractor and an appropriate Change Order shall be issued. If the payments then or thereafter due the Contractor are not sufficient to cover such amount, the Contractor shall pay the difference to the Owner.

13.2.6 The Contractor shall bear the cost of making good all work of the Owner or separate contractors destroyed or damaged by such correction or removal.

13.2.7 Nothing contained in this Paragraph 13.2 shall be construed to establish a period of limitation with respect to any other obligation which the Contractor might have under the Contract Documents, including Paragraph 4.5 hereof. The establishment of the time period of one year after the Date of Substantial Completion or such longer period of time as may be prescribed by law or by the terms of any warranty required by the Contract Documents relates only to the specific obligation of the Contractor to correct the Work, and has no relationship to the time within which his obligation to comply with the Contract Documents may be sought to be enforced, nor to the time within which proceedings may be commenced to establish the Contractor's liability with respect to his obligations other than specifically to correct the Work.

13.3 ACCEPTANCE OF DEFECTIVE OR NON-CONFORMING WORK

13.3.1 If the Owner prefers to accept defective or non-conforming Work, he may do so instead of requiring its removal and correction, in which case a Change Order will be issued to reflect a reduction in the Contract Sum where appropriate and equitable. Such adjustment shall be effected whether or not final payment has been made.

ARTICLE 14

TERMINATION OF THE CONTRACT

14.1 TERMINATION BY THE CONTRACTOR

14.1.1 If the Work is stopped for a period of thirty days under an order of any court or other public authority having jurisdiction, or as a result of an act of government, such as a declaration of a national emergency making materials unavailable, through no act or fault of the Contractor or a Subcontractor or their agents or employees or any other persons performing any of the Work under a contract with the Contractor, or if the Work should be stopped for a period of thirty days by the Contractor because the Architect has not issued a Certificate for Payment as provided in Paragraph 9.7 or because the Owner has not made payment thereon as provided in Paragraph 9.7, then the Contractor may, upon seven additional days' written notice to the Owner and the Architect, terminate the Contract and recover from the Owner payment for all Work executed and for any proven loss sustained upon any materials, equipment, tools, construction equipment and machinery, including reasonable profit and damages.

14.2 TERMINATION BY THE OWNER

14.2.1 If the Contractor is adjudged a bankrupt, or if he makes a general assignment for the benefit of his creditors, or if a receiver is appointed on account of his insolvency, or if he persistently or repeatedly refuses or fails, except in cases for which extension of time is provided, to supply enough properly skilled workmen or proper materials, or if he fails to make prompt payment to Subcontractors or for materials or labor, or persistently disregards laws, ordinances, rules, regulations or orders of any public authority having jurisdiction, or otherwise is guilty of a substantial violation of a provision of the Contract Documents, then the Owner, upon certification by the Architect that sufficient cause exists to justify such action, may, without prejudice to any right or remedy and after giving the Contractor and his surety, if any, seven days' written notice, terminate the employment of the Contractor and take possession of the site and of all materials, equipment, tools, construction equipment and machinery thereon owned by the Contractor and may finish the Work by whatever method he may deem expedient. In such case the Contractor shall not be entitled to receive any further payment until the Work is finished.

14.2.2 If the unpaid balance of the Contract Sum exceeds the costs of finishing the Work, including compensation for the Architect's additional services made necessary thereby, such excess shall be paid to the Contractor. If such costs exceed the unpaid balance, the Contractor shall pay the difference to the Owner. The amount to be paid to the Contractor or to the Owner, as the case may be, shall be certified by the Architect, upon application, in the manner provided in Paragraph 9.4, and this obligation for payment shall survive the termination of the Contract.

AIA® • © 1976 • THE AMERICAN INSTITUTE OF ARCHITECTS, 1735 NEW YORK AVENUE, N.W., WASHINGTON, D.C. 20006

FIG. 11.2 Design-build general conditions

THE ASSOCIATED GENERAL CONTRACTORS

GENERAL CONDITIONS FOR DESIGN-BUILD AGREEMENT BETWEEN OWNER AND CONTRACTOR

INSTRUCTIONS FOR CONTRACTOR

1. These documents are intended to be used with AGC Document No. 6a, Standard Form of Design-Build Agreement Between Owner and Contractor, Revised   , as the General Conditions. These conditions primarily govern the obligations of the Subcontractors and in addition establish the general procedures for the administration of construction.

2. In all cases your attorney should be consulted to advise you on their use and any modifications.

3. Nothing contained herein is intended to conflict with local, state or federal laws or regulations.

4. It is recommended all insurance matters be reviewed with your insurance consultant and carrier such as implications of errors and omission liability, completed operations, and waiver of subrogation.

5. Each article should be reviewed by the Contractor as to the applicability to a given project and contractual conditions.

6. Special conditions and terms for the project or the subcontracts should cover the following:

   — subcontractor retainages
   — payment schedules
   — insurance limits
   — builder's risk deductible, if any.

7. If the Owner does not provide Builder's Risk Insurance, Paragraph 12.2 will need to be modified.

Certain provisions of this document have been derived, with modifications, from the following document published by The American Institute of Architects: AIA Document A201, General Conditions, © 1976. Usage made of AIA language, with the permission of AIA, does not apply AIA endorsement or approval of this document. Further reproduction of copyrighted AIA materials without separate written permission from AIA is prohibited.

© 1977 Associated General Contractors of America

Reproduced by permission of The Associated General Contractors of America.

FIG. 11.2 (*continued*)

THE ASSOCIATED GENERAL CONTRACTORS

GENERAL CONDITIONS FOR DESIGN-BUILD AGREEMENT BETWEEN OWNER AND CONTRACTOR

TABLE OF CONTENTS

Certain provisions of this document have been derived, with modifications, from the following document published by The American Institute of Architects: AIA Document A201, General Conditions, © 1976, by The American Institute of Architects. Usage made of AIA language, with the permission of AIA, does not imply AIA endorsement or approval of this document. Further reproduction of copyrighted AIA materials without separate written permission AIA is prohibited.

© 1977 Associated General Contractors of America

FIG. 11.2 *(continued)*

ARTICLE I

CONTRACT DOCUMENTS

1.1 DEFINITIONS

1.1.1 THE CONTRACT DOCUMENTS

The Contract Documents consist of the Contract between the Contractor and the Subcontractor, the Conditions of the Contract (General, Supplementary and other Conditions), the Drawings (and Criteria if the drawings are not complete), the Specifications, all Addenda issued prior to execution of the Contract, and all Modifications issued after the execution of the contract. A Modification is (1) a written amendment to the Contract signed by both parties, (2) a Change Order, (3) a written interpretation issued by the Architect/Engineer pursuant to Subparagraph 3.2.2; or (4) a written order for a minor change in the Work issued on the Owner's behalf pursuant to Paragraph 13.4. In addition, the Subcontractor assumes toward the Contractor all the obligations and responsibilities which the Contractor assumes toward the Owner under the Agreement between the Owner and the Contractor. A copy of the pertinent parts of the Agreement will be made available on request. The Contract Documents do not include Bidding or Proposal Documents such as the Advertisement or Invitation To Bid, Requests for Proposals, sample forms, Subcontractors Bid or Proposal, or portions of Addenda relative to any of these, or any other documents other than those set forth in this subparagraph unless specifically set forth in the Agreement with the Subcontractor.

1.1.2 THE CONTRACT

The Contract Documents form the Contract with the Subcontractor. This Contract represents the entire and integrated agreement and supersedes all prior negotiations, representations, or agreements, either written or oral. The Contract may be amended or modified only by a Modification as defined in Subparagraph 1.1.1.

1.1.3 THE WORK

The Work comprises the completed construction performed by the Contractor with his own forces or, as to the Subcontractor, the completed construction required by a Subcontractor's Contract and includes all labor necessary to reproduce such construction required of the Contractor or a particular Subcontractor, and all materials and equipment incorporated or to be incorporated in such construction.

1.1.4 THE PROJECT

The Project is the total construction to be performed under the Agreement between the Owner and Contractor of which the Work is a part.

1.2 EXECUTION, CORRELATION AND INTENT

1.2.1 By executing his Agreement, each Subcontractor represents that he has visited the site, familiarized himself with the local conditions under which the Work is to be performed and correlated his observations with the requirements of the Contract Documents.

1.2.2 The intent of the Contract Documents is to include all items necessary for the proper execution and completion of the Work. The Contract Documents are complementary, and what is required by any one shall be as binding as if required by all. Work not covered in the Contract Documents will not be required unless it is consistent therewith and is reasonably inferable therefrom as being necessary to produce the intended results. Words and abbreviations in the Contract Documents which have well-known technical or trade meanings are used in accordance with such recognized meanings.

1.2.3 The organization of the Specifications into divisions, sections and articles, and the arrangements of Drawings shall not control the Contractor in dividing the Work among Subcontractors or in establishing the extent of Work to be performed by any trade.

1.3 OWNERSHIP AND USE OF DOCUMENTS

1.3.1 Unless otherwise provided in the Contract Documents, the Subcontractor will be furnished, free of charge, all

FIG. 11.2 *(continued)*

copies of Drawings and Specifications reasonably necessary for the execution of the Work.

1.3.2 All Drawings, Specifications and copies thereof furnished by the Contractor are and shall remain his property. They are to be used only with respect to this Project and are not to be used on any other project. With the exception of one contract set for each party, such documents are to be returned or suitably accounted for to the Contractor on request at the completion of the Work. Submission or distribution to meet official regulatory requirements or for other purposes in connection with the Project is not to be construed as publication in derogation of the Contractor's common law copyright or other reserved rights.

ARTICLE 2

OWNER

2.1 DEFINITION

2.1.1 The Owner is the person or entity identified as such in the Agreement between the Owner and Contractor and is referred to throughout the Contract Documents as if singular in number and masculine in gender. The term Owner means the Owner or his authorized representative.

2.2 INFORMATION AND SERVICES REQUIRED OF THE OWNER

2.2.1 The Owner will furnish all surveys describing the physical characteristics, legal limitations and utility locations for the site of the Project, and a legal description of the site.

2.2.2 Except as provided in Subparagraph 5.7.1 the Owner will secure and pay for necessary approvals, easements, assessments and charges required for the construction, use, or occupancy of permanent structures or permanent changes in existing facilities.

2.2.3 Information or services under the Owner's control will be furnished by the Owner with reasonable promptness to avoid delay in the orderly progress of the Work.

2.2.4 The Owner shall forward all instructions to the Subcontractors through the Contractor.

ARTICLE 3

ARCHITECT/ENGINEER

3.1 DEFINITION

3.1.1 The Architect/Engineer is the person lawfully licensed to practice architecture or engineering or an entity lawfully practicing architecture or engineering and identified as such in the Agreement between the Owner and Contractor and is referred to throughout the Contract Documents as if singular in number and masculine in gender. The term Architect/Engineer means the Architect/Engineer or his authorized representative.

3.1.2 Nothing contained in the Contract Documents shall create any contractual relationship between the Architect/Engineer and any Subcontractor.

3.2 ARCHITECT/ENGINEER'S DUTIES DURING CONSTRUCTION

3.2.1 The Architect/Engineer shall at all times have access to the Work wherever it is in preparation and progress. When directed by the Contractor, the Subcontractor shall provide facilities for such access so the Architect/Engineer may perform his functions under the Contract Documents.

3.2.2 The Architect/Engineer will be the interpreter of the requirements of the Drawings and Specifications. The Architect/Engineer will, within a reasonable time, render such interpretations as are necessary for the proper execution of the progress of the Work.

FIG. 11.2 *(continued)*

3.2.3 All interpretations of the Architect/Engineer shall be consistent with the intent of and reasonably inferable from the Contract Documents and will be in writing or in the form of drawings. All requests for interpretations shall be directed through the Contractor. Neither the Architect/Engineer nor the Contractor will be liable to the Subcontractor for the result of any interpretation or decision rendered in good faith in such capacity.

3.2.4 The Architect/Engineer's decisions in matters relating to artistic effect will be final if consistent with the intent of the Contract Documents.

3.2.5 The Architect/Engineer will have authority to reject Work which does not conform to the Contract Documents. Whenever, in his opinion, he considers it necessary or advisable for the implementation of the intent of the Contract Documents, he will have authority to require special inspection or testing of the Work in accordance with Subparagraph 8.7.2 whether or not such Work be then fabricated, installed or completed. However, neither the Architect/Engineer's authority to act under this Subparagraph 3.2.5, nor any decision made by him in good faith either to exercise or not to exercise such authority, shall give rise to any duty or responsibility of the Architect/Engineer to the Subcontractor, any Sub-Subcontractor, any of their agents or employees, or any other person performing any of the Work.

3.2.6 The Architect/Engineer will review and approve or take other appropriate action upon Subcontractor's submittals such as Shop Drawings, Product Data and Samples, but only for conformance with the design concept of the Work and with the information given in the Contract Documents. Such action shall be taken with reasonable promptness so as to cause no delay. The Architect/Engineer's approval of a specific item shall not indicate approval of an assembly of which the item is a component.

3.2.7 The Architect/Engineer along with the Contractor will conduct inspections to determine the dates of Substantial Completion and final completion.

3.2.8 The Architect/Engineer will communicate with the Subcontractors through the Contractor.

ARTICLE 4

CONTRACTOR

4.1 DEFINITION

4.1.1 The Contractor is the person or entity which has entered into an agreement with the Owner to design and construct the Project and is referred to throughout the Contract Documents as if singular in number and masculine in gender. The Contractor is authorized to enter into agreements with Subcontractors to perform the Work necessary to complete the Project and to perform some of the construction with his own forces. The term Contractor means the Contractor acting through his authorized representative.

4.2 ADMINISTRATION OF THE CONTRACT

4.2.1 The Contractor will provide the general administration of the Project as herein described.

4.2.2 The Contractor shall have the responsibility to supervise and coordinate the work of all Subcontractors.

4.2.3 The Contractor shall prepare and update all Construction Schedules and shall direct the Work with respect to such schedules.

4.2.4 The Contractor shall have authority to reject Work which does not conform to the Contract Documents and to require any Special Inspection and Testing in accordance with Subparagraph 8.7.2.

4.2.5 The Contractor will prepare and issue Change Orders to the Subcontractors in accordance with Article 13.

4.2.6 The Contractor along with the Architect/Engineer will conduct inspections to determine the dates of Substantial Completion and final completion, and will receive and review written warranties and related documents required by the Contract and assembled by the Subcontractor.

FIG. 11.2 *(continued)*

4.3 CONTRACTOR'S RIGHT TO STOP THE WORK

4.3.1 If the Subcontractor fails to correct defective Work as required by Paragraph 14.2 or persistently fails to carry out the Work in accordance with the Contract Documents, the Contractor may order the Subcontractor to stop the Work, or any portion thereof, until the cause for such order has been eliminated.

4.3.2 If the Subcontractor defaults or neglects to carry out the Work in accordance with the Contract Documents and fails within seven days after receipt of written notice from the Contractor to commence and continue correction of such default or neglect with diligence and promptness, the Contractor may, by written notice, and without prejudice to any other remedy he or the Owner may have, make good such deficiencies. In such case an appropriate Change Order shall be issued deducting from the payments then or thereafter due the Subcontractor the cost of correcting such deficiencies, including compensation for the Architect/Engineer's and Contractor's additional services made necessary by such default, neglect or failure.

ARTICLE 5

SUBCONTRACTORS

5.1 DEFINITION

5.1.1 A Subcontractor is the person or entity identified as such in the Agreement between the Contractor and a Subcontractor and is referred to throughout the Contract Documents as if singular in number and masculine in gender. The term Subcontractor means the Subcontractor or his authorized representative.

5.2 REVIEW OF CONTRACT DOCUMENTS

5.2.1 The Subcontractor shall carefully study and compare the Contract Documents and shall at once report to the Contractor any error, inconsistency or omission he may discover. The Subcontractor shall not be liable to the Owner, the Contractor or the Architect/Engineer for any damage resulting from any such errors, inconsistencies or omissions.

5.3 SUPERVISION AND CONSTRUCTION PROCEDURES

5.3.1 The Subcontractor shall supervise and direct the Work, using his best skill and attention. He shall be solely responsible for all construction means, methods, techniques, sequences and procedures and for coordinating all portions of the Work under the Contract subject to the overall coordination of the Contractor.

5.3.2 The Subcontractor shall be responsible to the Contractor for the acts and omissions of his employees and all his Sub-subcontractors and their agents and employees and other persons performing any of the Work under a contract with the Subcontractor.

5.3.3 Neither observations nor inspections, tests or approvals by persons other than the Subcontractor shall relieve the Subcontractor from his obligations to perform the Work in accordance with the Contract Documents.

5.4 LABOR AND MATERIALS

5.4.1 Unless otherwise specifically provided in the Contract Documents, the Subcontractor shall provide and pay for all labor, materials, equipment, tools, construction equipment and machinery, transportation, and other facilities and services necessary for the proper execution and completion of the Work.

5.4.2 The Subcontractor shall at all times enforce strict discipline and good order among his employees and shall not employ on the Work any unfit person or anyone not skilled in the task assigned to him.

5.5 WARRANTY

5.5.1 The Subcontractor warrants to the Owner and the Contractor that all materials and equipment furnished under this Contract will be new unless otherwise specified, and that all Work will be of good quality, free from faults and defects and in conformance with the Contract Documents. All Work not so conforming to these requirements, including substitutions

FIG. 11.2 (*continued*)

not properly approved and authorized, may be considered defective. If required by the Contractor, the Subcontractor shall furnish satisfactory evidence as to the kind and quality of materials and equipment. This warranty is not limited by the provisions of Paragraph 14.2.

5.6 TAXES

5.6.1 The Subcontractor shall pay all sales, consumer, use and other similar taxes for the Work or portions thereof provided by the Subcontractor which are legally enacted at the time bids or proposals are received, whether or not yet effective.

5.7 PERMITS, FEES AND NOTICES

5.7.1 Unless otherwise provided in the Contract Documents, the Subcontractor shall secure and pay for all permits, governmental fees, licenses and inspections necessary for the proper execution and completion of the Work, which are customarily secured after execution of the contract and which are legally required at the time bids or proposals are received.

5.7.2 The Subcontractor shall give all notices and comply with all laws, ordinances, rules, regulations and orders of any public authority bearing on the performance of the Work.

5.7.3 Unless otherwise provided in the Contract Documents, it is not the responsibility of the Subcontractor to make certain that the Contract Documents are in accordance with applicable laws, statutes, building codes and regulations. If the Subcontractor observes that any of the Contract Documents are at variance therewith in any respect, he shall promptly notify the Contractor in writing, and any necessary changes shall be by appropriate Modification.

5.7.4 If the Subcontractor performs any Work knowing it to be contrary to such laws, ordinances, rules and regulations, and without such notice to the Contractor, he shall assume full responsibility therefor and shall bear all costs attributable thereto.

5.8 ALLOWANCES

5.8.1 The Subcontractor shall include in the Contract Sum as defined in 10.1.1 all allowances stated in the Contract Documents. Items covered by these allowances shall be supplied for such amounts and by such persons as the Contractor may direct.

5.8.2 Unless otherwise provided in the Contract Documents:

.1 These allowances shall cover the cost to the Subcontractor, less any applicable trade discount, of the materials and equipment required by the allowance delivered at the site, and all applicable taxes;

.2 The Subcontractor's costs for unloading and handling on the site, labor, installation costs, overhead, profit and other expenses contemplated for the original allowance shall be included in the Contract Sum and not in the allowance;

.3 Whenever the cost is more than or less than the allowance, the Contract Sum shall be adjusted accordingly by Change Order, the amount of which will recognize changes, if any, in handling costs on the site, labor, installation costs, overhead, profit and other expenses.

5.9 SUPERINTENDENT

5.9.1 The Subcontractor shall employ a competent superintendent and necessary assistants who shall be in attendance at the Project site during the progress of the Work. The superintendent shall be satisfactory to the Contractor, and shall not be changed except with the consent of the Contractor, unless the superintendent proves to be unsatisfactory to the Subcontractor or ceases to be in his employ. The superintendent shall represent the Subcontractor and all communications given to the superintendent shall be as binding as if given to the Subcontractor. Important communications shall be confirmed in writing. Other communications shall be so confirmed on written request in each case.

FIG. 11.2 *(continued)*

5.10 PROGRESS SCHEDULE

5.10.1 The Subcontractor, immediately after being awarded the Contract, shall prepare and submit for the Contractor's information an estimated progress schedule for the Work. The progress schedule shall be related to the entire Project to the extent required by the Contract Documents and shall provide for expeditious and practicable execution of the Work. This schedule shall indicate the dates for the starting and completion of the various stages of construction, shall be revised as required by the conditions of the Work, and shall be subject to the Contractor's approval.

5.11 DRAWINGS AND SPECIFICATIONS AT THE SITE

5.11.1 The Subcontractor shall maintain at the site for the Contractor one copy of all Drawings, Specifications, Addenda, Change Orders and other Modifications, in good order and marked currently to record all changes made during construction. These Drawings, marked to record all changes during construction, and approved Shop Drawings, Product Data and Samples shall be delivered to the Contractor for the Owner upon completion of the Work.

5.12 SHOP DRAWINGS, PRODUCT DATA AND SAMPLES

5.12.1 Shop Drawings are drawings, diagrams, schedules and other data especially prepared for the Work by the Subcontractor or any Sub-subcontractor, manufacturer, supplier or distributor to illustrate some portion of the Work.

5.12.2 Product Data are illustrations, standard schedules, performance charts, instructions, brochures, diagrams and other information furnished by the Subcontractor to illustrate a material, product or system for some portion of the Work.

5.12.3 Samples are physical examples which illustrate materials, equipment or workmanship and establish standards by which the Work will be judged.

5.12.4 The Subcontractor shall review, approve and submit through the Contractor with reasonable promptness and in such sequence as to cause no delay in the Work or in the work of any separate contractor, all Shop Drawings, Product Data and Samples required by the Contract Documents.

5.12.5 By approving and submitting Shop Drawings, Product Data and Samples, the Subcontractor represents that he has determined and verified all materials, field measurements, and field construction criteria related thereto, or will do so, and that he has checked and coordinated the information contained within such submittals with the requirements of the Work and of the Contract Documents.

5.12.6 The Contractor, if he finds such submittals to be in order, will forward them to the Architect/Engineer. If the Contractor finds them not to be complete or in proper form, he may return them to the Subcontractor for correction or completion.

5.12.7 The Subcontractor shall not be relieved of responsibility for any deviation from the requirements of the Contract Documents by the Architect/Engineer's approval of Shop Drawings, Product Data or Samples under Subparagraph 3.2.6 unless the Subcontractor has specifically informed the Architect/Engineer and Contractor in writing of such deviation at the time of submission and the Architect/Engineer has given written approval to the specific deviation. Subcontractor shall not be relieved from responsibility for errors or omissions in the Shop Drawings, Product Data or Samples by the Architect/Engineer's approval thereof.

5.12.8 The Subcontractor shall direct specific attention, in writing or on resubmitted Shop Drawings, Product Data or Samples, to revisions other than those requested by the Architect/Engineer or Contractor on previous submittals.

5.12.9 No portion of the Work requiring submission of a Shop Drawing, Product Data or Sample shall be commenced until the submittal has been approved by the Architect/Engineer. All such portions of the Work shall be in accordance with approved submittals.

5.13 USE OF SITE

5.13.1 The Subcontractor shall confine operations at the site to areas designated by the Contractor, permitted by law, ordinances, permits and the Contract Documents and shall not unreasonably encumber the site with any materials or equipment.

FIG. 11.2 *(continued)*

5.14 CUTTING AND PATCHING OF WORK

5.14.1 The Subcontractor shall be responsible for all cutting, fitting or patching that may be required to complete the Work or to make its several parts fit together properly. He shall provide protection of existing Work as required.

5.14.2 The Subcontractor shall not damage or endanger any portion of the Work or the work of the Contractor or any separate contractors or subcontractors by cutting, patching or otherwise altering any work, or by excavation. The Subcontractor shall not cut or otherwise alter the work of the Contractor or any separate contractor except with the written consent of the Contractor and of such separate contractor. The Subcontractor shall not unreasonably withhold from the Contractor or any separate contractor his consent to cutting or otherwise altering the Work.

5.15 CLEANING UP

5.15.1 The Subcontractor at all times shall keep the premises free from accumulation of waste materials or rubbish caused by his operations. At the completion of the Work he shall remove all his waste materials and rubbish from and about the Project as well as all his tools, construction equipment, machinery and surplus materials.

5.15.2 If the Subcontractor fails to clean up, the Contractor may do so and the cost thereof shall be charged to the Subcontractor.

5.16 COMMUNICATIONS

5.16.1 The Subcontractor shall forward all communications to the Owner and Architect/Engineer through the Contractor.

5.17 ROYALTIES AND PATENTS

5.17.1 The Subcontractor shall pay all royalties and license fees. He shall defend all suits or claims for infringement of any patent rights and shall save the Owner and Contractor harmless from loss on account thereof, except that the Owner shall be responsible for all such loss when a particular design, process or the product of a particular manufacturer or manufacturers is specified, but if the Subcontractor has reason to believe that the design, process or product specified is an infringement of a patent, he shall be responsible for such loss unless he promptly gives such information to the Contractor.

5.18 INDEMNIFICATION

5.18.1 To the fullest extent permitted by law, the Subcontractor shall indemnify and hold harmless the Owner, the Contractor and the Architect/Engineer and their agents and employees from and against all claims, damages, losses, and expenses including but not limited to, attorneys' fees, arising out of or resulting from the performance of the Work, provided that any such claim, damage, loss or expense (1) is attributable to bodily injury, sickness, disease or death, or to injury to or destruction of tangible property (other than the Work itself) including the loss of use resulting therefrom, and (2) is caused in whole or in part by any negligent act or omission of the Subcontractor, any Sub-subcontractor, anyone directly or indirectly employed by any of them or anyone for whose acts any of them may be liable, regardless of whether or not it is caused in part by a party indemnified hereunder. Such obligation shall not be construed to negate, abridge or otherwise reduce any other right or obligation of indemnity which would otherwise exist as to any party or person described in this Paragraph 5.18.

5.18.2 In any and all claims against the Owner, the Contractor or the Architect/Engineer or any of their agents or employees by any employee of the Subcontractor, any Sub-subcontractor, anyone directly or indirectly employed by any of them or anyone for whose acts any of them may be liable, the indemnification obligation under this Paragraph 5.18 shall not be limited in any way by any limitation on the amount or type of damages, compensation or benefits payable by or for the Subcontractor or any Sub-subcontractor under workers' or workmen's compensation acts, disability benefit acts or other employee benefit acts.

5.18.3 The obligations of the Subcontractor under this Paragraph 5.18 shall not extend to the liability of the Architect/Engineer, his agents or employees arising out of (1) the preparation or approval of maps, drawings, opinions, reports, surveys, designs or specifications, or (2) the giving of or the failure to give directions or instruction by the Architect/Engineer, his agents or employees provided such giving or failure to give is the primary cause of the injury or damage.

FIG. 11.2 *(continued)*

ARTICLE 6

SUB-SUBCONTRACTORS

6.1 DEFINITION

6.1.1 A Sub-subcontractor is a person or entity who has a direct contract with a Subcontractor to perform any of the Work at the site. The term Sub-subcontractor is referred to throughout the Contract Documents as if singular in number and masculine in gender and means a Sub-subcontractor or his authorized representative.

6.2 AWARD OF SUB-SUBCONTRACTS AND OTHER CONTRACTS FOR PORTIONS OF THE WORK

6.2.1 Unless otherwise required by the Contract Documents or in the Bidding or Proposal Documents, the Subcontractor shall furnish to the Contractor in writing, for acceptance by the Contractor in writing, the names of the persons or entities (including those who are to furnish materials or equipment fabricated to a special design) proposed for each of the principal portions of the Work. The Contractor will promptly reply to the Subcontractor in writing if the Contractor, after due investigation, has reasonable objection to any such proposed person or entity. Failure of the Contractor to reply promptly shall constitute notice of no reasonable objection.

6.2.2 The Subcontractor shall not contract with any such proposed person or entity to whom the Contractor has made reasonable objection under the provisions of Subparagraph 6.2.1. The Subcontractor shall not be required to contract with anyone to whom he has a reasonable objection.

6.2.3 If the Contractor refuses to accept any person or entity on a list submitted by the Subcontractor in response to the requirements of the Contract Documents, the Subcontractor shall submit an acceptable substitute; however, no increase in the Contract Sum shall be allowed for any such substitution.

6.2.4 The Subcontractor shall make no substitution for any Sub-subcontractor, person or entity previously selected if the Contractor makes reasonable objection to such substitution.

6.3 SUB-SUBCONTRACTUAL RELATIONS

6.3.1 By an appropriate agreement, written where legally required for validity, the Subcontractor shall require each Sub-subcontractor, to the extent of the work to be performed by the Sub-subcontractor, to be bound to the Subcontractor by the terms of the Contract Documents and to assume toward the Subcontractor all the obligations and responsibilities which the Subcontractor, by these Documents, assumes toward the Owner, the Contractor, or the Architect/Engineer. Said agreement shall preserve and protect the rights of the Owner, the Contractor and the Architect/Engineer under the Contract Documents with respect to the Work to be performed by the Sub-subcontractor so that the subcontracting thereof will not prejudice such rights, and shall allow to the Sub-subcontractor, unless specifically provided otherwise in the Subcontractor—Sub-subcontractor agreement, the benefit of all rights, remedies and redress against the Subcontractor that the Subcontractor, by these Documents, has against the Contractor. Where appropriate, the Subcontractor shall require each Sub-subcontractor to enter into similar agreements with his Sub-subcontractors. The Subcontractor shall make available to each proposed Sub-subcontractor, prior to the execution of the Sub-subcontract, copies of the Contract Documents to which the Sub-subcontractor will be bound by this Paragraph 6.3, and shall identify to the Sub-subcontractor any terms and conditions of the proposed Sub-subcontract which may be at variance with the Contract Documents. Each Sub-subcontractor shall similarly make copies of such Documents available to his Sub-subcontractors.

ARTICLE 7

SEPARATE SUBCONTRACTS

7.1 MUTUAL RESPONSIBILITY OF SUBCONTRACTORS

7.1.1 The Subcontractor shall afford the Contractor and other Subcontractors reasonable opportunity for the introduc-

FIG. 11.2 *(continued)*

tion and storage of their materials and equipment and the execution of their work, and shall connect and coordinate his Work with others under the general direction of the Contractor.

7.1.2 If any part of the Subcontractor's Work depends, for proper execution or results, upon the Work of the Contractor or any separate Subcontractor, the Subcontractor shall, prior to proceeding with the Work, promptly report to the Contractor any apparent discrepancies or defects in such Work that render it unsuitable for such proper execution and results. Failure of the Subcontractor so to report shall constitute an acceptance of the other Subcontractor's or Contractor's Work as fit and proper to receive his Work, except as to defects which may subsequently become apparent in such work by others.

7.1.3 Any costs caused by defective or ill-timed work shall be borne by the party responsible therefor.

7.1.4 Should the Subcontractor wrongfully cause damage to the Work or property of the Owner or to other work on the site, the Subcontractor shall promptly remedy such damage as provided in Subparagraph 11.2.5.

7.1.5 Should the Subcontractor wrongfully cause damage to the work or property of any separate Subcontractor or other contractor, the Subcontractor shall, upon due notice, promptly attempt to settle with the other contractor by agreement, or otherwise resolve the dispute. If such separate contractor sues the Owner or the Contractor or initiates an arbitration proceeding against the Owner or Contractor on account of any damage alleged to have been caused by the Subcontractor, the Owner or Contractor shall notify the Subcontractor who shall defend such proceedings at the Subcontractor's expense, and if any judgment or award against the Owner or Contractor arises therefrom, the Subcontractor shall pay or satisfy it and shall reimburse the Owner or Contractor for all attorneys' fees and court or arbitration costs which the Owner or Contractor has incurred.

7.2 CONTRACTOR'S RIGHT TO CLEAN UP

7.2.1 If a dispute arises between the separate Subcontractors as to their responsibility for cleaning up as required by Paragraph 5.15, the Contractor may clean up and charge the cost thereof to the Subcontractors responsible therefor as the Contractor shall determine to be just.

ARTICLE 8

MISCELLANEOUS PROVISIONS

8.1 GOVERNING LAW

8.1.1 The Contract shall be governed by the law of the place where the Project is located.

8.2 SUCCESSORS AND ASSIGNS

8.2.1 The Contractor and the Subcontractor each binds himself, his partners, successors, assigns and legal representatives of such other party in respect to all covenants, agreements and obligations contained in the Contract Documents. Neither party to the Contract shall assign the Contract or sublet it as a whole without the written consent of the other.

8.3 WRITTEN NOTICE

8.3.1 Written notice shall be deemed to have been duly served if delivered in person to the individual or member of the firm or entity or to an officer of the corporation for whom it was intended, or if delivered at or sent by registered or certified mail to the last business address known to him who gives the notice.

8.4 CLAIMS FOR DAMAGES

8.4.1 Should either party to the subcontract agreements suffer injury or damage to person or property because of any act or omission of the other party or of any of his employees, agents or others for whose acts he is legally liable, claim shall be made in writing to such other party within a reasonable time after the first observance of such injury or damage.

FIG. 11.2 *(continued)*

8.5 PERFORMANCE BOND AND LABOR AND MATERIAL PAYMENT BOND

8.5.1 The Contractor shall have the right to require the Subcontractor to furnish bonds in a form and with a corporate surety acceptable to the Contractor covering the faithful performance of the Contract and the payment of all obligations arising thereunder if and as required in the Bidding or Proposal Documents or in the Contract Documents.

8.6 RIGHTS AND REMEDIES

8.6.1 The duties and obligations imposed by the Contract Documents and the rights and remedies available thereunder shall be in addition to and not a limitation of any duties, obligations, rights and remedies otherwise imposed or available by law.

8.6.2 No action or failure to act by the Contractor, Architect/Engineer or Subcontractor shall constitute a waiver of any right or duty afforded any of them under the Contract documents, nor shall any such action or failure to act constitute an approval of or acquiescence in any breach thereunder, except as may be specifically agreed in writing.

8.7 TESTS

8.7.1 If the Contract Documents, laws, ordinances, rules, regulations or orders of any public authority having jurisdiction require any portion of the Work to be inspected, tested or approved, the Subcontractor shall give the Architect/Engineer timely notice of its readiness so the Architect/Engineer and Contractor may observe such inspection, testing, or approval. The Subcontractor shall bear all costs of such inspections, tests or approvals unless otherwise provided.

8.7.2 If the Architect/Engineer or Contractor determines that any Work requires special inspection, testing or approval which Subparagraph 8.7.1 does not include, he will, through the Contractor, instruct the Subcontractor to order such special inspection, testing or approval and the Subcontractor shall give notice as in Subparagraph 8.7.1. If such special inspection or testing reveals a failure of the Work to comply with the requirements of the Contract Documents, the Subcontractor shall bear all costs thereof, including compensation for the Architect/Engineer's and Contractor's additional services made necessary by such failure. If the work complies, the Contractor shall bear such costs and an appropriate Change Order shall be issued.

8.7.3 Required certificates of inspection, testing or approval shall be secured by the Subcontractor and promptly delivered by him through the Contractor to the Architect/Engineer.

8.7.4 If the Architect/Engineer or Contractor is to observe the inspections, tests or approvals required by the Contract Documents, he will do so promptly and, where practicable, at the source of supply.

8.8 ARBITRATION

8.8.1 All claims, disputes and other matters in question arising out of, or relating to this Contract or the breach thereof, except as set forth in Subparagraph 3.2.4 with respect to the Architect/Engineer's decisions on matters relating to artistic effect, and except for claims which have been waived by the making or acceptance of final payment provided by Subparagraphs 10.8.4 and 10.8.5, shall be decided by arbitration in accordance with the Construction Industry Arbitration Rules of the American Arbitration Association then obtaining unless the parties mutually agree otherwise. This agreement to arbitrate shall be specifically enforceable under the prevailing arbitration law. The award rendered by the arbitrators shall be final, and judgment may be entered upon it in accordance with applicable law in any court having jurisdiction thereof.

8.8.2 Notice of the demand for arbitration shall be filed in writing with the other party to the Contract and with the American Arbitration Association. The demand for arbitration shall be made within a reasonable time after the claim, dispute or other matter in question has arisen, and in no event shall it be made after the date when institution of legal or equitable proceedings based on such claim, dispute or other matter in question would be barred by the applicable statute of limitations.

8.8.3 The Subcontractor shall carry on the Work and maintain the progress schedule during any arbitration proceedings, unless otherwise agreed by him and the Contractor in writing.

FIG. 11.2 *(continued)*

8.8.4 All claims which are related to or dependent upon each other shall be heard by the same arbitrator or arbitrators even though the parties are not the same unless a specific contract prohibits such consolidation.

ARTICLE 9

TIME

9.1 DEFINITIONS

9.1.1 Unless otherwise provided, the Contract Time is the period of time allotted in the Contract Documents for the Substantial Completion of the Work as defined in Subparagraph 9.1.3 including authorized adjustments thereto.

9.1.2 The date of commencement of the Work is the date established in a notice to proceed. If there is no notice to proceed, it shall be the date of the Subcontractor Agreement or such other date as may be established therein.

9.1.3 The Date of Substantial Completion of the Work or designated portion thereof is the Date certified by the Architect/Engineer when construction is sufficiently complete, in accordance with the Contract Documents, so the Owner can occupy or utilize the Work or designated portion thereof for the use for which it is intended.

9.1.4 The term day as used in the Contract Documents shall mean calendar day unless otherwise specifically designated.

9.2 PROGRESS AND COMPLETION

9.2.1 All time limits stated in the Contract Documents are of the essence of the Contract.

9.2.2 The Subcontractor shall begin the Work on the date of commencement as defined in Subparagraph 9.1.2. He shall carry the Work forward expeditiously with adequate forces and shall achieve Substantial Completion within the Contract Time.

9.3 DELAYS AND EXTENSIONS OF TIME

9.3.1 If the Subcontractor is delayed at any time in the progress of the Work by any act or neglect of the Owner, Contractor, or the Architect/Engineer, or by any employee of either, or by any separate contractor employed by the Owner, or by changes ordered in the Work, or by labor disputes, fire, unusual delay in transportation, adverse weather conditions not reasonably anticipatable, unavoidable casualties or any causes beyond the Subcontractor's control, or by delay authorized by the Owner or the Contractor pending arbitration, or by any other cause which the Contractor determines may justify the delay, then the Contract Time shall be extended by Change Order for such reasonable time as the Contractor may determine.

9.3.2 Any claim for extension of time shall be made in writing to the Contractor no more than twenty days after the commencement of the delay; otherwise it shall be waived. In the case of a continuing cause of delay only one claim is necessary. The Subcontractor shall provide an estimate of the probable effect of such delay on the progress of the Work.

9.3.3 If no agreement is made stating the dates upon which interpretations as set forth in Subparagraph 3.2.2 shall be furnished, then no claim for delay shall be allowed on account of failure to furnish such interpretations until fifteen days after written request is made for them, and not then unless such claim is reasonable.

9.3.4 It shall be recognized by the Subcontractor that he may reasonably anticipate that as the job progresses, the Contractor will be making changes in and updating Construction Schedules pursuant to the authority given him in Subparagraph 4.2.3. Therefore, no claim for an increase in the Contract Sum for either accelaration or delay will be allowed for extensions of time pursuant to this paragraph 9.3 or for other changes in Construction Schedules which are of the type ordinarily experienced in projects of similar size and complexity.

9.3.5 This Paragraph 9.3 does not exclude the recovery of damages for delay by either party under other provisions of the Contract Documents.

FIG. 11.2 *(continued)*

ARTICLE 10

PAYMENTS AND COMPLETION

10.1 CONTRACT SUM

10.1.1 The Contract Sum is stated in the agreement between the Contractor and the Subcontractor including adjustments thereto and is the total amount payable to the Subcontractor for the performance of the Work under the Contract Documents.

10.2 SCHEDULE OF VALUES

10.2.1 Before the first Application for Payment, the Subcontractor shall submit to the Contractor a schedule of values allocated to the various portions of the Work prepared in such form and supported by such data to substantiate its accuracy as the Contractor may require. This schedule, unless objected to by the Contractor, shall be used only as a basis for the Subcontractor's Applications for Payment.

10.3 APPLICATIONS FOR PAYMENT

10.3.1 At least ten days before the date for each progress payment established in the Subcontractor's Agreement, the Subcontractor shall submit to the Contractor an itemized Application for Payment, notarized if required, supported by such data substantiating the Subcontractor's right to payment as the Owner or the Contractor may require, and reflecting such retainage as provided in the Contractor's Subcontract.

10.3.2 Unless otherwise provided in the Contract Documents, payments will be made on account of materials or equipment not incorporated in the Work but delivered and suitably stored at the site and, if approved in advance by the Contractor, payments may similarly be made for materials or equipment stored at some other location agreed upon in writing. Payments made for materials or equipment stored on or off the site shall be conditioned upon submission by the Subcontractor of bills of sale or such other procedures satisfactory to the Contractor to establish the Owner's title to such materials or equipment or otherwise protect the Owner's interest, including applicable insurance and transportation to the site for those materials and equipment stored off the site.

10.3.3 The Subcontractor warrants that title to all Work, materials and equipment covered by an Application for Payment will pass to the Owner either by incorporation in the construction or upon the receipt of payment by the Subcontractor, whichever occurs first, free and clear of all liens, claims, security interests or encumbrances, hereinafter referred to in this Article 10 as "liens"; and that no Work, materials or equipment covered by an Application for Payment will have been acquired by the Subcontractor, or by any other person performing the Work at the site or furnishing materials and equipment for the Project, subject to an agreement under which an interest therein or an encumbrance thereon is retained by the seller or otherwise imposed by the Subcontractor or such other person. All Subcontractors and Sub-subcontractors agree that title will so pass upon their receipt of payment from the Subcontractor.

10.4 PROGRESS PAYMENTS

10.4.1 If the Subcontractor has made Application for Payment as above, the Contractor will, with reasonable promptness but not more than seven days after the receipt of payment from the Owner make payment in accordance with the subcontract.

10.4.2 No approval of an application for a progress payment, nor any progress payment, nor any partial or entire use or occupancy of the Project by the Owner, shall constitute an acceptance of any Work not in accordance with the Contract Documents.

10.4.3 The Subcontractor shall promptly pay each Sub-subcontractor upon receipt of payment out of the amount paid to the Subcontractor on account of such Sub-subcontractor's Work, the amount to which said Sub-subcontractor is entitled, reflecting the percentage actually retainea, if any, from payments to the Subcontractor on account of such Sub-subcontractor's Work. The Subcontractor shall, by an appropriate agreement with each Sub-subcontractor, also require each Sub-subcontractor to make payments to his subcontractors in a similar manner.

FIG. 11.2 *(continued)*

10.5 PAYMENTS WITHHELD

10.5.1 The Contractor may decline to approve an Application for Payment if in his opinion the Application is not adequately supported. If the Subcontractor and Contractor cannot agree on a revised amount, the Contractor shall process the Application for the amount he deems appropriate. The Contractor may also decline to approve any Applications for Payment or, because of subsequently discovered evidence or subsequent inspections, he may nullify in whole or in part any approval previously made to such extent as may be necessary in his opinion because of:

.1 defective work not remedied;

.2 third party claims filed or reasonable evidence indicating probable filing of such claims;

.3 failure of the Subcontractor to make payments properly to Sub-subcontrators or for labor, materials or equipment;

.4 reasonable evidence that the Work cannot be completed for the unpaid balance of the Contract Sum;

.5 damage to the Contractor, the Owner or another contractor working at the Project;

.6 reasonable evidence that the Work will not be completed within the Contract time; or

.7 persistent failure to carry out the Work in accordance with the Contract Documents.

10.5.2 When the above grounds in Subparagraph 10.5.1 are removed, payment shall be made for amounts withheld because of them.

10.6 FAILURE OF PAYMENT

10.6.1 If the Subcontractor is not paid within seven days after payment to the Contractor by the Owner, then the Subcontractor may, upon seven days additional written notice to the Contractor, stop the work until payment of the amount owing has been received. The Contract Sum shall be increased by the amount of the Subcontractor's reasonable costs of shut-down, delay and start-up, which shall be effected by appropriate Change Order in accordance with Paragraph 13.3.

10.7 SUBSTANTIAL COMPLETION

10.7.1 Warranties required by the Contract Documents shall commence on the Date of Substantial Completion of the Project as established by the Certificate of Substantial Completion between the Contractor and Owner.

10.8 FINAL COMPLETION AND FINAL PAYMENT

10.8.1 Upon receipt of written notice that the Work is ready for final inspection and acceptance and upon receipt of a final Application for Payment, the Architect/Engineer and Contractor will promptly make such inspection and, when they find the Work acceptable under the Contract Documents and the Contract fully performed, the Contractor will promptly approve final payment.

10.8.2 Neither the final payment nor the remaining retained percentage shall become due until the Subcontractor submits to the Contractor (1) an affidavit that all payrolls, bills for materials and equipment, and other indebtedness connected with the Work for which the Owner or his property might in any way be responsible, have been paid or otherwise satisfied, (2) consent of surety, if any, to final payment and (3), if required by the Owner, other data establishing payment or satisfaction of all such obligations, such as receipts, releases and waivers of liens arising out of the Contract, to the extent and in such form as may be designated by the Owner. If any Sub-subcontractor refuses to furnish a release or waiver required by the Owner or Contractor, the Subcontractor may furnish a bond satisfactory to the Owner and Contractor to indemnify them against any such lien. If any such lien remains unsatisfied after all payments are made, the Subcontractor shall refund to the Owner or Contractor all moneys that the latter may be compelled to pay in discharging such lien, including all costs and reasonable attorneys' fees.

FIG. 11.2 *(continued)*

10.8.3 If, after Substantial Completion of the Work, final completion thereof is materially delayed through no fault of the Subcontractor or by the issuance of Change Orders affecting final completion, and the Contractor so confirms, the Owner or Contractor shall, upon certification by the Contractor, and without terminating the Contract, make payment of the balance due for that portion of the Work fully completed and accepted. If the remaining balance for Work not fully completed or corrected is less than the retainage stipulated in the Contract Documents, and if bonds have been furnished as provided in Paragraph 8.5, the written consent of the surety to the payment of the balance due for that portion of the Work fully completed and accepted shall be submitted by the Subcontractor to the Contractor prior to such payment. Such payment shall be made under the terms and conditions governing final payment, except that it shall not constitute a waiver of claims.

10.8.4 The making of final payment shall constitute a waiver of all claims by the Contractor except those arising from:

.1 unsettled liens;

.2 faulty or defective Work appearing after Substantial Completion;

.3 failure of the Work to comply with the requirements of the Contract Documents; or

.4 terms of any special warranties required by the Contract Documents.

10.8.5 The acceptance of final payment shall constitute a waiver of all claims by the Subcontractor except those previously made in writing and identified by the Subcontractor as unsettled at the time of the Final Application for Payment.

ARTICLE 11

PROTECTION OF PERSONS AND PROPERTY

11.1 SAFETY PRECAUTIONS AND PROGRAMS

11.1.1 The Subcontractor shall be responsible for initiating, maintaining and supervising all safety precautions and programs in connection with the Work.

11.1.2 If the Subcontractor fails to maintain the safety precautions required by law or directed by the Contractor, the Contractor may take such steps as necessary and charge the Subcontractor therefor.

11.1.3 The failure of the Contractor to take any such action shall not relieve the Subcontractor of his obligations in Subparagraph 11.1.1.

11.2 SAFETY OF PERSONS AND PROPERTY

11.2.1 The Subcontractor shall take all reasonable precautions for the safety of, and shall provide all reasonable protection to prevent damage, injury or loss to:

.1 all employees on the Work and all other persons who may be affected thereby;

.2 all the Work and all materials and equipment to be incorporated therein, whether in storage on or off the site, under the care, custody or control of the Subcontractor or any of his Sub-subcontractors or Subsubsubcontractors; and

.3 other property at the site or adjacent thereto, including trees, shrubs, lawns, walks, pavements, roadways, structures and utilities not designated for removal , relocation or replacement in the course of construction.

FIG. 11.2 *(continued)*

11.2.2 The Subcontractor shall give all notices and comply with all applicable laws, ordinances, rules, regulations and lawful orders of any public authority bearing on the safety of persons or property or their protection from damage, injury or loss.

11.2.3 The Subcontractor shall erect and maintain, as required by existing conditions and progress of the Work, all reasonable safeguards for safety and protection, including posting danger signs and other warnings against hazards, promulgating safety regulations and notifying owners and users of adjacent utilities. If the Subcontractor fails to so comply he shall, at the direction of the Contractor, remove all forces from the Project without cost or loss to the Owner or Contractor, until he is in compliance.

11.2.4 When the use or storage of explosives or other hazardous materials or equipment is necessary for the execution of the Work the Subcontractor shall exercise the utmost care and shall carry on such activities under the supervision of properly qualified personnel.

11.2.5 The Subcontractor shall promptly remedy all damage or loss (other than damage or loss insured under Paragraphs 12.2 and 12.3) to any property referred to in Clauses 11.2.1.2 and 11.2.1.3 caused in whole or in part by the Subcontractor, his Sub-contractors; his Sub-subcontractor, or anyone directly or indirectly employed by any of them or by anyone for whose acts any of them may be liable and for which the Subcontractor is responsible under Clauses 11.2.1.2 and 11.2.1.3, except damage or loss attributable to the acts or omissions of the Owner or Architect/Engineer or anyone directly or indirectly employed by either of them or by anyone for whose acts either of them may be liable, and not attributable to the fault of negligence of the Subcontractor. The foregoing obligations of the Subcontractor are in addition to his obligation under Paragraph 5.17.

11.2.6 The Subcontractor shall designate a responsible member of his organization at the site whose duty shall be the prevention of accidents. This person shall be the Subcontractor's superintendent unless otherwise designated by the Subcontractor in writing to the Contractor.

11.2.7 The Subcontractor shall not load or permit any part of the Work to be loaded so as to endanger its safety.

11.3 EMERGENCIES

11.3.1 In any emergency affecting the safety of persons or property, the Subcontractor shall act, at his discretion, to prevent threatened damage, injury or loss. Any additional compensation or extension of time claimed by the Subcontractor on account of emergency work shall be determined as provided in Article 13 for Changes in the Work.

ARTICLE 12

INSURANCE

12.1 SUBCONTRACTOR'S LIABILITY INSURANCE

12.1.1 The Subcontractor shall purchase and maintain such insurance as will protect him from claims set forth below which may arise out of or result from the Subcontractor's operations under the Contract, whether such operations be by himself or by any of his Sub-subcontractors or by anyone directly or indirectly employed by any of them, or by anyone for whose acts any of them may be liable:

.1 claims under workers' compensation, disability benefit and other similar employee benefit acts which are applicable to the work to be peformed;

.2 claims for damages because of bodily injury, occupational sickness or disease, or death of his employees under any employer's liability law including, if applicable, those required under maritime or admiralty law for wages, maintenance, and cure;

.3 claims for damages because of bodily injury, sickness or disease, or death of any person other than his employees;

FIG. 11.2 *(continued)*

.4 claims for damages insured by usual personal injury liability coverage which are sustained (1) by any person as a result of an offense directly or indirectly related to the employment of such person by the Subcontractor, or (2) by any other person;

.5 claims for damages other than to the work itself because of injury to or destruction of tangible property, including loss of use resulting therefrom; and

.6 claims for damages because of bodily injury or death of any person or property damage arising out of the ownership, maintenance or use of any motor vehicle.

12.1.2 The insurance required by Subparagraph 12.1.1 shall be written for not less than any limits of liability specified in the Contract Documents, or required by law, whichever is greater.

12.1.3 The insurance required by Subparagraph 12.1.1 shall include premises-operations (including explosion, collapse and underground coverage), elevators, independent contractors, products and completed operations, and contractual liability insurance (on a "blanket basis" designating all written contracts), all including broad form property damage coverage. Liability insurance may be arranged under Comprehensive General Liability policies for the full limits required or by a combination of underlying policies for lesser limits with the remaining limits provided by an Excess or Umbrella Liability Policy.

12.1.4 The foregoing policies shall contain a provision that coverages afforded under the policies will not be cancelled or not renewed until at least sixty days' prior written notice has been given to the Contractor. Certificates of Insurance acceptable to the Contractor shall be filed with the Contractor prior to commencement of the Work. Upon request, the Subcontractor shall allow the Contractor to examine the actual policies.

12.2 PROPERTY INSURANCE

12.2.1 Unless otherwise provided, the Owner will purchase and maintain property insurance upon the entire Work at the site to the full insurable value thereof. This insurance shall include the interests of the Owners, the Contractor, the Subcontractors, and Sub-subcontractors in the Work and shall insure against the perils of fire and extended coverage, and shall include "all risk" insurance for physical loss or damage.

12.2.2 The Owner will effect and maintain such boiler and machinery insurance as may be necessary and/or required by law. This insurance shall include the interest of the Owner, the Contractor, the Subcontractors, and Sub-subcontractors in the Work.

12.2.3 Any loss insured under Paragraph 12.2 and 12.3 is to be adjusted with the Owner and Construction Manager and made payable to the Owner and Construction Manager as trustees for the insureds, as their interests may appear, subject to the requirements of any applicable mortgagee clause.

12.2.4 The Owner, the Contractor, the Architect/Engineer, the Subcontractors, and the Sub-subcontractors waive all rights against each other and any other contractor or subcontractor engaged in the project for damages caused by fire or other perils to the extent covered by insurance provided under Paragraphs 12.2 and 12.3, or any other property or consequential loss insurance applicable to the Project, equipment used in the Project, or adjacent structures, except such rights as they may have to the proceeds of such insurance. If any policy of insurance requires an endorsement to maintain coverage with such waivers, the owner of such policy will cause the policy to be so endorsed. The Owner will require, by appropriate agreement, written where legally required for validity, similar waivers in favor of the Subcontractors and Sub-subcontractors by any separate contractor and his subcontractors.

12.2.5 The Owner and Contractor shall deposit in a separate account any money received as trustees, and shall distribute it in accordance with such agreement as the parties in interest may reach, or in accordance with an award by arbitration in which case the procedure shall be as provided in Paragraph 8.8. If after such loss no special agreement is made, replacement of damaged work shall be covered by an appropriate Change Order.

12.2.6 The Owner and Contractor as trustees shall have power to adjust and settle any loss with the insurers unless one

FIG. 11.2 *(continued)*

of the parties in interest shall object in writing within five days after the occurrence of loss to the Owner's and Contractor's exercise of this power, and if such objection be made, arbitrators shall be chosen as provided in Paragraph 8.8. The Owner and Contractor as trustees shall, in that case, make settlement with the insurers in accordance with the directions of such arbitrators. If distribution of the insurance proceeds by arbitration is required, the arbitrators will direct such distribution.

12.2.7 If the Owner finds it necessary to occupy or use a portion or portions of the Work prior to Substantial Completion thereof, such occupancy shall not commence prior to a time mutually agreed to by the Owner and Contractor and to which the insurance company or companies providing the property insurance have consented by endorsement to the policy or policies. This insurance shall not be cancelled or lapsed on account of such partial occupancy. Consent of the Contractor and of the insurance company or companies to such occupancy or use shall not be unreasonably withheld.

ARTICLE 13

CHANGES IN THE WORK

13.1 CHANGE ORDERS

13.1.1 A Change Order is a written order to the Subcontractor signed by the Contractor, issued after the execution of the Contract, authorizing a Change in the Work or an adjustment in the Contract Sum or the Contract Time. The Contract Sum and the Contract Time may be changed only by Change Order. A Change Order signed by the Subcontractor indicates his agreement therewith, including the adjustment in the Contract Sum or the Contract Time.

13.1.2 The Contractor, without invalidating the Contract, may order Changes in the Work within the general scope of the Contract consisting of additions, deletions or other revisions, the Contract Sum and the Contract Time being adjusted accordingly. All such changes in the Work shall be authorized by Change Order, and shall be performed under the applicable conditions of the Contract Documents.

13.1.3 The cost or credit to the Contractor resulting from a Change in the Work shall be determined in one or more of the following ways:

.1 by mutual acceptance of a lump sum properly itemized and supported by sufficient substantiating data to permit evaluation; or

.2 by unit prices stated in the Contract Documents or subsequently agreed upon; or

.3 by cost to be determined in a manner agreed upon by the parties and a mutually acceptable fixed or percentage fee; or

.4 by the method provided in Subparagraph 13.1.4.

13.1.4 If none of the methods set forth in Clauses 13.1.3.1, 13.1.3.2 or 13.1.3.3 is agreed upon, the Subcontractor, provided he receives a written order signed by the Contractor, shall promptly proceed with the Work involved. The cost of such Work shall be determined by the Contractor on the basis of the reasonable expenditures and savings of those performing the Work attributable to the change, including, in the case of an increase in the Contract Sum, a reasonable allowance for overhead and profit. In such case, and also under Clauses 13.1.3.3 and 13.1.3.4 above, the Subcontractor shall keep and present, in such form as the Contractor may prescribe, an itemized accounting together with appropriate supporting data for inclusion in a Change Order. Unless otherwise provided in the Contract Documents, cost shall be limited to the following: cost of materials, including sales tax and cost of delivery; cost of labor, including social security, old age and unemployment insurance, and fringe benefits required by agreement or custom; workers' or workmen's compensation insurance; bond premiums; rental value of equipment and machinery; and the additional costs of supervision and field office personnel directly attributable to the change. Pending final determination of cost, payments on account shall be made as determined by the Contractor. The amount of credit to be allowed by the Subcontractor for any deletion or change which results in a net decrease in the Contract Sum will be the amount of the actual net cost as confirmed by the Contractor.

FIG. 11.2 *(continued)*

When both additions and credits covering related Work or substitutions are involved in any one change, the allowance for overhead and profit shall be figured on the basis of the net increase, if any, with respect to that change.

13.1.5 If unit prices are stated in the Contract Documents or subsequently agreed upon, and if the quantities originally contemplated are so changed in a proposed Change Order that application of the agreed unit prices to the quantities of Work proposed will cause substantial inequity to the Owner, the Contractor, or the Subcontractor, the applicable unit prices shall be equitably adjusted.

13.2 CONCEALED CONDITIONS

13.2.1 Should concealed conditions encountered in the performance of the Work below the surface of the ground or should concealed or unknown conditions in an existing structure be at variance with the conditions indicated by the Contract Documents or should unknown physical conditions below the surface of the ground or should concealed or unknown conditions in an existing structure of an unusual nature, differing materially from those ordinarily encountered and generally recognized as inherent in work of the character provided for in this Contract, be encountered, the Contract Sum shall be equitably adjusted by Change Order upon claim by either party made within twenty days after the first observance of the conditions.

13.3 CLAIMS FOR ADDITIONAL COST

13.3.1 If the Subcontractor wishes to make a claim for an increase in the Contract Sum, he shall give the Contractor written notice thereof within twenty days after the occurrence of the event giving rise to such claim. This notice shall be given by the Subcontractor before proceeding to execute the Work, except in an emergency endangering life or property in which case the Subcontractor shall proceed in accordance with Paragraph 11.3. No such claim shall be valid unless so made. Any change in the Contract Sum resulting from such claim shall be authorized by Change Order.

13.3.2 If the Subcontractor claims that additional cost is involved because of, but not limited to, (1) any written interpretation issued pursuant to Subparagraph 3.2.2, (2) any order by the Contractor to stop the Work pursuant to Paragraph 4.3 where the Subcontractor was not at fault, or (3) any written order for a minor change in the Work issued pursuant to Paragraph 13.4, the Subcontractor shall make such claim as provided in Subparagraph 13.3.1.

13.4 MINOR CHANGES IN THE WORK

13.4.1 The Architect/Engineer will have authority to order through the Contractor minor changes in the Work not involving an adjustment in the Contract Sum or an extension of the Contract Time and not inconsistent with the intent of the Contract Documents. Such changes shall be effected by written order and such changes shall be binding on the Owner, the Contractor and the Subcontractor. The Subcontractor shall carry out such written orders promptly.

ARTICLE 14

UNCOVERING AND CORRECTION OF WORK

14.1 UNCOVERING OF WORK

14.1.1 If any portion of the Work should be covered contrary to the request of the Contractor or Architect/Engineer, or to requirements specifically expressed in the Contract Documents, it must, if required in writing by the Contractor, be uncovered for their observation and replaced, at the Subcontractor's expense.

14.1.2 If any other portion of the Work has been covered which the Contractor or the Architect/Engineer has not specifically requested to observe prior to being covered, the Architect/Engineer or Contractor may request to see such Work and it shall be uncovered by the Subcontractor. If such Work be found in accordance with the Contract Documents, the cost of uncovering and replacement shall, by appropriate Change Order, be charged to the Contractor. If such Work be found not in accordance with the Contract Documents, the Subcontractor shall pay such costs unless it be found that this condition was caused by a separate subcontractor employed as provided in Article 7, and in that event the separate subcontractor shall be responsible for the payment of such costs.

FIG. 11.2 *(continued)*

14.2 CORRECTION OF WORK

14.2.1 The Subcontractor shall promptly correct all Work rejected by the Architect/Engineer or the Contractor as defective or a failing to conform to the Contract Documents whether observed before or after Substantial Completion and whether or not fabricated, installed or completed. The Subcontractor shall bear all costs of correcting such rejected Work, including compensation for the Architect/Engineer's and/or Contractor's additional services made necessary thereby.

14.2.2 If, within one year after the Date of Substantial Completion of Work or designated portion thereof, or within one year after acceptance by the Owner of designated equipment or within such longer period of time as may be prescribed by law or by the terms of any applicable special warranty required by the Contract Documents, any of the Work is found to be defective or not in accordance with the Contract Documents, the Subcontractor shall correct it promptly after receipt of a written notice from the Owner or Contractor to do so unless the Owner or Contractor has previously given the Subcontractor a written acceptance of such condition. This obligation shall survive the termination of the Contract. The Owner or Contractor shall give such notice promptly after discovery of the condition.

14.2.3 The Subcontractor shall remove from the site all portions of the Work which are defective or non-conforming and which have not been corrected under Subparagraphs 14.2.1, 5.5.1 and 14.2.2 unless removal has been waived by the Owner.

14.2.4 If the Subcontractor fails to correct defective or non-conforming Work as provided in Subparagraphs 5.5.1, 14.2.1 and 14.2.2, the Owner or Contractor may correct it in accordance with Subparagraph 4.3.2.

14.2.5 If the Subcontractor does not proceed with the correction of such defective or non-conforming Work within a reasonable time fixed by written notice from the Contractor, the Owner or Contractor may remove it and may store the materials or equipment at the expense of the Subcontractor. If the Subcontractor does not pay the cost of such removal and storage within ten days thereafter, the Owner or Contractor may upon ten additional days' written notice sell such Work at auction or at private sale and shall account for the net proceeds thereof, after deducting all the costs that should have been borne by the Subcontractor, including compensation for the Contractor's additional services made necessary thereby. If such proceeds of sale do not cover all costs which the Subcontractor and an approriate Change Order shall be issued. If the payments then or thereafter due the Subcontractor are not sufficient to cover such amount, the Subcontractor shall pay the difference to the Owner or Contractor.

14.2.6 The Subcontractor shall bear the cost of making good all work of the Contractor, other Subcontractors other contractors destroyed or damaged by such removal or correction.

14.3 ACCEPTANCE OF DEFECTIVE OR NONCONFORMING WORK

14.3.1 If the Owner or Contractor prefers to accept defective or non-conforming Work, he may do so instead of requiring its removal and correction, in which case a Change Order will be issued to reflect reduction in the Contract Sum where appropriate and equitable. Such adjustment shall be effected whether or not final payment has been made.

ARTICLE 15

TERMINATION OF THE CONTRACT

15.1 TERMINATION BY THE SUBCONTRACTOR

15.1.1 If the Work should be stopped for a period of thirty days by the Subcontractor because of failure to receive payment in accordance with the Contract, then the Subcontractor may, upon seven additional days' written notice to the Contractor, terminate the Contract and recover from the Contractor payment for all Work executed and for any proven loss sustained upon any materials, equipment, tools, construction equipment and machinery, including reasonable profit and damages.

FIG. 11.2 *(continued)*

15.2 TERMINATION BY THE CONTRACTOR

15.2.1 If the Subcontractor is adjudged a bankrupt, or if he makes a general assignment for the benefit of his creditors, or if a receiver is appointed on account of his insolvency, or if he persistently or repeatedly refuses or fails, except in cases for which extension of time is provided, to supply enough properly skilled workmen or proper materials, or if he fails to make prompt payment to Sub-subcontractors or for materials or labor, or persistently disregards laws, ordinances, rules, regulations or orders of any public authority having jurisdiction, or otherwise is guilty of a substantial violation of a provision of the Contract Documents, then the Contractor may, without prejudice to any right or remedy and after giving the Subcontractor and his surety, if any, seven days' written notice, terminate the employment of the Subcontractor and take possession of the site and of all materials, equipment, tools, construction equipment and machinery thereon owned by the Subcontractor and may finish the Work by whatever method he may deem expedient. In such case the Subcontractor shall not be entitled to receive any further payment until the Work is finished.

15.2.2 If the unpaid balance of the Contract Sum exceeds the cost of finishing the Work, including compensation for the Contractor's additional services made necessary thereby, such excess shall be paid to the Subcontractor. If such costs exceed the unpaid balance, the Subcontractor shall pay the difference to the Contractor.

15.2.3 The Contractor has the option to terminate his Agreement with the Owner if the Work is stopped for a period of thirty days under an order of any court or other public authority having jurisdiction, or as a result of an act of government, such as a declaration of a national emergency making materials unavailable, or from the failure of the Owner to make payment. If the Contractor exercises such option he may then terminate the contracts with the Subcontractors and the Subcontractors shall be entitled to recover such amounts for his proven losses as the Contractor may be able to recover from the Owner.

tions. Ownership of the documents should be defined in this area, and statements regarding copies to be furnished may also be found here.

An important element of this portion of the conditions is the statement precluding any contractual relationship between the subcontractor or sub-subcontractors and the owner or the design professional. Such a statement is recommended as an attempt to avoid possible misunderstandings that could arise since all of these parties are mentioned in the general conditions.

11.4.2 DUTIES AND RESPONSIBILITIES OF PARTIES

The second area of information contained within the general conditions is of extreme importance. This concerns the duties and responsibilities of those who are party to the contract, namely the owner and the contractor. This section should first of all define the owner and the contractor under the terms of the agreement. This is particularly important if the owner is a public body. In such cases, the general conditions may also address the subject of the owner's representative or contracting officer. The duties and responsibilities of the owner are fairly limited in scope and relate primarily to the furnishing of information and the prompt payment of approved pay requests. A statement regarding the obligation of the owner to perform assigned responsibilities with reasonable promptness should be included. This will permit action to be taken by the contractor against any unnecessary delays.

In most contracts, this section of the general conditions will also address the owner's rights. This will include the owner's right to stop the work for just cause and to carry out the work upon the default or negligence of the contractor.

Considerably more time and attention is devoted to the duties and responsibilities of the contractor. Normal information will include the responsibility for supervising and executing the work, furnishing necessary permits and fees, paying applicable taxes, supplying such items as progress schedules and payment schedules, reviewing and accepting the contract documents, supplying required warranties, and conforming to other items that may be included. "Other items" may include the requirement to furnish competent supervision for the project or to include all cash allowances. Cash allowances are pre-set amounts of money specified in the documents to cover certain items of the work. Two of the more common instances when an allowance is used would be for finish hardware in building construction or for face brick.

A controversial item under the contractor's duties is the matter of indemnification. Such a provision will normally require the contractor to hold harmless the owner and the design professional to the extent permitted by law. This means that the contractor shall protect and defend them from claims and expenses coming out of the performance of the construction. This particular provision has been the subject of much intra-industry debate and is generally regarded as the prime reason that one of the standard versions of the general conditions was revised and reissued a few years ago.

An imbalanced listing of duties and responsibilities to be placed upon the contractor should not be interpreted as taking unfair advantage of that party to the contract. It must be remembered that the general conditions as well as the other contract documents are written primarily to ensure that the owner will receive that which is being paid for and to govern and control the actions of the contractor in supplying the project.

11.4.3 ROLE OF DESIGN PROFESSIONAL

It was stated earlier that the general conditions cover those who are affected by the contract. Such is the case with the design professional who, though not signatory to the agreement, is still discussed in the general conditions. This discussion will include the design professional's right of access to the project site, the fact that the design professional is the owner's representative, and information regarding the design professional's authority to make certain decisions. The courts have handed down rulings that have somewhat limited the authority of the design professional to make decisions. Generally, the design professional may make decisions pertaining to the intent of the drawings with respect to the design. In the case of the specifications, however, the courts are prone to base decisions on what has actually been said, rather than permit the design professional to interpret.

Care is usually taken in this portion of the general conditions to define and delimit the design professional's duties with respect to the progress of the work. Specifically, the responsibility for conformance with the intent of the design documents remains that of the contractor and will not be assumed by the design professional. In an attempt to emphasize this condition, some general conditions state that the design professional will "observe" the work on behalf of the owner, but that such activity will not constitute "inspection" or "supervision."

Processing of shop drawings may also be covered in this portion of the general conditions. The designer's role in this activity is explained in the section on approval procedures in Chapter 18.

11.4.4 ROLE OF SUBCONTRACTORS

The portion of the general conditions dealing with subcontractors will normally include the definition of a subcontractor, any listing or approval methods with respect to subcontractors, their relationship to the general contractor under the contract, and provisions or limitations with respect to payment.

The approval of subcontractors by the owner and the design professional is worthy of special notice. The contractor is precluded from awarding a subcontract to anyone that the owner or the design professional objects to for valid reasons. This raises the possibility of affecting the general contractor's status as low bidder. It also creates the uncomfortable situation of causing ill feelings on the part of the rejected subcontractor.

This portion of the general conditions repeats again the absence of any contractual relationship between the owner and the subcontractor.

11.4.5 WORK BY OWNER

Unless there is legislation forbidding it, the general conditions will reserve for the owner the right to award separate contracts on portions of the work. Some states have enacted laws that require a single contract to be awarded, in which case this provision would be deleted from the general conditions or modified by the supplemental conditions.

Also covered in this part of the general conditions would be the mutual responsibility of the different contractors on the project. This would include cooperation with one another as well as coordination of the various parts of the contract. It is normal for this portion of the general conditions to also reserve for owners the right to do work with their own forces. The latter provision should be examined carefully on each project for possible conflicts between union and open shop labor forces.

11.4.6 TIME CONSIDERATIONS

The time aspect of the contract portion of the general conditions is one that is often the subject of litigation. For this reason, particular care should be given to its preparation. This section covers such items as dates of commencement, contract time allowed for completion, delays and extensions of time, and the formal procedure that shall be followed to implement each of these provisions. Clear distinction should be made between delays that are unavoidable and therefore a basis for time extension, and delays that the contractor should plan for and for which no extension will be granted.

11.4.7 PAYMENTS

The payments part of the general conditions will include coverage of the procedures to be followed for the contractor to request periodic payments under the contract, the time intervals when these requests should be submitted and to whom, the timing and procedure for the design professional's approval or modification of these requests, and the conditions under which the owner is to make payment. The percentage of retained payments is also covered, along with any information related to any decrease in the percentage as certain stages of the work are completed, and the timing and conditions under which the retained percentages will be paid. This section may also include provisions for the contractor to charge the owner interest on any approved payments that become past due. This particular provision is one of fairly recent origin and is still not widely enforced by many contractors.

Completion of the project is a topic that should be thoroughly defined. Final payment and payment of all retained percentages is dependent upon this stage being reached. The general conditions should state what is an acceptable level of completion, whether it consists of total completion or only that affording the owner beneficial use of the project. If it is a highway project, completion to the point of permitting full traffic may be accepted, with shoulders and medians still being in the stage of needing minor work. For a building project, beneficial occupancy may be used as the criterion. Such definitions are also necessary in order to apply any bonus or penalty clauses.

11.4.8 INSURANCE AND BONDS

Insurance coverage and bonds that are to be furnished should be described in detail. Each type of insurance that the contractor is to carry should be listed, along with the amount of coverage for each type. This section will also include information regarding submission of proof of coverage, time limitations and procedures to be followed in case of cancellation, and to whom notice is to be given. Such information shall be furnished for insurance carried by the owner as well as insurance to be carried by the contractor.

Information regarding bonds for the project should include each type of bond that is to be furnished along with the amount for each. The general conditions also include statements as to who shall be responsible for the payment of bond premiums, and how notification is to be furnished. If the owner wishes to specify a particular bonding company, this information should be listed in the special conditions. It should be stated whether the bonding company's standard form of bond will be acceptable or whether some other form will be required.

11.4.9 CHANGES

Changes in the work provisions consist primarily of a statement reserving for the owner the right to order changes in the work. Information regarding the issuance of change orders as well as any limitations on the scope of the ordered changes should be included. The limitations provision is sometimes expressed as a percentage of the base contract. This part of the general conditions will also address the situation regarding the issuance of change orders relative to any claims that the contractor may make for additional costs. Of importance in this connection is the establishing of realistic time limitations for this eventuality. Reference should include a definite procedure that is to be followed in submitting such claims. Litigation can often be avoided by a careful preparation of this provision.

11.4.10 ARBITRATION

In an increasing number of instances, construction contracts contain a provision for the arbitration of disputes that arise in connection with the contract. This provision has as its purpose the avoidance of the high costs and lengthy delays that many feel accompany litigation. This section of the general conditions should clearly state the conditions under which arbitration is indicated and the time limits attendant to the arbitration process. Information relative to the rules of arbitration that will apply should also be given. This matter is often covered by referring to a particular arbitration association and saying that the rules and procedures of that association shall apply.

All parties should check to see that all contracts contain the same provisions regarding arbitration. This means that the contract between the owner and the design professional, between the owner and the contractor, and between the contractor and the subcontractors should contain similar arbitration provisions. If this recommendation is not followed, some parties may find themselves obligated to arbitration by one contract and have no alternative to litigation on the other contract. Such a situation can prove costly to the parties and can also involve considerable delay in settlement.

11.4.11 TERMINATION

The general conditions should contain information regarding termination of the contract by either the owner or the contractor. The owner is usually permitted to terminate the contract upon failure to perform by the contractor. Notice provisions as well as time limitation in connection with the notice should be covered. The owner is also normally permitted to terminate the contract with no prejudice to the contractor if it is in the owner's best interest to do so. If this provision is included, information should be presented as to payment of the contractor for work already performed and payment to the contractor for lost revenue.

The conditions under which the contractor has the right to terminate the contract or to stop work should be clearly defined. This section would include a listing of those reasons for contract termination and the provisions made for determination of payment due to the contractor. Among the reasons given for the contractor to terminate the contract are the owner's failure to pay promptly, halting of the work by a court order or public body, or unwarranted delays on the part of the design professional. Again, information should be given covering time limitation of notice, form of notice, and method of determining the final amount of payment due the contractor.

11.4.12 OTHER PROVISIONS

There are several other items that may be included within the general conditions. These may involve such items as patents and royalties, and ownership of drawings. These may be included when the owner or contracting authority finds it advantageous to control these conditions in all or most of their contracts. On the other hand if it is a unique situation, they would more properly be included within the supplemental or special conditions.

12 SUPPLEMENTARY CONDITIONS

FIGURES

12.1 DEFINITION

Supplementary conditions are similar to the general conditions in that they both present information relative to the conditions of the agreement. Supplementary conditions are also known as "special conditions," "supplemental general conditions," or "supplemental conditions." These terms are used interchangeably to refer to the same document. The supplementary conditions cover provisions of the contract that are of a general nature but are not covered or included in the general conditions.

12.2 REASONS FOR USE

When a standardized version of the general conditions is used, it may not include all the conditions of the contract needed for a particular project. It may also be true that some of the conditions contained within the standardized version are not desired for the particular project. Supplementary conditions are indicated for any of the following reasons:

> To make modifications to a provision within the general conditions. These modifications may cover a wide range of topics, such as the amount of insurance, number of copies of drawings to be furnished, or who is to secure the necessary permits.
>
> To effect deletions from the general conditions. There may be provisions in the general conditions that do not apply to a particular project, or provisions that contain cost factors that the owner does not desire. Examples would be if a project does not include a particular type of work or if the owner elects not to require any contract bonds. In such cases the appropriate portions of the general conditions would be deleted since they no longer apply.
>
> To make additions to the general conditions. Many owners, especially governmental agencies, require that certain provisions be included in the contract. If the design firm is more accustomed to doing private work, the general conditions normally used would not include such provisions. These added requirements would be included in the supplementary conditions. Additions to the supplementary conditions may also replace other provisions that have been deleted.

12.3 RELATIONSHIP TO GENERAL CONDITIONS

Since the supplementary conditions change or modify the general conditions, they are to be considered part of the general conditions. Therefore they apply to the contract and are included as part of all areas where the general conditions are included by reference. Even though such a relationship is understood and implied, it is highly desirable to include a statement in the contract documents to this effect.

By virtue of the fact that the supplementary conditions make changes in the general conditions, wherever there are conflicts between the two documents, the provisions of the supplementary conditions shall govern. In order to emphasize the close relationship between the supplementary conditions and the general conditions, the general should be followed immediately by the supplementary conditions when they are bound with the specifications within the cover.

12.4 USE OF CHECKLIST

The number of items that may dictate the necessity for supplemental conditions is as high as the number of items in the construction project. Any part of the work may be an area in which the owner or the design professional wishes to introduce special requirements. While such a condition precludes any checklist from being complete, the use of a checklist is still recommended. Even though a checklist cannot contain all the possible items, its use will serve to at least diminish the number of errors of omission.

12.5 GOVERNMENTAL PROVISIONS

As previously mentioned, the specific requirements of various governmental agencies often necessitate the use of supplementary conditions. The following are typical of such requirements:

> Davis-Bacon Act. This legislation requires that prevailing area wages be paid on contracts over a stated amount when federal funds are used, in whole or in part. While it is not uncommon for the wage rate schedule to be included with the bidding documents, the schedule is just as likely to be included by reference, with the bidders expected to secure the information from the office of the owner or the design professional, or from the particular agency involved. Many states have enacted companion legislation referred to as "little Davis-Bacon bills."

> Buy America Act. Some contracts may carry the provision that certain materials or equipment must be of United States origin or manufacture. Such a requirement will be covered within the supplementary conditions. This section will establish the requirement and the items to which the requirement will apply. Such a provision will also usually include procedures to be followed for obtaining waivers from the requirement.

> State preference requirements. Many contracts awarded by state agencies contain requirements for preference to be given to in-state firms and materials when awarding subcontracts and material orders. Through a system of award points, the use of preferred subcontractors and material suppliers may also allow a firm to be adjudged the successful bidder even if their bid is not the lowest one submitted. The purpose of such a provision in the supplementary conditions is to favor local firms and products.

> Value engineering. An increasing number of contracts contain provisions for the contractor to suggest changes from the original design that may lead to cost savings for the owner. Under the value engineering provisions, the contractor is permitted to share in any savings that are realized. The procedure to be followed and the apportioning of any savings between the owner and the contractor would be covered in the supplementary conditions.

12.6 PROCEDURE TO MODIFY STANDARD FORM

As discussed earlier in this chapter, supplementary conditions are used to make modifications, deletions, or additions to the general conditions. Figure 12.1 shows the AIA version of the Federal Supplementary Conditions. Note that Article 15 deals with modifications while Articles 16 and 17 are devoted to additional conditions. Other examples are the following:

> General Conditions, Article 13, Paragraph A, after the words "state government," insert "and Federal Government."
> General Conditions, Article 17, Paragraph C, add new subparagraph "3)" as follows: "3) To any preference priority or allocation order duly issued by the Federal Government."

(text continues on page 270)

FIG. 12.1 Federal Supplementary conditions

THE AMERICAN INSTITUTE OF ARCHITECTS

AIA Document A201/SC

Federal Supplementary Conditions of the Contract For Construction

THIS DOCUMENT HAS IMPORTANT LEGAL CONSEQUENCES; CONSULTATION WITH AN ATTORNEY IS ENCOURAGED WITH RESPECT TO ITS COMPLETION OR MODIFICATION

1977 EDITION

TABLE OF ARTICLES

This document may be used for U.S. Department of Health, Education and Welfare Federally Assisted Construction Projects.

This document has been reproduced with the permission of the American Institute of Architects under application number 79082. Further reproduction, in part or in whole, is not authorized. Because AIA documents are revised from time to time, users should ascertain from AIA the current edition of the document reproduced above.

FIG. 12.1 *(continued)*

SUPPLEMENTARY CONDITIONS OF THE CONTRACT FOR CONSTRUCTION

ARTICLE 15

MODIFICATIONS OF THE GENERAL CONDITIONS

15.1 MODIFICATION OF PARAGRAPH 1.2, EXECUTION, CORRELATION AND INTENT

15.1.1 Add the following to Subparagraph 1.2.1:

Such Documents shall be enumerated in the Owner-Contractor Agreement.

15.2 MODIFICATION OF PARAGRAPH 4.8, ALLOWANCES

Substitute the following for Subparagraph 4.8.1:

4.8.1 The Contractor shall include in the Contract Sum all allowances stated in the Contract Documents. Items covered by these allowances shall be supplied for such amounts as the Owner may direct and by such persons as the Owner shall have determined by competitive bidding through public advertising. The Contractor shall purchase the items covered by these allowances by the award of Subcontract to the lowest responsive and responsible bidder.

15.3 MODIFICATION OF PARAGRAPH 5.2, AWARD OF SUBCONTRACTS AND OTHER CONTRACTS FOR PORTIONS OF THE WORK

Substitute the following for Subparagraph 5.2.2:

5.2.2 The Contractor shall not contract with any person or entity declared ineligible under Federal laws or regulations from participating in Federally assisted construction projects or to whom the Owner or the Architect has made reasonable objection under the provisions of Subparagraph 5.2.1. The Contractor shall not be required to contract with anyone to whom he has a reasonable objection.

15.4 MODIFICATION OF PARAGRAPH 7.5, PERFORMANCE BOND AND LABOR AND MATERIAL PAYMENT BOND

Substitute the following for Subparagraph 7.5.1:

7.5.1 The Contractor shall furnish a Performance Bond in an amount equal to one hundred percent (100%) of the Contract Sum as security for the faithful performance of this Contract and also a Labor and Material Payment Bond in an amount not less than one hundred percent (100%) of the Contract Sum or in a penal sum not less than that prescribed by State, Territorial or local law, as security for the payment of persons performing labor on the Project under this Contract and furnishing materials in connection with this Contract. The Performance Bond and the Labor and Material Payment Bond may be in one or in separate instruments in accordance with local law and shall be delivered to the Owner not later than the date of execution of the Contract.

15.5 MODIFICATION OF PARAGRAPH 10.1, SAFETY PRECAUTIONS AND PROGRAMS

Add the following Subparagraph 10.1.2:

10.1.2 Attention is invited to the regulations issued by the Secretary of Labor pursuant to Section 107 of the Contract Work Hours and Safety Standards Act (40 U.S.C. 333) entitled "Safety and Health Regulations for Construction" (29 CFR Part 1926). The Contractor shall be required to comply with those regulations to the extent that any resulting Contract involves construction.

15.6 MODIFICATION OF ARTICLE 11, INSURANCE

Refer to Article 17 for the limits of and any modifications to the insurance requirements of this Article.

ARTICLE 16

ADDITIONAL CONDITIONS

16.1 SUBSTITUTION OF MATERIALS AND EQUIPMENT

16.1.1 Whenever a material, article or piece of equipment is identified on the Drawings or in the Specifications by reference to manufacturers' or vendors' names, trade names, catalog numbers, or the like, it is so identified for the purpose of establishing a standard, and any material, article, or piece of equipment of other manufacturers or vendors which will perform adequately the duties imposed by the general design will be considered equally acceptable provided the material, article, or piece of equipment so proposed is, in the opinion of the Architect, of equal substance, appearance and function. It shall not be purchased or installed by the Contractor without the Architect's written approval.

16.2 FEDERAL INSPECTION

16.2.1 The authorized representatives and agents of the Federal Government shall be permitted to inspect all Work, materials, payrolls, records of personnel, invoices of materials, and other relevant data and records.

16.3 LANDS AND RIGHTS-OF-WAY

16.3.1 Prior to the start of construction, the Owner shall obtain all lands and rights-of-way necessary for the execution and completion of Work to be performed under this Contract.

16.4 EQUAL OPPORTUNITY

16.4.1 During the performance of this Contract the Contractor agrees as follows:

.1 The Contractor will not discriminate against any employee or applicant for employment because of race, color, sex, religion, national origin or age. The

FIG. 12.1 *(continued)*

Contractor will take affirmative action to insure that applicants are employed and that employees are treated during employment without regard to their race, color, sex, religion, national origin or age. Such action shall include, but not be limited to, the following: employment, upgrading, demotion or transfer; recruitment or recruitment advertising; layoff or termination; rates of pay or other forms of compensation; and selection for training, including apprenticeship. The Contractor agrees to post in conspicuous places, available to employees and applicants for employment, notices to be provided by an appropriate agency of the Federal Government setting forth the requirements of these nondiscrimination provisions.

.2 The Contractor will state, in all solicitations or advertisements for employees placed by or on behalf of the Contractor, that all qualified applicants will receive consideration for employment without regard to race, color, sex, religion, national origin or age.

.3 The Contractor will send to each labor union or representative of workers with which he has a collective bargaining agreement or other contract or understanding a notice to be provided by the Owner, advising the labor union or workers' representative of the Contractor's commitments under Section 202 of Executive Order No. 11246 of September 24, 1965 as amended, and shall post copies of the notice in conspicuous places available to employees and applicants for employment.

.4 The Contractor will comply with all provisions of Executive Order No. 11246 of September 24, 1965, as amended, and of the rules, regulations and relevant orders of the Secretary of Labor.

.5 The Contractor will furnish all information and reports required by Executive Order No. 11246 of September 24, 1965 as amended, and by the rules, regulations and orders of the Secretary of Labor, or pursuant thereto, and shall permit access to his books, records and accounts by an appropriate agency of the Federal Government and by the Secretary of Labor for purposes of investigation to ascertain compliance with such rules, regulations and orders.

.6 In the event of the Contractor's noncompliance with the Equal Opportunity conditions of this Contract or with any of such rules, regulations or orders, this Contract may be cancelled, terminated or suspended in whole or in part, and the Contractor may be declared ineligible for further Government contracts or Federally assisted contracts in accordance with procedures authorized in Executive Order No. 11246 of September 24, 1965, as amended, and such other sanctions may be imposed and remedies invoked as provided in said Executive Order, or by rule, regulation or order of the Secretary of Labor, or as otherwise provided by law.

.7 The Contractor will include all clauses 16.4.1.1 to 16.4.1.7 inclusive in every Subcontract or purchase order unless exempted by rules, regulations or orders of the Secretary of Labor issued pursuant to Section 204 of Executive Order 11246 of September 24, 1965, as amended, so that such provisions will be binding upon each Subcontractor or vendor. The Contractor will take such action with respect to any Subcontractor or vendor as the appropriate agency of the Federal Government may direct as a means of enforcing such provisions, including sanctions for noncompliance, provided, however, that in the event the Contractor becomes involved in, or is threatened with, litigation with a Subcontractor or vendor as a result of such direction by the appropriate agency of the Federal Government, the Contractor may request the United States to enter into such litigation to protect the interests of the United States.

16.4.2 Exemptions to the requirements of the above Equal Opportunity conditions are construction Contracts and Subcontracts not exceeding $10,000, and Contracts and Subcontracts with regard to Work performed outside the United States by employees who were not recruited in the United States.

16.4.3 Unless otherwise provided, the above Equal Opportunity provisions are not required to be inserted in sub-subcontracts except for Sub-subcontracts involving the performance of construction Work at the site of construction, in which case the provisions must be inserted in all Sub-subcontracts.

16.5 CERTIFICATION OF NONSEGREGATED FACILITIES

(Applicable to Contracts and Subcontracts exceeding $10,000 which are not exempt from the provisions of Paragraph 16.4, "Equal Opportunity" of this Article 16.)

16.5.1 By entering into an agreement related to the Work described in the Contract Documents the Contractor or Subcontractor certifies that he does not maintain or provide for his employees any segregated facilities at any of his establishments, and that he does not permit his employees to perform their services at any location under his control where segregated facilities are maintained. The Contractor or Subcontractor further certifies that he will not maintain or provide for his employees any segregated facilities at any of his establishments and that he will not permit his employees to perform their services at any location under his control where segregated facilities are maintained. The Contractor or Subcontractor agrees that a breach of this certification is a violation of Paragraph 16.4, "Equal Opportunity." As used herein, the term "segregated facilities" means any waiting rooms, work areas, rest rooms and washrooms, restaurants and other eating areas, time clocks, locker rooms and other storage or dressing areas, parking lots, drinking fountains, recreation or entertainment areas, transportation, and housing facilities provided for employees on the basis of race, color, religion, age or national origin, because of habit, local custom, or otherwise. The Contractor further agrees that (except where he has obtained identical certifications from proposed Subcontractors for specific time peri-

FIG. 12.1 *(continued)*

ods) he will obtain identical certifications from proposed Subcontractors prior to the award of Subcontracts exceeding $10,000 which are not exempt from the provisions of Paragraph 16.4 "Equal Opportunity;" that he will retain such certifications in his files; and that he will forward the following notice to such proposed Subcontractors (except where the proposed Subcontractors have submitted identical certifications for specific time periods):

"NOTICE TO PROSPECTIVE SUBCONTRACTORS OF REQUIREMENT FOR CERTIFICATION OF NONSEGREGATED FACILITIES

A Certification of Nonsegregated Facilities, as required by the May 9, 1967, order (32 Federal Register 7439, May 19, 1967) on Elimination of Segregated Facilities, by the Secretary of Labor, must be submitted prior to the award of a Subcontract exceeding $10,000 which is not exempt from the provisions of Paragraph 16.4 'Equal Opportunity.' The Certification may be submitted either for each Subcontract or for all Subcontracts during a period, i.e. quarterly, semiannually or annually."

16.5.2 The penalty for making false statements in Certifications required by Paragraph 16.5.1 is prescribed in 18 USC 1001.

16.6 MINIMUM WAGES

16.6.1 All mechanics and laborers employed or working directly upon the site of the Work shall be paid unconditionally and not less often than once a week, and without subsequent deduction or rebate on any account [except such payroll deductions as are permitted by the Copeland Regulations (29 CFR Part 3)], the full amounts due at time of payment computed at wage rates not less than the aggregate of the basic hourly rates and the rates of payments, contributions, or costs for any fringe benefits contained in the wage determination decision of the Secretary of Labor, which is attached hereto and made a part hereof, regardless of any contractual relationship which may be alleged to exist between the Contractor or Subcontractor and such laborers and mechanics, and the wage decision shall be posted by the Contractor at the site of the Work in a prominent place where it can easily be seen by the workers.

16.6.2 The Owner shall require that any class of laborers or mechanics, including apprentices and trainees, which is not listed in the wage determination and which is to be employed under the Contract shall be classified or reclassified conformably to the wage determination and a report of the action taken shall be sent to the appropriate Federal agency. If the interested parties cannot agree on the proper classification or reclassification of a particular class of laborers or mechanics, including apprentices and trainees, to be used, the Owner shall submit the question together with his recommendation through the appropriate Federal agency to the Secretary of Labor for final determination.

16.6.3 The Owner shall require, whenever the minimum wage rate prescribed in the Contract for a class of laborers or mechanics includes a fringe benefit which is not expressed as an hourly wage rate and the Contractor is obliged to pay a cash equivalent of such a fringe benefit, an hourly cash equivalent thereof to be established. In the event interested parties cannot agree upon a cash equivalent of the fringe benefit, the question, accompanied by the recommendation of the Owner shall be referred to the Secretary of Labor for determination.

16.6.4 If the Contractor does not make payments to a trustee or other third person, he may consider as part of the wages of any laborer or mechanic the amount of any costs reasonably anticipated in providing benefits under a plan or program of a type expressly listed in the wage determination decision of the Secretary of Labor which is a part of this Contract: provided, however, the Secretary of Labor has found, upon the written request of the Contractor, that the applicable standards of the Davis-Bacon Act have been met. The Secretary of Labor may require the Contractor to set aside in a separate account assets for the meeting of obligations under the plan or program.

16.7 PAYROLLS AND BASIC RECORDS

16.7.1 The Contractor shall maintain payrolls and basic records relating thereto during the course of the Work and shall preserve them for a period of three years thereafter for all laborers and mechanics, including apprentices and trainees, working at the site of the Work. Such records shall contain the name and address of each employee, his correct classification, rate of pay (including rates of contributions for, or costs assumed to provide, fringe benefits), daily and weekly number of hours worked, deductions made and actual wages paid. Whenever the Contractor has obtained approval from the Secretary of Labor as provided in Paragraph 16.6.4 the Contractor shall maintain records which show that the commitment to provide such benefits is enforceable, that the plan or program is financially responsible, and that the plan or program has been communicated in writing to the laborers or mechanics affected, and records which show the costs anticipated or the actual cost incurred in providing such benefits.

16.7.2 The Contractor shall submit weekly a copy of all payrolls to the Owner. The prime Contractor shall be responsible for the submission of copies of payrolls of all Subcontractors. Each such copy shall be accompanied by a statement signed by the Contractor indicating that the payrolls are correct and complete, that the wage rates contained therein are not less than those determined by the Secretary of Labor, and that the classification set forth for each laborer or mechanic, including apprentices and trainees, conform to the work he performed. Submission of the "Weekly Statement of Compliance" required under this Contract and the Copeland Regulations of the Secretary of Labor (29 CFR Part 3) shall satisfy the requirement for submission of the above statement. The Contractor shall submit also a copy of any approval by the Secretary of Labor with respect to fringe benefits which is required by Paragraph 16.6.4.

FIG. 12.1 *(continued)*

16.7.3 Contractors employing apprentices or trainees under approved programs shall include a notation on the first weekly certified payrolls submitted to the Owner that their employment is pursuant to an approved program and shall identify the program.

16.7.4 The Contractor will make the records required under the labor standards clauses of the Contract available for inspection by authorized representatives of the Owner, the appropriate Federal agency and the U.S. Department of Labor, and shall permit such representatives to interview employees during working hours on the job.

16.8 APPRENTICES AND TRAINEES

16.8.1 Apprentices will be permitted to work at less than the predetermined rate for the work they perform when they are employed and individually registered in a bona fide apprenticeship program registered with the U.S. Department of Labor, Employment Training Administration (formerly Manpower), Bureau of Apprenticeship and Training, or with a State apprenticeship agency recognized by the Bureau, or if a person is employed in his first 90 days of probationary employment as an apprentice in such an apprenticeship program who is not individually registered in the program but who has been certified by the Bureau of Apprenticeship and Training or a State apprenticeship agency (where appropriate) to be eligible for probationary employment as an apprentice. The allowable ratio of apprentices to journeymen in any craft classification shall not be greater than the ratio permitted to the Contractor as to his entire work force under the registered program. Any employee listed on a payroll at an apprentice wage rate, who is not a trainee as defined in Subparagraph 16.8.2 of this Paragraph or is not registered or otherwise employed as stated above, shall be paid the wage rate determined by the Secretary of Labor for the classification of work he actually performed. The Contractor will be required to furnish to the Owner or a representative of the U.S. Department of Labor written evidence of the registration of his program and apprentices, as well as the appropriate ratios and wage rates (expressed in percentages of the journeyman hourly rates), for the area of construction prior to using any apprentices on the Contract work. The wage rate paid apprentices shall be not less than the appropriate percentage of the journeyman's rate contained in the applicable wage determination.

16.8.2 Trainees will not be permitted to work at less than the predetermined rate for the work performed unless they are employed pursuant to and individually registered in a program which has received prior approval, evidenced by formal certification, by the U.S. Department of Labor, Employment Training Administration (formerly Manpower), Bureau of Apprenticeship and Training. The ratio of trainees to journeymen shall not be greater than permitted under the plan approved by the Bureau of Apprenticeship and Training. Every trainee must be paid not less than the rate specified in the approved program for his level of progress. Any employee listed in the payroll at a trainee rate who is not registered and participating in a training plan approved by the Bureau of Apprenticeship and Training shall be paid not less than the wage rate determined by the Secretary of Labor for the classification of work he actually performed. The Contractor will be required to furnish to the Owner or a representative of the U.S. Department of Labor, written evidence of the certification of his program, the registration of the trainees, and the ratios and wage rates prescribed in that program. In the event the Bureau of Apprenticeship and Training withdraws approval of a training program, the Contractor will no longer be permitted to utilize trainees at less than the applicable predetermined rate for the work performed until an acceptable program is approved.

16.8.3 The utilization of apprentices, trainees and journeymen under this part shall be in conformity with the equal employment opportunity requirements of Executive Order 11246, as amended, and 29 CFR Part 30.

16.9 COMPLIANCE WITH COPELAND REGULATIONS

16.9.1 The Contractor shall comply with the Copeland Regulations of the Secretary of Labor (29 CFR Part 3), which are incorporated herein by reference. In addition, the Weekly Statement of Compliance required by these regulations shall also contain a statement that the fringe benefits paid are equal to or greater than those set forth in the minimum wage decision.

16.10 CONTRACT WORK HOURS AND SAFETY STANDARDS ACT—OVERTIME COMPENSATION AND SAFETY STANDARDS (40 USC 327-330)

16.10.1 The Contractor shall not require or permit any laborer or mechanic, including apprentices and trainees, in any work week in which he is employed on any work under this Contract to work in excess of 8 hours in any calendar day or in excess of 40 hours in such work week on work subject to the provisions of the Contract Work Hours and Safety Standards Act unless such laborer or mechanic, including apprentices and trainees, receives compensation at a rate not less than one and one-half times his basic rate of pay for all such hours worked in excess of 8 hours in any calendar day or in excess of 40 hours in such work week, whichever is the greater number of overtime hours. The "basic rate of pay" as used in this provision shall be the amount paid per hour, exclusive of the Contractor's contribution or cost for fringe benefits, and any cash payment made in lieu of providing fringe benefits, or the basic hourly rate contained in the wage determination, whichever is greater.

16.10.2 In the event of any violation of the provisions of Subparagraph 16.10.1 the Contractor shall be liable to any affected employee for any amounts due, and to the United States for liquidated damages. Such liquidated damages shall be computed with respect to each individual laborer or mechanic, including apprentices and trainees, employed in violation of the provisions of Subparagraph 16.10.1, in the sum of $10 for each calendar day on which such employee

FIG. 12.1 *(continued)*

was required or permitted to be employed on such work in excess of 8 hours or in excess of the standard work week of 40 hours without payment of the overtime wages required by Subparagraph 16.10.1.

16.10.3 The Contractor shall not require or permit any laborer or mechanic, including apprentices and trainees, employed in the performance of this Contract to work in surroundings or under conditions which are unsanitary, hazardous, or dangerous to his health as determined under construction safety and health standards promulgated by the Secretary of Labor by regulation (29 CFR Part 1926, 36 FR 7340, April 17, 1971) pursuant to Section 107 of the Contract Work Hours and Safety Standards Act.

16.11 WITHHOLDING OF FUNDS

16.11.1 The Owner may withhold or cause to be withheld from the Contractor as much of the accrued payments or advances as may be considered necessary (a) to pay the laborers and mechanics, including apprentices and trainees, employed by the Contractor or any Subcontractor on the Work the full amount of wages required by the Contract, and (b) to satisfy any liability of any Contractor for liquidated damages under Paragraph 16.10 hereof entitled "Contract Work Hours and Safety Standards Act—Overtime Compensation and Safety Standards (40 USC 327-330)."

16.11.2 If the Contractor or any Subcontractor fails to pay any laborer or mechanic, including apprentices and trainees, employed or working on the site of the Work, all or part of the wages required by the Contract, the Owner may, after written notice to the prime Contractor take such action as may be necessary to cause suspension of any further payments or advances until such violations have ceased.

16.12 SUBCONTRACTS

16.12.1 The Contractor will insert in all Subcontracts, Paragraphs 16.6 through 16.13 inclusive, entitled: "Minimum Wages," "Payrolls and Basic Records," "Apprentices and Trainees," "Compliance with Copeland Regulations," "Contract Work Hours and Safety Standards Act—Overtime Compensation and Safety Standards (40 USC 327-330)," "Withholding of Funds," "Subcontracts," and "Contract Termination—Debarment," and shall further require all Subcontractors to incorporate physically these same Paragraphs in all Sub-subcontracts.

16.12.2 The term "Contractor" as used in such Paragraphs in any Subcontract shall be deemed to refer to the Subcontractor except when the phrase "prime Contractor" is used.

16.13 CONTRACT TERMINATION—DEBARMENT

16.13.1 A breach of Paragraphs 16.6 through 16.13 inclusive, respectively entitled "Minimum Wages," "Payrolls and Basic Records," "Apprentices and Trainees," "Compliance with Copeland Regulations," "Contract Work Hours and Safety Standards (40 USC 327-330)," "Withholding of Funds," and "Subcontracts" may be grounds for termination of the Contract and for debarment as provided in 29 CFR 5.6.

16.14 USE AND OCCUPANCY OF PROJECT PRIOR TO ACCEPTANCE BY THE OWNER

16.14.1 The Contractor agrees to use and occupancy of a portion or unit of the Project before formal acceptance by the Owner under the following conditions:

.1 A Certificate of Substantial Completion shall be prepared and executed as provided in Subparagraph 9.8.1 of the accompanying General Conditions of the Contract for Construction, except that when, in the opinion of the Architect, the Contractor is chargeable with unwarranted delay in completing Work or other Contract requirements the signature of the Contractor will not be required. The Certificate of Substantial Completion shall be accompanied by a written endorsement of the Contractor's insurance carrier and surety permitting occupancy by the Owner during the remaining period of Project Work.

.2 Occupancy by the Owner shall not be construed by the Contractor as being an acceptance of that part of the Project to be occupied.

.3 The Contractor shall not be held responsible for any damage to the occupied part of the Project resulting from the Owner's occupancy.

.4 Occupancy by the Owner shall not be deemed to constitute a waiver of existing claims in behalf of the Owner or Contractor against each other.

.5 If the Project consists of more than one building, and one of the buildings is to be occupied, the Owner, prior to occupancy of that building, shall secure permanent property insurance on the building to be occupied and necessary permits which may be required for use and occupancy.

16.14.2 With the exception of Clause 16.14.1.5, use and occupancy by the Owner prior to Project acceptance does not relieve the Contractor of his responsibility to maintain all insurance and bonds required of the Contractor under the Contract until the Project is completed and accepted by the Owner.

FIG. 12.1 *(continued)*

ARTICLE 17

INSURANCE REQUIREMENTS

17.1 The insurance required by Subparagraph 11.1.1 shall be written for not less than the following:

1. Workers' Compensation
   - (a) State — Statutory
   - (b) Applicable Federal (e.g., Longshoremen's) — Statutory
   - (c) Employer's Liability — $__________

2. Comprehensive General Liability
   (Including Premises—Operations; Independent Contractor's Protective; Products and Completed Operation Broad Form Property Damage)
   - (a) Bodily Injury
     - (1) Each Occurrence — $__________
     - (2) Annual Aggregate — $__________
   - (b) Property Damage
     - (1) Each Occurence — $__________
     - (2) Annual Aggregate — $__________

4. Personal Injury
   Annual Aggregate — $__________

5. Completed Operations and Products Liability shall be maintained for ________( ) years after final payment.

6. Property Damage Liability Insurance shall include coverage for the following hazards:
   - ☐ X
   - ☐ C
   - ☐ U

7. Comprehensive Automobile Liability
   - (a) Bodily Injury
     - (1) Each Person — $__________
     - (2) Each Occurrence — $__________
   - (b) Property Damage
     - (1) Each Occurrence — $__________

8. If an exposure exists, Aircraft Liability (owned and non-owned) and Watercraft Liability (owned and non-owned), with limits approved by the Owner shall be provided.

9. The Contractor shall carry insurance in addition to that specifically named above as follows:

Coverage	*Amount*
	$__________

17.2 Revise the first sentence of Subparagraph 11.3.1 as follows:

11.3.1 Until the Work is completed and accepted by the Owner, the ________________* shall purchase and maintain property insurance upon the entire Work at the site to the full insurable value thereof.

*(Insert Owner or Contractor. Occasionally the Owner may require the Contractor to furnish the property insurance.)

As mentioned earlier, there is an almost infinite list of items that may necessitate the supplementary conditions, with corresponding examples for each. Figure 12.1 and the above two citations have been given for illustration purposes.

DIVISION VI

CONSTRUCTION SPECIFICATIONS

13 INTRODUCTION TO SPECIFICATIONS

13.1 DEFINITIONS

The dictionary defines specifications as "a detailed precise presentation of something . . . " Specifications may be further defined as written explanations delimiting the work to be undertaken and as written instructions regarding materials and methods that are to be used to accomplish the work of the project. Yet another definition is that they are written instructions to be used in conjunction with the drawings so that together they fully describe and define the work that is to be accomplished, along with the methods and quality that will be required. Specifications are one of the contract documents for a construction project and as such should be prepared with as much care to detail as the drawings and any of the other contract documents.

13.2 RELATIONSHIP TO DRAWINGS

Like most of the other construction contract documents, the construction specifications are not intended to be used independently. The specifications in particular are very closely related to the drawings in their preparation and in their use. They are prepared as a supplement to the drawings and are intended to complement the information that the drawings provide.

Since the drawings and the specifications together present the total information regarding the project, there should be some guidelines as to what information goes in each. In general, it can be stated that the specifications should include all material that is impossible or impractical to show on the drawings. The drawings will usually be the first choice for the inclusion of material because of the following:

> There is less chance for misunderstanding. It is difficult to misinterpret a drawing, while any combination of words may be susceptible to a number of interpretations.
>
> Less time and effort required. Because the graphic form of presentation utilizes a universal language, it provides greater clarity with a minimum effort required. What can be accomplished through a small drawing may require several pages of written material to convey the same information. This is reinforced by the saying, "one picture is worth a thousand words."

In connection with the foregoing conditions, the drawings will normally be used to provide a physical description of the project. This will include plans and elevations as well as larger scaled details pertaining to specific parts of the project. For a building project, the drawings would show the location of doors and windows, room arrangements, framing plans and details, as well as the location of equipment, fixtures, finishes, materials, etc. For a civil project they would include profiles, details of spillways, curb and gutter details, bridge framing plans, etc. In both types of projects, site and grading information would be illustrated in the drawings.

The specifications are usually the preferred source of information regarding such items as finish and gauge of materials, erection and installation methods, quality of material required, quality of workmanship and how it will be measured, tests and inspections, and cash allowances and unit costs.

13.2.1 CONFLICT PROVISIONS

Since the drawings and the specifications are both used to present information regarding the project, it is almost inevitable that there will be occasions when the two documents conflict. Although in most cases the practice within the construction industry is that the specifications shall govern over the drawings in cases of conflict, it is dangerous to automatically assume this. The specifications should instead contain provisions covering this situation, and should also contain language connecting them to the drawings. An example of this provision might be as follows: "Anything shown on the drawings but not mentioned in the specifications, or mentioned in the specifications but not shown on the drawings, shall be deemed to have been mentioned and/or shown in both." This should be construed not as an attempt on the part of the design professional to avoid proper responsibility but rather as a recognition of the joint dependency of the two documents.

13.3 PURPOSE OF SPECIFICATIONS

The construction specifications serve a number of purposes. First, they are a guide to the bidders in the preparation of cost estimates upon which their proposals are based. This is accomplished by giving information regarding quality of material that will be acceptable, particular methods that will be required, etc. Second, they are a guide to the contractor during the construction phase. They present information that is helpful regarding erection and installation as well as the ordering of material and the performance of work. Third, they form part of the agreement between the owner and the contractor. Since they are usually included as part of the agreement by reference, they actually make up part of the contract.

13.4 PREPARED BY WHOM

Construction specifications are usually prepared by one or several members of the design firm that has been retained by the owner. While one person or department may have the assigned responsibility for their preparation, all members of the design firm who are connected with the project will have some influence on the specifications, and many will make contributions to them. The structural engineer will make decisions during the design of the structure that must be reflected in the specifications. The draftsman who places certain notes on the margins of the original tracings is actually making minor design decisions that will be incorporated into the specifications. Even the owner, when making decisions regarding the financial aspects of the project, will be contributing to information that will go into the specifications.

13.5 QUALIFICATIONS FOR SPECIFICATION WRITER

In most cases, however, a single individual or a small group of individuals has the responsibility of producing the formal set of specifications. It is important that these people be well qualified for the assigned task. A mistake is often made in many design offices when they assign the responsibility for specification preparation to a young and relatively inexperienced architect or engineer. The quality of document produced usually reflects this inexperience and may lead to a lowering of quality in the final project. Because of the importance of the specifications in the overall construction process, care should be taken to ensure that the specification writer is highly knowledgeable and experienced and in addition has a high degree of interest in the preparation of specifications.

A qualified specification writer should possess the following minimum qualifications:

A thorough knowledge of the construction process. In many instances this can mean understanding area practices in the location of the particular project in addition to an overall knowledge of the industry. Since the award of subcontracts is based upon the specifications, they should reflect the manner in which work is performed by different crafts in that area.

A good understanding of construction materials. This will include the capabilities of different materials to perform under varying conditions and how such materials may be incorporated with other materials, as well important points in inspection and testing.

A familiarity with construction methods. If possible, this understanding should be based on field experience. As a minimum, the specification writer should have spent a great deal of time observing field construction methods. Many a specification has been written by a person who, upon being questioned regarding a proposed method, hasn't the faintest idea how the work is to be accomplished.

An understanding of the design process. Since the specifications are an extension of the design in that they contribute to effecting the construction, the writer of the specifications should understand what the designer is trying to accomplish. As an absolute minimum, specification writers should have an empathy for the design process so that they are not working at cross purposes with designers.

A high ability with written language. The specification writer communicates only through a written form and must thus possess the ability to express ideas and to present information in a clear, concise manner.

A sense of fairness. This is a qualification that is often overlooked. Specifications can be written to give an unfair advantage to particular parties through the seemingly innocent selection of words and phrases. It should be the specification writer's responsibility to avoid the deliberate or inadvertant occurrence of such a situation.

13.6 USERS OF SPECIFICATIONS

The construction specifications are used by a large number of persons and firms. During the bidding process, estimators for the general contractor will "take off" the entire set of specifications, meaning that they will review the requirements pertaining to a given branch of the work. In addition, all subcontractors and material suppliers planning to submit a quotation to the general contractor will review those parts of the specifications that apply to the particular branches of the work with which they are involved. Although they do not normally do so, the subcontractors and material suppliers would also be well advised to check the general conditions since the provisions contained therein will have an effect upon the subcontracts and purchase orders issued by the general contractor.

During the construction stage of a project, the specifications will be used by another varied group of individuals. This will include the project manager, project superintendent, crew foremen, individual workers, inspectors, schedulers, cost control personnel, shop drawing checkers, as well as the many persons involved in the fabrication and installation of various elements of the work. Prior to the start of construction, the specifications will be reviewed by planning and zoning departments, the officials who issue proper permits, and the various governmental agencies that may be concerned with environmental impact, employment practices, etc.

13.7 PRIMARY BENEFIT

In spite of the large number and variety of people who will be using a set of construction specifications, they are primarily prepared to supply information to the contractor for the benefit of the owner. They are written to ensure that the owner receives the amount and quality of service that has been contracted for. Even though most owners are not technically qualified to review the specifications and in most instances do not even read them, they still remain the primary beneficiaries of their impact.

13.8 TYPES OF PROVISIONS

Construction specifications contain two types of requirements or provisions, which may be identified as general and specific. The general provisions are those which pertain to the work as a whole or to all the work within one section of the specifications. If they apply to the entire project, they may be grouped at the beginning of the set of specifications as a separate section or division. If they apply only to the work of one particular trade section, they will usually be placed at the beginning of that section of the specifications.

Specific provisions are sometimes referred to as technical provisions. These are provisions or statements spelling out the technical requirements of the project and normally having restricted application, usually to only a limited and specific part of the project.

13.9 ASSIGNMENT OF RESPONSIBILITY

The requirements of the project are enumerated and explained through the use of these two types of provisions. Also of importance is the fact that the provisions will make it possible and practicable to assign the responsibility for each of these provisions to the proper party. This is considered vital under the contract system of construction. Thoughtful preparation of the specifications can greatly enhance this process.

14

SOURCES FOR SPECIFICATION INFORMATION

14.1 USE OF STANDARD FORMS

Although originality and innovation are highly desirable and much admired in the practice of architecture and engineering, the preparation of specifications is one area where such qualities are not advised. It is preferable that the specifications be presented in standard forms and language so that they can be readily understood. It should also be remembered that many of the persons using a set of specifications may have received less formal education than the person who has prepared the specifications. The specification writer's purpose should be to present information in as clear a manner as possible, not to try to impress the reader.

14.2 REFERENCE SOURCES

Any large construction project involves the collection of a tremendous amount of knowledge and information. It is not practical for a specification writer to remember all of this material or even to be in total possession of it. For this reason, the preparation of a set of specifications will involve considerable research as well as extensive use of references. A competent specification writer will quickly compile an extensive personal library and will at all times have ready access to much additional reference information. Although remembering information may not be important or even desirable, remembering where to get that information is.

There are many sources of information available to the specification writer. These sources include specification standards published by professional and industry associations, recommended specification requirements disseminated by the materials manufacturers, and guide specifications published by governmental agencies as well as some professional associations. In addition, most municipal, county, and state codes contain information that is helpful to the specification writer. Efforts should be made to become familiar with the type of material available from each source so that little time will be lost in securing that information when it is needed.

14.3 CODES AS REFERENCES

Building, plumbing, and electrical codes exist in most large cities. County codes are in existence in some areas, and an increasing number of states are enacting state codes. The law generally requires that construction in a given area conform with applicable codes and ordinances. It is imperative that the specification writer be aware of the requirements of the code that impinges upon the project and insure that the specifications are prepared in conformance with them.

There may be occasions when codes will overlap. In such a case, it should be remembered that the code requirement that is the highest or most demanding will prevail. If the project is located in an area where there is no local code in force, the design professional is not given "free rein." Under such circumstances, "accepted standards of good practice" must be observed. In building construction this can best be accomplished by using one of the so-called model codes that have been published. These codes have national distribution and use, and some municipalities have adopted them for local use.

Except as a base for doing work in accordance with accepted standards of good practice, these model codes have no force unless adopted by legislation.

Specific requirements from these codes can be used, and their use and adoption, in whole or in part, is strongly encouraged by the various associations. The list of model codes includes the following:

The Uniform Building Code, produced by the International Conference of Building Officials.

The National Building Code, produced by the American Insurance Association.

The Southern Standard Building Code, produced by the Southern Building Code Congress.

The Basic Building Code, produced by the Building Officials and Code Administrators International, Inc.

The National Plumbing Code, produced by the American Society of Mechanical Engineers and the American Public Health Association.

The National Electrical Code, produced by the American Insurance Association.

The Life Safety Code, produced by the National Fire Protection Association International.

14.3.1 CODE DIFFERENCES

Differences exist between requirements contained within the various "model" codes, just as they exist between codes for different political divisions. Some of these differences may reflect local needs and requirements, while others may exist because of political or parochial interests. Permitting a multistory office building to be constructed of timber in the Northwest is a reflection of local preference and interest. Since the source of codes is the police power of the state to safeguard the health and safety of the public, codes will no doubt continue to reflect local concerns and welfare.

14.4 SPECIFICATION STANDARDS

Some sources of information have been widely accepted as standards for construction specifications. The use of such standards can result in the saving of much time and space since the specification writer can include a particular requirement by reference. One of the better known is the ASTM Standards, produced by the American Society for Testing and Materials. This organization, which has been in existence since the beginning of this century, is highly respected for its thoroughness and fairness. An example of using an ASTM requirement by reference would be "concrete aggregate shall conform to ASTM C33." While the actual wording of the requirement may take many paragraphs, or even pages, it may be included in the specifications by reference with a single sentence.

Other organizations produce information regarding the preparation of specifications that is used as standards in the construction industry. Examples of these standards would include the following:

Standard Grading and Dressing Rules for Lumber, produced by the West Coast Lumbermen's Association.

American National Standards Institute

Specifications for the Design, Fabrication and Erection of Structural Steel for Buildings, produced by the American Institute of Steel Construction.

14.5 AGENCY STANDARDS

Several branches and agencies of the federal government publish standards and guide specifications that are recommended for use in projects done for them. Some of these "suggestions" are in fact absolute requirements and must be followed. Some of these requirements pertain to materials and methods, while others pertain to employment practices or wage rates. These requirements are contained within such documents as Military Specifications, Federal Specifications, Corps of Engineers Regulations, Federal Highway Administration Guidelines, Federal Housing Administration Minimum Property Standards, and many more. The specification writer should check the requirements of each governmental project to ensure that all applicable standards have been considered and included. While some particular requirements may find adoption in a small number of private projects, this is seldom the case.

14.6 MANUFACTURERS' SPECIFICATIONS

Most major manufacturers of construction material publish recommended specifications that they encourage the design professionals to adopt. While this represents an important information source for the specification writer and can be very helpful, care must be exercised in their use since they represent a parochial interest on the part of the particular manufacturer. One metal deck manufacturer presented suggested specifications for a metal deck that included a requirement for section modulus. Examination disclosed the fact that no other company marketed a metal deck that could meet this section modulus. As a result, adoption of this suggested specification would restrict the bidding to only one supplier, even though many other decks were actually equivalent.

Material manufacturers produce considerable technical literature to inform all persons in the industry about their product. Included within this literature are the manufacturer's recommended specifications along with much additional technical information and promotional material. The F. W. Dodge Corporation compiles much of this technical literature into bound volumes known as Sweet's Files. These represent a convenient source of information for the specification writer. Different sets are published covering differing parts of the construction industry. They are furnished without charge to some firms, depending upon the firm's annual volume of business.

Many manufacturers are members of an industry association that is composed of firms producing allied products. Examples are the Metal Building Manufacturers Association, National Ready Mixed Concrete Association, and Valve Manufacturers Association. Many of these industry associations publish standards and suggested specifications for use in connection with their members' products. These are usually more general than those published by an individual firm and leave the bidding open to a greater number of competitors. Such associations are quite willing to furnish free copies to specification writers in the hope that it will increase the business volume of their member firms.

14.7 GUIDE SPECIFICATIONS

There are several guide specifications that can be purchased. These are essentially outline specifications with blank spaces to be filled in with information peculiar to a particular project. Some of these are published by professional associations such as the American Institute of Architects, while others have been written by individuals as a business venture. Although such guide specifications may be helpful to the specification writer, the experienced individual will quickly outgrow the need for them.

14.8 RESPONSIBILITY FOR USE

Any one or all of the aforementioned sources may prove to be very helpful to a specification writer and to that extent their use should be encouraged. However, the individual specification writer must keep in mind that the responsibility for the project specifications cannot be shifted to someone else merely because other sources have been used as references. For this reason, any quoted or copied material should be carefully reviewed to ensure that the information contained is in agreement with what the specification writer wishes to present.

15 PROJECT SPECIFICATIONS

15.1 WHEN WRITTEN

The design professional will prepare a set of preliminary drawings before proceeding with the final or working drawings. In conjunction with the preliminary drawings, an outline set of specifications will be written. These will contain only the basic information regarding materials and methods for the project. Information related to contract conditions, etc. will be included only if it is an unusual or drastic requirement. As work begins and progresses on the working drawings, work will also proceed on the project specifications. These are the final specifications containing all requirements for the project that cannot be shown on the drawings.

15.2 PERFORMANCE SPECIFICATIONS

There are two basic approaches or philosophies in the writing of specifications for a construction project. The first is the performance specification, also known as the results system, and the second is the methods system. The specification writer using the performance method describes in detail the required performance or service characteristics of the finished product or system. When using this method, it is important that the end results be susceptible to the proper tests or measurements to ensure compliance.

Under the performance method of specification writing, the method that is to be used to achieve the desired results is left to the discretion of the contractor. Any method that will satisfactorily realize the desired results may be used, with the contractor normally required to furnish a guarantee regarding the results. The methods of testing and/or measurement that will be used to evaluate the results should be clearly spelled out in the specifications so as to avoid any surprises for the contractor, and also to allow the contractor to better select the method to be used.

15.2.1 ADVANTAGES

The performance approach to specification writing provides for wide competition among products and systems. It places the expertise of individual contractors on a premium basis in that the wider their knowledge and experience, the greater their chances for success. On the other hand, the specification writer who utilizes the performance approach to specifications does not need to have as extensive a knowledge of products and systems as would be required under the methods system.

15.2.2 SPECIFYING TEST METHODS

One of the major difficulties encountered in the use of the performance type of specifications is devising adequate testing and measuring systems by which the results can be evaluated. The testing methods specified should be short term so that completion of other parts of the project will not be delayed. If it is specified that a heating system shall be capable of maintaining a temperature of 70 degrees Fahrenheit at a level 30 inches above the floor when the outside temperature is − 20 degrees Fahrenheit, it is not practical to wait until the outside temperature drops to the specified level to measure the results. Rather, a testing method should be specified that will permit evaluation of the heating system whenever it is installed, even if it is in the middle of the summer. Performance specifications are widely used in connection with equipment and machinery requirements and many systems of testing and measurement exist in this area.

15.2.3 USING AN ORDERLY SYSTEM

When using the performance method, the specification writer should avoid drastic variations from performance standards of most products. Variation from these standards will result in increased prices to the owner, sometimes for only a minimal change in quality. It is also important for the specification writer to present the performance specification information in an orderly manner. This should consist of:

1. A general description of the product or system contemplated.
2. The requirements pertaining to design and installation.
3. The conditions under which the product or system is expected to operate.
4. The requirements regarding performance (these must be specific).
5. Detailed information regarding tests and measurements that will be used to evaluate the performance.
6. Requirements regarding any guarantees or warranties that must be furnished.

By following such a method of presenting the performance specification information, it will be easier to guard against errors of omission, and it will also be easier for the contractor to make a bid.

15.3 METHODS SYSTEM SPECIFICATIONS

The other basic approach to specification writing is called the methods system. It is also referred to as prescription specifications, the materials and workmanship method, or descriptive specifications. The specification writer using this system describes in detail the materials that are to be used and the procedure that is to be followed in incorporating these materials into the final project. If the material is specified by brand or manufacturer's name, much space and time can be saved in preparing the specifications. In addition, the specification writer has the advantage of knowing ahead of time the performance characteristics of the material or system because of its past performance on other projects.

15.3.1 WHEN USED

The methods approach to specification writing is used when the design professional determines that it is not practical or possible to evaluate the adequacy of the material or system through a series of tests or measurements. This is also true when an inspection of the outside appearance of work will not necessarily disclose defects. This may well be the case in connection with brick masonry, concrete, or structural steel. In such cases the material would be specified according to established standards and the method of installation or erection would be specified in order to guard against deficiencies.

15.3.2 ACCEPTANCE OF RESPONSIBILITY

The specification writer who uses the methods approach to specifications must be prepared to accept more responsibility than would be the case if the performance system were used. Since both the materials and methods to be used are specified in detail, it would be extremely unfair to force a contractor who has complied with the specifications to be responsible for performance. This holds true even when the brand name approach is used for specifying materials. Since the particular brand of product has been selected by the specification writer, it follows that the specification writer is the one who must warrant its performance.

15.3.3 REALISTIC REQUIREMENTS

Since the method to be used is described in detail under the methods system of specifications, it behooves the specification writer to make sure that the requirements are realistic. Specifications have been written for fairly standard projects with such requirements as reinforcing steel placement to the nearest 1/32 of an inch, soil compaction to an unreasonable value, and many other impractical requirements that far exceed the workers' abilities. When using the methods system of specifications, the specification writer must be fully acquainted with both materials and field methods. Lack of this knowledge may result in impractical requirements that in turn lead to additional costs for the owner.

15.4 OPEN VS. CLOSED SPECIFICATIONS

Specifications can be written in two ways: They may be open to all qualified bidders or they may be available to only one firm. The first situation is referred to as "open" specifications and the second is called "closed" specifications. For construction projects where public funds are being used, the closed specification is illegal. The result of a closed specification is to eliminate competition, even if the closed specification is used for only one product for the project. Since this is contrary to the public interest, legislation has been enacted forbidding its use on public works. On the other hand, open specifications make it possible for the greatest number of firms or manufacturers to compete for the contract for that branch of the work.

15.4.1 APPROVED EQUAL PHRASE

A common approach to specifying materials for a project, which is a combination of the two systems mentioned above, is called a "restricted" specification. This is most often used when the brand name approach is used in specifying materials. Under this system, several brand names (three is common) will be listed followed by the phrase "or approved equal." The specification writer must exercise care to ensure that all brands listed are in fact equal and that an unfair advantage is not granted to one of the bidders. The phrase "or approved equal" is interpreted to mean approved as equal by the design professional. The contractor does not have the authority to establish equality of products under such a system. When listing brand names, however, the specification writer must accept any one of them as meeting the specification requirements without any partiality or evidence of preference. If a contractor feels that another brand name should be added to the list already given in the specifications, the request must be made during the bidding stage. If the design professional is in agreement, an addendum will be issued to all bidders.

15.4.2 DISADVANTAGES

The use of a restricted specification does not violate the law prohibiting closed specifications on public projects. Specification writers may occasionally encounter some difficulty in finding products that are equal and that meet their approval. If this is the case, it would probably be wise to consider using a performance type of specification for that part of the work.

If private funds are being used for a project, the specification writer is permitted to write a closed specification. There may be situations where only one manufacturer's product will give the desired results, in which case a closed specification can be used. Such practice is not normally advised, however, since the supplier, being aware of the absence of any competition, will be strongly tempted to increase the price above that which would be quoted competitively.

15.5 EMERGING FORMS

More recently developed forms of construction contracts, such as design-build and fast-track, have caused changes in the specifications for projects under such contract systems. If projects under a design-build form of contract are on a negotiated basis, formal specifications may vary from being quite sketchy to being practically nonexistent. Such a situation becomes even more prevalent under a cost plus form of contract.

When competitive bids are taken on a design-build contract, the specifications in most cases are of the performance type. Specifications for such contracts generally specify the owner's wishes and the successful bidder is required to produce a design that will meet the stated needs. As part of this design, a fairly complete set of specifications may be produced.

ORGANIZING THE SPECIFICATIONS 16

16.1 NEED FOR ORGANIZATION

The specifications contain a large body of information that must be used by a number of different persons. That information covers many areas, each area being used by different people at different times to assist them in meeting their responsibilities. The absence of organization in the presentation of the specification material can create a chaotic condition and make work more difficult for each person involved. If there is a lack of organization, greater periods of time are required to locate the necessary information, and the risk increases that some of the information will be missed entirely. On the other hand, if the information is arranged in an orderly and logical fashion, the use of the specifications is facilitated and each person's job is made easier. As a result, the contractor is better able to prepare a bid, the field personnel are better able to perform their activities, and the owner receives a better project at a more favorable price.

16.1.1 CHECKING FOR OMISSIONS

The first person who must be concerned with the organization of the specifications is the specification writer. One of the specification writer's major concerns is to guard against the omission of any necessary information. While the contractor is responsible for constructing the project, this responsibility does not extend to items that are not specifically required. Although the owner may have desired an exposed aggregate finish for a particular part of the concrete work in the project, and although the design professional was in agreement and intended to include it in the requirements, such intent must be included within the specifications for the contractor to include it in the bid. Good organization is the most effective means of guarding against the omission of necessary information. In addition, the specification writer should strive to facilitate the use of the specification information by all those involved with the project. Again, good organization is the most effective means of accomplishing this goal.

16.1.2 EASE OF REFERENCE

From the day a project is put out for bids until the day it is completed, there is a great deal of correspondence generated. Much of this correspondence pertains to questions and decisions related to the contract documents. It is helpful if the person writing the letter can refer to certain items within the specifications in such a way as to reduce misunderstandings. Good organization will include a numbering system that makes this possible. The same thing is true with regard to the award of subcontracts and material purchase orders. If these can be prepared by referring to parts of the specifications by number, it will eliminate the necessity of extended descriptive texts. Such a procedure will also lessen the possibility of omissions or overlaps in the award of subcontracts and purchase orders.

16.2 STANDARD FORMAT

Each design office develops its own "standard" format for specifications. This format evolves as a result of the experience and preferences of those involved. As the users of the specifications become more familiar with the format of that particular firm, their efficiency and understanding improves. However, construction firms are increasingly operating over wider geographic areas, many of them on a national scale. Even though the specifications from each design firm are prepared in accordance with their own standard format, the existence of so many standards will inevitably lead to confusion and a reduction in the effectiveness of communication. For these reasons, it is highly desirable for all design firms to use a single standard for specification format.

16.2.1 TRADE SECTIONS

Many experienced specification writers are now utilizing the Construction Specification Institute's three-part approach to the trade sections of the specifications. The first part may be described as the "general" section and includes information regarding the scope of the work, submittals for approval, delivery (including storage and handling), job conditions, and guarantees. The second part deals with the product itself and includes a description of the material or equipment. Examples would include the mix requirements for concrete or the fabrication requirements for structural steel. The third part deals with the execution of construction. This part would describe quality of workmanship, installation or erection procedures, housekeeping, and any applicable provisions for testing and inspection.

The above is an example of a standardized approach to the writing of the specifications for trade sections. It can readily be seen that adherence to such a format would be conducive to the improvement of the communication process. If such standardization can then be expanded to include all parts of the specifications, even greater benefits can be realized.

16.3 CSI FORMAT

A standardized format that is increasingly being followed by specification writers is that presented by the Construction Specifications Institute (CSI). CSI has as a major goal the improvement of the quality of construction specifications. One aspect of their efforts is to provide an approach to uniformity in specification writing by furnishing a standard arrangement of specification material.

16.3.1 DIVISIONS

Under the CSI format the project specifications are divided into 17 broad areas. These areas are referred to as "divisions." Each division is derived from an interrelationship of material, trade, function, or space. The CSI format has found widespread acceptance within the construction industry. Proponents claim that it leads to increased bidding accuracy, greater project control for the contractor, as well as assisting the design professional in producing better and more complete specifications. The major avenue of benefit stems from the use of the uniform format. The 17 divisions as presented by CSI are as follows:

0. Bidding and Contract Requirements
1. General Requirements
2. Site Work
3. Concrete
4. Masonry
5. Metals
6. Wood and Plastics
7. Thermal and Moisture Protection
8. Doors and Windows

9. Finishes
10. Specialties
11. Equipment
12. Furnishings
13. Special Construction
14. Conveying Systems
15. Mechanical
16. Electrical

16.3.2 SECTIONS

Under the CSI format, each division contains a varying number of "sections." Sections are classified as "broadscope" or "narrowscope" title headings. The narrowscope headings detail the subject matter covered by each broadscope title, while the broadscope titles list the various areas of information under a particular division. For example, the broadscope titles under Division 5—Metals are:

05100 Structural Metal Framing
05200 Metal Joists
05300 Metal Decking
05400 Cold-Formed Metal Framing
05500 Metal Fabrication
05700 Ornamental Metal
05800 Expansion Control
05900 Metal Finishes

Each broadscope title includes a number of narrowscope titles under it. For example, 05500—Metal Fabrication includes:

05501 Anchor Bolts
05502 Expansion Bolts
05510 Metal Stairs
05515 Ladders
05520 Handrails and Railings
05521 Pipe and Tube Railings
05530 Gratings and Floor Plates
05540 Castings
05550 Custom Enclosures
05551 Heat-Cooling Unit Enclosures

Such a system offers great flexibility to the specification writer. A simple project may make use of only broadscope titles, while a highly complex project may use the broadscope titles and also a large number of the narrowscope titles. The CSI format assigns a specific location to each section. Thus users of the specifications can be assured of ease in locating specific information.

16.4 OTHER STANDARD FORMATS

There are other organizations and associations that have developed standard formats for specifications. The American Association of State Highway and Transportation Officials (AASHTO) has developed guide specifications for highway construction as well as for major structures and bridges. Their highway construction specification format contains the following seven sections:

100 General Provisions
200 Earthwork
300 Base Courses
400 Bituminous Pavements
500 Rigid Pavement
600 Miscellaneous Construction
700 Materials Details

Each section includes a varying number of subsections. Section 200—Earthwork contains the following:

201 Clearing and Grubbing
202 Removal of Structures and Obstructions
203 Excavation and Embankment
204 Subgrade Preparation
205 Pre-watering of Excavation Areas
206 Overhaul
207 Structure Excavation for Conduits and Minor Structures
208 Water Pollution Control

Each subsection may in turn contain a number of sub-subsections as conditions necessitate. As an example subsection 204 may include the following:

204.01 Description
204.02 Construction Requirements
204.03 Method of Measurement
204.04 Basis of Payment

The above examples illustrate the flexibility inherent within a standardized specification format. At the same time, the efficiency and accuracy of both the preparation and the use of the specifications can be increased.

17 SPECIFICATION WRITING

17.1 PROCEDURES PRIOR TO WRITING

There are several approaches that can be used in the actual production of a set of construction specifications. Regardless of which approach is used, however, the writing of the specifications should be preceded by a thorough study and review of the drawings. The specification writer should constantly remember that the specifications must be coordinated with the drawings and that they should generally contain only information that cannot be better shown on the drawings. The specification writer should follow an outline in the preparation of the specifications. This may be an outline prepared for the particular project, or a standard published checklist.

Before any writing is begun, all notes generated by the designers and draftsmen should be collected. These notes represent decisions made by various members of the firm during the production of the drawings. Throughout this process, decisions have been made regarding the selection of materials and other matters, all of which should be reflected in the specifications.

Design offices use different methods to record these pieces of information. Some rely on notes placed on the margin of the tracings, while others may use card files or loose-leaf notebooks. Any of these systems can be satisfactory if they are properly monitored and supervised to ensure completeness.

17.2 CUT AND PASTE METHOD

One of the more widely used methods of writing and assembling a set of specifications is referred to as "cut and paste." The specification writer selects a set of specifications from a previous project of a similar nature. These are reviewed to delete those portions that are not applicable to the current project and new parts are written where necessary. The review must be a thorough one to make certain that extraneous requirements are not retained. It is also important to follow an outline to guard against errors of omission.

After the new set has been completely assembled, it will have to be retyped to present a finished appearance. Many offices will make two copies to be used in the preparation of cut and paste specifications for future projects. This method of specification preparation has the advantage of saving time and contributing to uniformity in the specifications produced by the firm.

17.3 USE OF STANDARD SHEETS

Some design offices make use of a system of standard paragraphs or sheets of specifications. When it is time to write the specifications the proper paragraphs or sheets are pulled from the files and assembled. Again, use of an outline is strongly recommended to guard against omissions or extraneous material.

Some firms place information into a computer and are thus able to assemble a set of specifications with a minimum of effort and time. In addition, computerized specifications with editing features and modern word processing equipment are widely available. It seems certain that this approach will be used by an increasing number of firms.

17.4 GUIDE SPECIFICATIONS

As previously mentioned, there are several sources for guide specifications. Some smaller design firms prefer to make use of these in order to save time and money in specification preparation. Guide specifications present work sheets containing information that is standard in nature, with blank spaces that are to be filled in with information specific to the current project. Their use should be limited to projects that are relatively routine in nature since they do not afford sufficient flexibility for custom construction.

17.5 SPECIFICATION LANGUAGE

Most of us have had the experience of having our words come back to us in a less than satisfactory manner. Our reaction may vary from "I didn't say that!" or "I didn't mean to say that," to "You misunderstood me. That was not my intention." The misunderstanding may be the result of a failure in either oral or written communication.

Since the specifications are normally included by reference as part of the contract, they are regarded as a legal document. Because they are a legal document, they may be subject to interpretation by someone other than the person who wrote them. Accordingly, the specification writer should exercise great care in the selection of words and phrases to minimize misunderstandings as much as possible. As previously mentioned, many of the persons who will use the specifications may have received less formal education than the specification writer, and this should be kept in mind as the specifications are prepared.

Because of licensing procedures that are in effect, it is usually assumed that the design professional possesses the necessary technical ability to design the project at hand. This technical ability will be of diminished value, however, if the ideas and instructions cannot be communicated in a clearly understood manner. Wise and careful selection of specification language can improve this communication process and thus contribute to the quality of the final project.

17.5.1 LANGUAGE GUIDELINES

Most specification writers try to follow the guidelines of the four C's in specification preparation. These guidelines state that the specifications should be clear, concise, complete, and correct. In order for this to be more easily accomplished, the specification writer should make use of clear concise language, with an increase in the use of accepted trade terms, and a corresponding decrease in the use of legal phraseology. Relatively short sentences are preferred over long involved sentences in which the meaning may be unclear to the reader. The value of a set of specifications is determined by how easily they are understood rather than how impressive the writing style is.

17.5.2 PUNCTUATION

It would be impractical to write a set of specifications without using any punctuation marks except periods. It is desirable, however, to reduce the use of punctuation marks to a minimum. This is especially true with respect to commas. Commas within a sentence may sometimes confuse a reader or create a situation where two meanings or interpretations are possible. It is usually better to make two or more sentences rather than one long complicated sentence requiring a number of commas or semicolons.

17.5.3 ABBREVIATIONS

The extensive use of abbreviations in specification writing is a practice that should be discouraged. Abbreviations that may be clear and well known to the specification writer or the design office may prove to be a source of puzzlement to some of the users of the specifications. Care should be exercised to include only those abbreviations that are generally well known within the industry or that are used by the general public. Some examples of abbreviations that can be safely used are as follows:

sq. ft. square feet
in. inch
psi pounds per square inch
mph miles per hour
gpm gallons per minute
rpm revolutions per minute
o.c. on center
c.c. center to center
ASTM American Society for Testing and Materials
Btu British thermal unit
AISC American Institute of Steel Construction

Abbreviations that should be generally avoided are those that are the creation of the particular specification writer or those whose usage is so restricted that their meaning will not be known by most of the specification users. Wise use of abbreviations leads to a more condensed set of specifications with no loss of clarity for the reader. It can save time in writing, typing, and reading of the specifications.

17.5.4 CAPITALIZATION

Capital letters are used in specification writing in accordance with the general rules of grammar and when using certain abbreviations, such as ASTM. However, capital letters are also used when presenting or referring to the following:

Those who are party to the contract, such as Owner, Contractor, Architect, Engineer

The contract documents, such as General Conditions of the Contract, Specifications, Drawings, Addendum.

Portions of the project, such as Ramp, Foyer, Kitchen, Spillway.

Grades of material, such as Select Structural, Heavy Duty, B Intermediate.

17.5.5 GRAMMATICAL MOOD

Most of the technical provisions of the specifications consist of orders and instructions that are being given to the contractor. As a result, the use of the imperative or indicative mood is appropriate. The use of either of these moods, rather than the usual moods of traditional sentence structure, will reduce the number of words necessary to convey the information. When these moods are used properly, no decrease in understanding should result. Alternating between the two may reduce the monotony of the specifications and may add to their clarity. An example of how the two moods may be used is as follows:

Standard: The Contractor shall apply two coats of paint "C" to the exposed surfaces.

Imperative: Apply two coats of paint "C" to exposed surfaces.

Indicative: Two coats of paint "C" shall be applied to exposed surfaces.

When using the imperative mood, the verb will be the first word in the sentence; when using the indicative mood, traditional sentence structure will be more closely followed. The imperative mood implies the words "the Contractor shall," while the indicative mood makes use of the word "shall" and implies "the Contractor." Both forms present complete information with no difference in clarity between them.

17.5.6 EXAMPLES OF MISUSE

There are numerous words that are misused quite often in specification writing and this misuse may lead to a misinterpretation of the specification writer's intent. While the following list is not all-inclusive, it is sufficient to convey the idea:

all: In most cases this word is unnecessary and can be eliminated. "Store steel joists on wooden pallets" reads as well as "store all steel joists on wooden pallets."

and/or: This is a holdover from the time when legal phraseology was commonly used in specifications. It should be replaced with "or." Instead of "approval shall be secured from the Architect and/or the Engineer," it would be better to specify who will give the approval. For example, "approval shall be secured from the Architect."

any: This word implies that the contractor has a choice in the selection of a particular provision, when that may not be the intention.

either: This word is often used when the meaning of the requirement requires the use of "both." "Install signs on either side of door" does not convey the same instructions as "install signs on both sides of door." If no choice or option is intended, "both" should be used.

every: This falls in the same category as "all" and can usually be eliminated.

1st class: Except when this is part of a formal classification standard, there is absolutely no measurable meaning attached to this phrase. It is virtually impossible to find a specification that calls for "second class" workmanship. Use instead those characteristics of material or workmanship that can be measured.

good: This is too indefinite when used in an attempt to define quality. See comments regarding "first class."

must: Use of this word implies a degree of obligation. The specification writer is advised to substitute "shall."

same: This word should not be used as a pronoun. It is better to structure the sentence so that the noun is mentioned in a direct manner.

said: This is often misused as an adjective. Instead of saying "said work shall be replaced" it would be better to say "defective work shall be replaced."

shall: This word is a command and is to be used in connection with instructions to the contractor. It should never be used when referring to acts of the owner or the design professional.

will: This word implies information regarding something that will happen. It is used in connection with the acts and duties of the owner and the design professional. It should not be used in connection with the duties of the contractor.

which: Representative of several pronouns used in specification writing. While their use cannot be entirely eliminated, the specification writer is usually better advised to repeat the noun rather than use a pronoun. Repeating the noun will lead to a clearer specification and reduce the chances for misunderstanding.

17.5.7 CLARITY

The specification writer should be on guard against the use of unnecessary words and phrases that do not contribute to a clearer understanding on the part of the reader. Many of these have been used in the past and have acquired a certain degree of respectability because of their age. Care should be exercised to eliminate any unnecessary verbiage. This would include possible "streamlining" of the specifications by omitting "the," "and," or "an" whenever possible.

17.6 INFORMATION FURNISHED BY OWNER

Although the specification writer is traditionally a member of the design professional team whose primary responsibility is to an owner, it is important that the specifications not be shaded to present any unfair advantages to the owner at the expense of the contractor. The specifications should present information in a straightforward manner, and no attempt should be made to impose unfair responsibilities upon any of the parties. One area where this principle is often violated is in the matter of information furnished by the owner. The owner should properly assume total responsibility for the accuracy of this information. Instead, many specifications require the contractor to verify this information and then to accept the responsibility for it.

In a similar manner, the specifications may require contractors to certify that they have examined the site prior to submitting the bid and that they accept responsibility for all site conditions, both evident and hidden. This may even extend to the accuracy of soil borings that have been taken at the owner's direction. It is unreasonable to require the contractor to determine the validity of such information during the bidding stage, since the time and expense of such an undertaking is not practical prior to contract letting.

17.7 UNUSUAL REQUIREMENTS

Specifications that are to be fair should warn the bidders of any unusual or difficult requirements. It should be the intent of the specifications to present information in sufficient quantity and with enough clarity to permit the bidder to prepare a responsive proposal. If a project contains requirements that are sufficiently out of the ordinary to affect the contractor's proposed cost, failure to so inform the bidder may place the contractor in a position of financial jeopardy. This situation presents no advantage to an owner since the affected contractor will be strongly tempted to "cheapen" the project in an attempt to protect the firm's potential profit. The specification writer should ensure that the bidders are made aware of any unusual or particularly stringent requirements by "flagging" or emphasizing them in the specifications.

17.8 ESCAPE CLAUSES

In an attempt to protect the owner's interests, some specification writers make use of exculpatory clauses, also referred to as "escape" or "weasel" clauses. These are clauses that shift the responsibility from the responsible party to the contractor. They thus enable either the owner or the design professional to avoid responsibility. The author recalls one specification that changed the standard provision for tying the plans and specifications together into an excape clause par excellence. The modified provision read as follows, with the escape provision being italicized here but not in the actual specifications:

> It is the intention that the plans and specifications shall be considered together as presenting complete information regarding the work required. Anything shown on the drawings but not mentioned in the specifications, or mentioned in the specifications but not shown on the drawings, *or not shown or mentioned in either but deemed necessary for the satisfactory completion of the project,* shall be deemed to have been mentioned or shown in both.

Such an approach is patently unfair to the contractor and should be strongly discouraged by any reputable firm. Concealing such a requirement in an otherwise reasonable provision makes it even more reprehensible and the design firm and its specification writer could well be censured by the appropriate professional societies.

17.9 REPRODUCTION

There are several ways in which the original typed copy of the specifications can be reproduced for distribution. On a small project where either a negotiated contract or restricted bidding will be used, the design office may elect to make a number of carbon copies. While the present technology of reproduction makes this impractical, it is still possible and may be used on occasion. Some offices still prefer to make use of either mimeograph or ditto processes for reproduction of specifications. Others have the specifications copied by the same firm and process that is used to duplicate the drawings. There are also firms utilizing the offset printing process using paper masters.

Many firms have a copy machine in their own office and use their own personnel to run off the necessary copies. There are a variety of machines on the market that produce a high quality copy at a relatively low cost. For specifications, it is preferable to use a machine that feeds the original copy in automatically and can be set to produce multiple copies. Some of the machines will also collate the copies, thus saving many hours of time for clerical staff.

Whatever method is used, it should produce clear copies to avoid errors of interpretation due to fuzzy reproduction methods. Some grades of paper will accept reproduction on both sides, while other papers should receive copy on one side only. It does neither the owner nor the design professional any good to create a good design and a high quality set of specifications if the quality of reproduction is so poor that they are difficult to read.

17.10 DISTRIBUTION

The number of sets of specifications needed is determined by the size of the project, whether the contract is to be negotiated or awarded as a result of competitive bidding, the type of project (which in turn determines the number of subcontractors), and the number of copies needed for permit authorities. For bidding purposes, drawings and specifications are generally furnished to each general construction firm indicating an intention to bid. Bidding documents are also placed in various plan rooms in the area or areas where there are potential bidders. Most of these documents are returned by the unsuccessful bidders prior to the award of the contract in order to recover their deposit. These sets may then be used for distribution during the construction stage. The following may be considered a representative listing of who will require sets of specifications and drawings during the construction stage:

One set retained by the office of the design professional. Some offices follow a practice of holding two sets, one the official file copy, and the other a working set for the office.

One set for the owner. Although the owner does not own the documents, a set is usually provided as a matter of courtesy.

Six sets for the general contractor. Although the size and complexity of the project may require that more than six sets be furnished, six is usually considered the minimum number. Some specifications also set this as a limiting number, with the contractor required to pay the reproduction charges for all sets over that number. The general contractor will normally maintain one set in the home office, at least one set in the field office, and as many sets as necessary for field operations. The general contractor also takes responsibility for seeing that the subcontractors and material suppliers have that portion of the documents that is necessary for them to meet their obligations.

One or more sets for each permit or approving authority. For some building projects, this may involve only one set to the city building department. In projects affected by a number of governmental regulations, this number may climb to 20 or more.

One set to the lending institution. If the owner has secured financing for the project, whoever is furnishing the financing, such as a bank, building and loan, or insurance firm, will normally require a set of documents.

DIVISION 

CONSTRUCTION DRAWINGS

18 TYPES OF DRAWINGS

FIGURES

18.1 PRE-DRAWING ACTIVITIES

Design professionals have the need to communicate the information regarding their design to the construction personnel who will be bidding and constructing the project. Information is conveyed by many documents, but the major vehicles are the drawings and the specifications. The specifications contain the written material and the drawings are graphic presentations regarding the project.

In a well-organized design office the preparation of the drawings is preceded by the writing of a comprehensive program. This program should state the needs, functions, and parameters of the proposed project. The preparation of a proper program may entail a considerable amount of research and study. Such items as client needs, cost and financing, site conditions, availability of materials and personnel, and project objectives should all be taken into account. This process will normally involve constant consultation with the owner to ensure that the design is proceeding in accordance with the owner's wishes. During the phase of program analysis the proposed points of the program are discussed in a critical fashion. Various parts are analyzed to determine if they meet the stated needs, with changes made where necessary. The end result of this process is program confirmation. At this point the design firm is ready to proceed with the drawings.

18.2 PRELIMINARY DRAWINGS

The first drawing stage involves schematic or outline drawings in which the requirements detailed in the program are translated into graphic form. These are often single line drawings with a minimum of dimensioning information furnished. The preparation of the single line drawings may be preceded by a diagram of functional relationships (DFR). This is a graphic analysis of various components or areas of the project and provides a means of determining their relationships with one another. The DFR is used quite widely on architectural projects but can also be easily adapted to civil works. The single line drawings are then prepared so as to reflect the conditions defined and determined by the DFR. It is fairly common practice to supplement the schematic drawings with sketches or renderings to further enlighten the client. Small scale models may also be used to advantage. The design professional should remember that most clients are not well versed in reading and interpreting drawings; therefore anything that can contribute to a more complete understanding should be considered.

18.2.1 DESIGN DEVELOPMENT

Schematic drawings are prepared by the design office to present a proposed solution based upon the confirmed program. Presentation is made to the owner and approval is requested. There may be occasions when a number of possible solutions, or partis, are presented for the owner's selection. Drawings at this stage of the design development are meant to let the client know how the project will look in accordance with the proposed solution. They should also allow the owner to determine how the project will operate. If it is a building project, the owner will want to determine space relationships and the flow of people or material. For a dam project, flood gate operation, location of diversion tunnels, and other similar information is critical.

18.2.2 INFORMATION TO BE INCLUDED

The schematic drawings are preliminary in nature and should not be confused with the first stage of the working drawings. Preliminary drawings may contain notes indicating choice of material. Sketches may indicate how the project will look if these materials are used. Dimensions are given in rough or approximate form so that the general scope of the project can be defined. Large scale details would be supplied at this stage only in the case of unusual requirements or conditions. Preliminary drawings should be developed to such a degree of completeness that the owner is able to visualize the total project. They should also be complete enough to allow the preparation of a realistic budget estimate.

18.2.3 APPROVAL PROCESS

Preliminary drawings should always be submitted to the owner for approval. The approval process should be formalized to the extent that it includes the affixing of authorized signatures. Many design offices use a stamp that is placed on each drawing submitted. This stamp includes a qualifying statement regarding the approval as well as a place for the approving signature. Such a procedure is recommended because it shows that the owner has examined the preliminary drawings and is granting approval and permission to develop the working drawings in accordance with them. Failure to follow such a procedure has sometimes led to disapproval of the working drawings, with a subsequent financial loss for the design professional.

18.3 WORKING DRAWINGS

The next stage involves the preparation of the working drawings. These are based upon the design presented in the preliminary drawings that have been approved by the owner. In the working drawings the approved design is not changed but is further developed and defined. The preliminary drawings provided a design concept for the owner's consideration while the working drawings will serve as a basis for the actual construction. The working drawings, together with the specifications, present a complete picture of the proposed project along with the information and instructions necessary for its construction. The preliminary drawings contain only those dimensions needed to convey the scope of the project to the owner. The working drawings contain detailed dimensioning information that establishes sizes, relationships, and location of all elements of the project. This will also include larger scale details to further explain a particular or unusual aspect of the project.

18.3.1 HOW THEY ARE USED AND BY WHOM

The working drawings are used by several groups of people in connection with a project. The drawings should be developed to such an extent that the requirements of each group are met. One of the first formal users of the working drawings will the proper permit authority. This may be a city building department or it may be several agencies of the federal government. The authority involved will be checking the working drawings to make certain that all requirements have been met. Such checks may range from examining structural adequacy to determining if environmental considerations have been satisfied. Usually at least two sets of the working drawings are submitted to each affected authority. When the determination has been made to grant approval, both sets are marked accordingly, with the authority retaining one set and the other set being returned to the owner or the design professional.

The next group to make use of the working drawings will be the general and specialty contractors who have elected to prepare proposals on the project. The working drawings should permit these persons to determine quantities of materials and labor that will be required for construction. Such examinations are very thorough and it is often during this stage that errors and discrepancies within the working drawings, or between the working drawings and the specifications, are discovered. The reporting of these to the design professional will be the instigation of correcting and clarifying addenda.

The culmination of the bidding stage is the award of a contract or contracts. Although the working drawings are not in active use at this point, they are involved since they will be incorporated into the agreement by reference. It is at this time that the owner, the design professional, and the contractor should each retain a set of working drawings as a "set of record." This will serve as a reliable reference to define the project for which the bid was submitted. Such a practice will aid in arriving at an equitable settlement in the case of any claims for extra work. It is not unusual for drawings to be changed (either expanded or corrected) during the construction stage. It is good practice for each party to retain a set of record to establish the basis for the original contract.

During the construction stage, the working drawings will be used by a large number of people. Included in this group would be schedulers, purchasing agents, project supervisors and managers, foremen, craftsmen, material suppliers, and inspectors. Each will be securing information from the working drawings to meet their individual responsibilities. The drawings thus must present details, dimensions, methods information, locations, and a host of other required facts. Some of the persons using the working drawings are "old hands" while others will be relative newcomers. It is better to prepare the working drawings so they can be interpreted by those with less experience than to assume only the experts will be involved.

There may be occasions when a lending institution is involved with financing of the construction project. When this happens, the institution will often require a set of working drawings for their records. In most instances, the drawings would not be put into actual use by the lending firm except in cases involving dispute. They may then be referred to as settlements are discussed, or in the case of litigation.

18.3.2 LEGAL REQUIREMENTS

Each sheet of the working drawings should carry the seal of the design professional who bears the responsibility for its preparation. Site drawings are often sealed by a land surveyor and general project drawings carry the seal of the engineer or architect, depending upon whether it is a building or civil project. Supporting drawings in the set may be sealed by a variety of design professionals, such as the landscape architect, mechanical or electrical engineer, and structural engineer, as well as by such specialists as a kitchen consultant. Each drawing should contain a title block, which gives such information as the name of the project, the number of the sheet, date of the drawing (along with provisions for dates of revisions), the type of information on the sheet (framing plan, interior details, etc.), and the name of the design professional. Some firms also include space for the name of the draftsman and the name of the person who performed the final check of the drawings.

18.3.3 ORGANIZING DRAWINGS FOR A BUILDING PROJECT

It is desirable that the sheets of the working drawings be arranged as nearly as possible in the order in which construction will take place. While it is virtually impossible to follow this exactly, due to some mixing of sheet contents, it should serve as a desirable guideline. For a building construction project, the following is a representative arrangement of working drawings:

1. Title sheet. Although the use of such a sheet is optional, it is considered by many offices to lend clarity and understanding to the working drawings. A title sheet would include the name of the project and its location, the name of the owner, the name of the design firm or firms, and a listing of all the drawing sheets and their contents. The last is especially helpful since it serves as a ready index or table of contents for the user, and it also permits an easy check to make sure that the particular drawings constitute a full set.
2. Site plan. This may be a single or several drawing sheets. Both the size and complexity of the property may dictate the need for a number of drawings in order to convey the required information. Site drawings will generally present:
   a) The location of the building on the property. This will include dimensions and angles to given reference points. The elevation of one of the floor levels will also be given in reference to an established bench mark.
   b) The grading plan for the site. This will cover both existing and proposed elevations. Such information may be indicated by contour lines or by showing existing and proposed elevations at different locations on the site. These locations may make use of a grid system and may also indicate corners of the proposed building.
   c) The proposed site improvements. Examples would include utilities, parking lots, sidewalks, signs, flagpoles, and trees to be retained or removed. Existing utilities as well as proposed extensions of utilities should be shown.
   d) Construction details of some of the site improvements are commonly shown on the site drawings. This is particularly true if the improvements are relatively standard in nature and small in number. However, if the site improvements are complex and extensive, details of their construction may be located on later drawings.
3. Floor plans. A plan drawing of each floor will be given, starting with the lowest level. These are actually plan sections showing location and arrangement of the various spaces. The plans will show location of window and door openings as well as horizontal dimensions of each space. Thickness of walls is indicated and wall material is partially indicated. Each space is identified by title. In addition to intermediate dimensions for each space and area, exterior dimensions are given for portions of the structure as well as the overall structure. The location of sections through the floor plans will also be indicated.

4. Exterior elevations. Scale drawings of each of the exterior elevations of the building are shown. These are orthographic rather than perspective drawings. They show the fenestration, which is the location of doors and windows, as well as all other exterior openings. Dashed lines will give information regarding the elevation of footings and below grade floor levels. Reference lines to the side of the elevation drawings will also give dimensions from floor to floor, elevations of floor levels, and height of walls. Elevation drawings may also be used to give keys to type and style of windows and doors. The type and extent of exterior finish material for the walls of the building are also shown. In addition, the location of wall sections taken will be shown as well as the location of control joints.
5. Roof plan. The roof plan shows the arrangement of the roof of the building from a viewpoint above the roof level. This view includes the location of roof drains, vent pipes, ventilators, power fans, and equipment that is located on the roof. It will also show the arrangement of any roof drainage scheme with the attendant ridges, slopes, and valleys. All projections through the roof will be located and identified. Parapet walls or gravel stops will be shown, including the location and extent of each. The roof plan will also show the location of any curbs and expansion joints. Details and sections explaining the above items will also generally be included on the sheets containing the roof plans.
6. Wall sections, transverse and longitudinal sections. The section drawings represent vertical cuts through the building. They are used to further explain the structural arrangements of the building. The locations of the wall sections are shown on the exterior elevations, while the locations of the transverse and longitudinal sections are shown on the floor plans. The wall sections supply information regarding the structure of the walls as well as information regarding window and door openings, and connection of the walls to the floor and roof levels. Dimensions are given for each wythe or material within the wall and how they are to be interconnected. Examples include masonry bonding, lintels, sill conditions, and fascia or eaves conditions. The transverse and longitudinal sections are cut through the entire building in opposing directions and serve to further explain the overall structure of the project. They are often drawn at a scale similar to the exterior elevations or floor plans, which is a smaller scale than that normally used for the wall sections.
7. Interior elevations and construction details. This next series of sheets may well be a mixture of information. Interior elevations may be shown to further define wall treatments or the positioning of equipment. Along with interior elevations, reflected ceiling plans may be included to cover particularly complex arrangements of tile, fixtures, trim, or diffusers and speakers. The construction details are usually drawn to larger scales and cover a variety of conditions needing more explanation than that conveyed by the smaller scale drawings. Each is given a descriptive and numerical designation, with the scale indicated. These sheets often include numerous schedules that are used to present infor-

mation in a concise manner. Typical use of schedules would include the following:

a) Window schedule. This would include information for each window designation regarding window size, type, material, strength of glass, sill material, lintel size or designation, and any special requirements.

b) Door schedule. This schedule would list door openings by designations given on plan and exterior and interior elevation drawings. Each listed door type would then cover door size, style of frame required, lintel size or designation, hardware requirements, type of door (flush, paneled, etc.), and information on such items as ventilators, grilles, or glass.

c) Lintel schedule. The size and length of lintels over door and window openings are given. A single-line sketch, such as

may often be included to show how lintel members are arranged. Size of bearing plates will also be given if they are required.

d) Room finish schedule. Each space is listed by number and title (ex., 103–Gymnasium), with information regarding requirements of base, flooring material, wall and ceiling finish, and remarks covering special requirements.

e) Although the four schedules mentioned above are the most common used on the architectural drawings, other schedules can be introduced for as many items as the design office deems desirable. An example of a joist schedule is shown in Fig. 18.1.

8. Footings and foundations. This series of drawings contains information relating to the substructure of the building. The first drawing will be a plan of the footings, with information regarding size, reinforcing, and elevations. Layout of foundation walls as well as information regarding slabs on grade will also be included. Sheets in this series will include sections of the footings and foundation walls. In this connection, schedules presenting information on footings and grade beams are often used. The drawings may also include elevations of foundation walls clarifying such information as masonry ledges, cutouts, or changes in top or bottom elevations. Some offices prefer to position this series of drawings after the site information sheets.

9. Structural drawings. This series of drawings covers floor and roof framing plans along with all necessary details. Schedules may be used for such items as columns, beams, and floors. This is more likely if reinforced concrete is the main structural material. These drawings will normally be prepared by or under the supervision of a structural engineer. As a result, they usually are sealed by the engineer, not the architect.

FIG. 18.1 Joist schedule

JOIST SCHEDULE					
			Reinforcing		
Mark	Width	Depth	Bot.	Top	Remarks
J-1	5	14 + 3	2#5	2#5 2#5	Supply hooks on 2-#5 top bars
J-2	5	14 + 3	2#5	2#5	
J-3	VARIES	14 + 3	5#5	5#5	Supply hooks on 3-#5 top bars
J-4	5	12 + 3	2#5	---	
J-5	9	14 + 3	2#5	2#5 2#5	
J-6	6	16 + 3	2#6	2#6 2#6	
J-7	6	14 + 3	2#5	2#5	2-#5 top bars trussed
J-8	7	14 + 3	2#5	2#5	
J-9	5	14 + 3	2#5	3#5	
J-10	9	14 + 3	3#5	3#5 3#5	Supply hooks on 3-#5 top bars
J-11	5	12 + 3	2#5	2#5 2#5	
J-12	5	14 + 3	2#5	---	

10. Mechanical and electrical. The last sheets cover information pertaining to the heating and air conditioning aspects of the building, as well as the plumbing and electrical. These may be prepared by one firm, but more often they are prepared by two. These sheets contain plan layouts and all the necessary details. Isometric drawings are often used in this series to add clarity to the design.

18.3.4 ORGANIZING DRAWINGS FOR A CIVIL PROJECT

As with the drawings for a building construction project, those for an engineering project are arranged as closely as possible in the order of construction. The following represents the order of drawings for a road project:

1. Cover sheet. The first sheet will normally contain the title of the project and the project number, a location map of the project, an index of symbols used on the drawings, and a listing of the drawings included.
2. Estimated quantities. In most road projects, this series of sheets will contain a detailed presentation of estimated quantities. Since most projects of this type are publicly financed, they are bid competitively. Road projects are usually bid under the unit price system based upon the engineer's estimate of quantities.
3. Plans and profiles. Plan drawings of the proposed road with related profiles are included here. These sheets will normally present the road plans on the top half of the sheet with the related profiles directly below. Profiles are longitudinal sections. They are usually drawn to two different scales. The horizontal scale will be the same as that used on the road plans, but the vertical scale will be larger for the purpose of clarity. Station marks are used to relate the plans to the profiles.

4. Cross sections and details. This series presents cross sections of the roadway and any necessary information on such items.

18.3.5 ORGANIZING DRAWINGS FOR OTHER PROJECTS

Engineering projects can include greatly varied construction, and the type and order of drawings will vary with each class of project. Although the general order of drawings is similar for most building projects, the order of drawings for engineering projects depends upon the type of project. This may be best illustrated by presenting a typical list of drawings for a bridge project:

Cover sheet
Plans and elevations
Notes and quantities
Boring logs
River stages
Footing layout
Pier structure and details
Abutments and details
Pile load tests
Framing plans
Girder elevations
Bridge deck information
Roadway geometry
Expansion joints
Bearing conditions
Electrical
Railings and guard rails

Drawing requirements for dams and large treatment plants would generate similarly lengthy lists.

18.4 SHOP DRAWINGS

Much of the material and equipment to be incorporated into the construction project requires drawings that are needed for fabrication and installation purposes. The working drawings present the design requirements and intent for the project but are usually not sufficient for fabrication purposes. For example, the structural drawings may call for a certain size of steel beam to frame between two steel columns. The center to center of column dimension will be given along with information regarding connection methods. This information presented in the working drawings must be expanded for fabrication and installation purposes. This would involve determination of the cut length for the steel beam, and the size, length, and location of welds or the size of connection angles along with size and location of bolt holes. This example illustrates that the shop drawings are actually an extension and further development of the working drawings. The shop drawings do not change the intent of the working drawings but instead make it possible for the intent to be realized.

18.4.1
PREPARED BY WHOM

Shop drawings are prepared by the subcontractor or material supplier responsible for the particular branch of the work. If a subcontractor orders material or awards a sub-subcontract, the shop drawings will be prepared by firms from this third level of contract. Some shop drawings are prepared by the general contractor. These usually cover field-related activities such as falsework and cofferdams. Subcontractors and material suppliers will furnish shop drawings in such areas as structural steel, millwork, toilet partitions, metal windows and doors, ductwork and piping, pumps, and valves.

18.4.2
APPROVAL PROCEDURE

Since the shop drawings are an extension or further development of the working drawings, they are subject to review by the design professional. Subcontractors and material suppliers first submit shop drawings to the general contractor. Under the terms of most contracts, the general contractor has the obligation to examine the shop drawings for errors. Following this check, the shop drawings are then submitted to the design professional. At this point they are checked for conformance with the intent of the design drawings. It should be understood that the check performed by the design professional is not all-inclusive. It covers only conformance with the intent of the design drawings and does not extend to such areas as dimensions, number of units, or construction methods. If the design drawings require a structural member of 16′–8½″ in length and the shop drawings designate this same member as 18′–16½″ in length, it is not the responsibility of the design professional to catch this error. Neither is it the responsibility of the design professional to uncover the error of listing a wrong model or size of an item unless it is proposed as a substitution and brought to the design professional's attention. Approval of the shop drawings by the design professional is a restricted approval. It means only that they have been checked against the intent of the design drawings. Such approval does not relieve the contractor from conforming with specific requirements of the design drawings and is not to be considered an excuse for errors or omissions.

18.4.3
TIME FOR APPROVAL

Since fabrication of the material or equipment cannot be started until the shop drawings have been checked, timely approval is of prime importance. The time required for approval will vary with the size and complexity of the work covered. It can also be influenced by how busy the various firms are with other projects. It is advisable to maintain a record of when each set of shop drawings is submitted. If they have not been returned within a reasonable period of time, a telephone call should be made to check on progress. Many construction firms follow the practice of including approval of the shop drawings as specific items in the project progress schedule. This serves as a communication of allowable time for approval and permits assignment of responsibility for delay.

18.4.4
HOW THEY ARE USED

A sufficient number of copies of the shop drawings should be submitted to permit each party involved to retain a copy. As an example, reinforcing steel would usually require a minimum of four copies, one each for the structural engineer, the architect, the general contractor, and the re-bar material supplier.

The shop drawings are then used in connection with the working drawings to fabricate the material and install or erect it into the project. Coordination between the two sets of drawings is necessary since neither one presents all of the necessary information.

18.5 AS-BUILT DRAWINGS

Some construction contracts contain a provision for the owner to be furnished "as-built" drawings. These drawings are not as comprehensive as the working drawings in that details and much other information are not included. As-built drawings are often "location" drawings to which the owner may refer for maintenance and repairs. Working drawings may show the proposed locations for water lines that are part of an irrigation system for a golf course, as-built drawings may show the actual location where lines were installed. A similar situation may be the chases and raceways for services in the floors of a rental office building. Details of the construction will not normally be shown except where there was variation from the working drawings.

18.5.1 PREPARATION AND COST

Although as-built drawings can be furnished by the design professional, the general contractor, or the subcontractor, in the majority of cases they are the responsibility of the general contractor. Under this condition, general contractors may delegate responsibility to the appropriate subcontractors, or they may retain the design professional to prepare the drawings.

The requirement for as-built drawings should be included in the bidding documents. They represent a definite cost to the contractor, and the owner should expect to pay that cost. If they have not been mentioned in the bidding documents, or if the decision to require them is made after the contract is signed, this additional service and cost can be covered by a change order.

19 PLAN READING AND INTERPRETATION

FIGURES

19.1 INTRODUCTION

As mentioned before, the proverb "one picture is worth a thousand words" is particularly applicable to construction contract documents. The design professional uses graphical representation as the first choice to present information, and only when that is no longer practical is the written word used. The design professional has the need and obligation to translate the design of the project into instructions and information that will permit the design to become a physical reality. Graphical representation accomplishes this goal with the least chance of misunderstanding.

19.1.1 DRAWINGS

The design professional presents in graphic form information relative to the design in a document known as the "drawings," "plans," or "blueprints." All three terms are used interchangeably to mean the same thing. The term "blueprints" is derived from the system of reproduction that resulted in white lines on a blue background. This is a wet process that is not presently in widespread use. Most printing of drawings is now done by the "ozalid" method or a similar dry process. The dry process produces prints having either black or blue lines on a white background.

19.2 TYPES OF VIEWS

There are three basic types of views that are used for working drawings: plan views, elevations, and isometric drawings. Each of these may be drawn to a small scale to present the entire scope of the project, or at least a large part of the project. Examples might be the entire floor plan of a building project or the elevation of a bridge. These same types of views may be drawn at a much larger scale to present more detailed information regarding a specific part of the project. Such drawings are referred to as "details."

19.2.1 PLAN VIEWS

Plan views are those drawings in which the graphic information is presented as projected on a horizontal plane. Most plan views are taken as if the project were viewed directly from above. In some instances the point of viewing is assumed to be over the project. The plan view of a highway or the roof plan of a building are examples of this. In other instances, a horizontal plane is passed through the project. An example of this is the floor plan of a building. Such views are commonly referred to as "plan sections." One special use of a plan view is a reflected ceiling plan. This view, which is presented as if it were viewed on a mirror, is used to locate various ceiling fixtures.

19.2.2 ELEVATIONS

Elevations are views that present the graphic information on a vertical plane. If the point of view is outside of the project or structure, the views are called exterior elevations. If the views are of the interior vertical surfaces of a project, they are called interior elevations. In cases where the vertical plane is "passed through" the project, the views are referred to as "sections." Longitudinal sections are those taken along the main or long axis of the project or building, while those taken along the short axis are referred to as either cross or transverse sections. Sections are often used to further explain the structure or organization of a project.

19.2.3 ISOMETRIC DRAWINGS

An isometric drawing is one in which all three surfaces are equally inclined to the drawing plane, with parallel lines being drawn parallel and in their true length. This is in contrast to a perspective drawing in which parallel lines appear to converge, and distances or lengths are foreshortened. Isometric drawings are often used to further explain parts of a structure, a piece of equipment, or an assembly. Piping diagrams are often shown by an isometric drawing.

19.3 SYMBOLS

Simply stated, a set of working drawings is a collection of graphic representations of the proposed project. This representation makes use of a number of symbols to convey the necessary information. Many symbols are in common use within the industry and a familiarity with them will make it easier to interpret the drawings. Ease of interpretation will also be enhanced if such symbols are used by the design professional.

One of the uses of symbols is to indicate the type of material when shown in section, the section being either horizontal or vertical. Although there are many such symbols—one for each material—some of the more common and widely used ones are shown here.

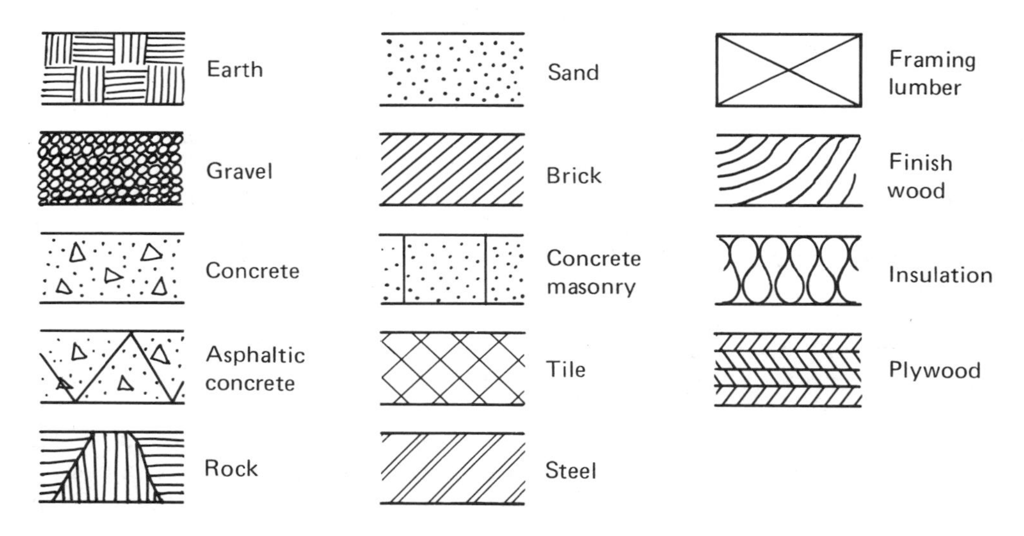

There are also many symbols used to represent various fixtures and equipment. A small sample is presented here.

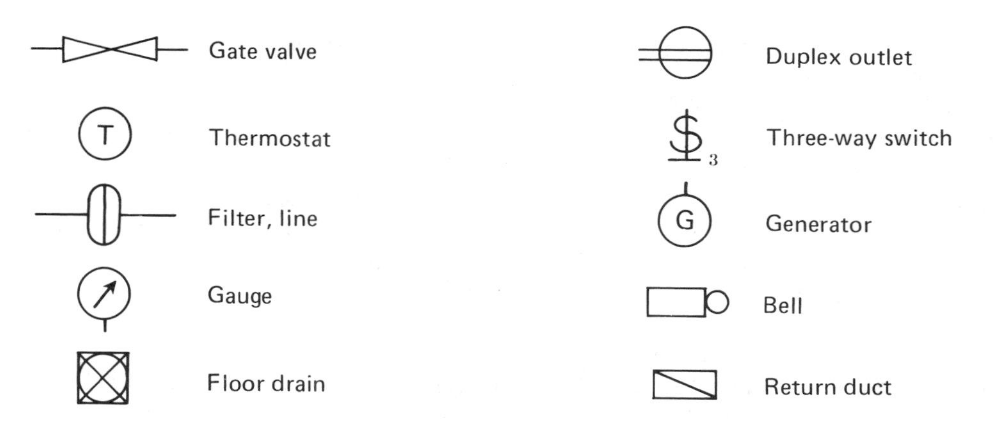

19.4 ABBREVIATIONS

Abbreviations and symbols are both used on drawings to save space and drafting time. As long as only well-known standards are used, there should be no decrease in understanding. A recommended practice followed by many firms is to furnish a key to symbols and abbreviations on one of the first drawings. A small number of widely accepted abbreviations is presented here:

alum	aluminum
AIA	American Institute of Architects
ASTM	American Society for Testing and Materials
AT	asphalt tile
bbl	barrel
BM	bench mark
CI	cast iron
C.C.	center to center
CMU	concrete masonry unit
cu yd	cubic yard
D & M	dressed and matched
ea	each
exp jt	expansion joint
fpm	feet per minute
ftg	footing
HM	hollow metal
ID	inside diameter
inv	invert
jt	joint
kwh	kilowatt-hour
mh	manhole
o.c.	on center
OD	outside diameter
d	penny
pl	plate
psi	pounds per square inch
R/S	reinforcing steel
rpm	revolutions per minute
specs	specifications
sq ft	square feet
SP	station point
str	structural
M	thousand
k	thousand pounds

T & G tongue and groove
vol volume
WC water closet
WS weather stripping
wt weight
w/o without
yd yard

19.5 INTERPRETATION

Only time and experience will make it easier for an individual to interpret a set of drawings. As familiarity with the various symbols and abbreviations is gained, the user will be able to more rapidly and accurately "read" a set of plans. The following section presents some fairly easy examples of drawing interpretation.

19.5.1 SECTIONED PLAN VIEWS

Example A in Fig. 19.1 represents a portion of a sectioned floor plan. The plan shows type of wall construction, dimensions of spaces and openings, location of equipment, designation and direction of elevations, names of spaces, and indications for details and sections.

The exterior wall is a nominal 1′–0″ thick and is constructed of 4″ face brick and 8″ concrete block. Space number 6 is a clerical room and is 14′–0″ wide. The room contains a cabinet whose elevation is shown on sheet number 14, elevation number 2. The window jamb is detailed on sheet number 6, detail number 8. The interior wall and connecting construction will be further explained by section 9 on sheet number 8. The door to the clerical room is of type 3, and further information regarding this door will be given under that number in the door schedule. Located in the 1′–4″ wall next to door number 3 is a fire extinguisher cabinet, designated by FEC. An elevation view of the exterior wall will be found on sheet 6 and is noted as elevation 3. The direction of sight is shown by the arrow at the end of the line.

19.5.2 ELEVATION DRAWINGS

Example B in Fig. 19.1 shows a portion of an exterior elevation of a building. Although most material is identified by notes, an effort is also made to indicate the material appearance graphically. It is obvious, for example, that the metal panel material above the windows has a vertical character from either bends or joints. The window glass is noted by "gl" and the operating direction of the windows is shown by a dashed line. The elevation of the finish floor level is given as a reference by 0′–0″. The gym floor is then noted as being 2′–8″ lower. Some sets of plans will use actual elevation of a floor based on an established bench mark, such as 79′–8¼″. The approximate finished grade level is indicated. The foundation wall and footing are shown in dashed lines since the earth would hide them from direct view. The rowlock course of brick serving as a window sill is noted along with the location of a control joint at the left end of the windows. A section has been indicated through the exterior wall and windows. This is a vertical section and can be found on sheet number 8 as section 5.

FIG. 19.1 Drawing examples A and B

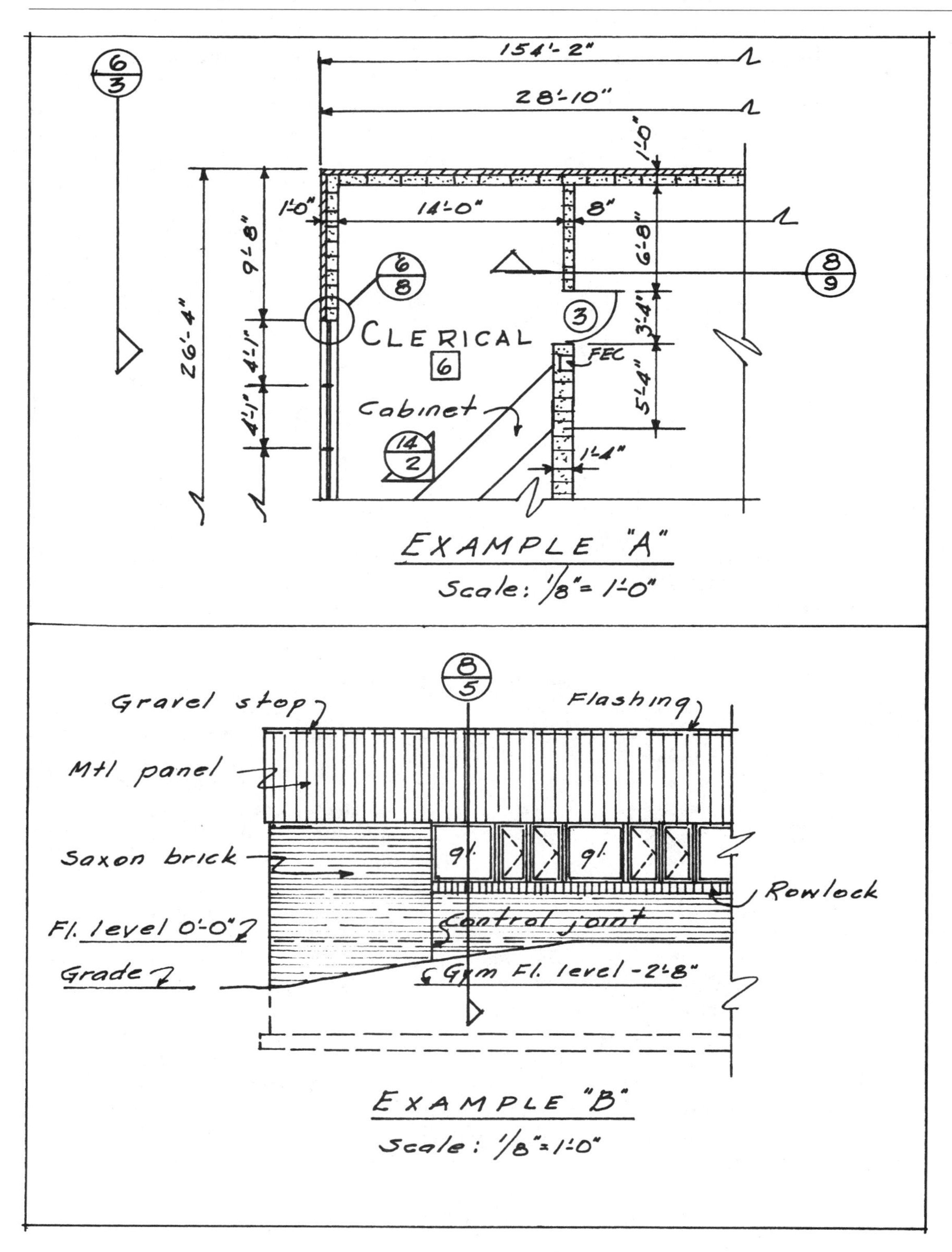

19.5.3 SITE DRAWINGS

A portion of a site plan is shown by Example C in Fig. 19.2. It shows a corner of a site with a portion of the structure that is to be built. The property lines are shown along with the curb line of the street. One corner of the building is located with respect to the property lines. The existing contours are shown by dashed lines, while proposed contours are represented by solid lines. This is the more common practice but some offices use the reverse of this designation. A contour line connects points of equal elevation. As additional information, the elevation is given for the finish grade at the corners of the building. All elevations are referenced to a bench mark located on the top of the fire hydrant located outside the property line.

Information is presented concerning trees; some are to be removed, while others will remain. The approximate location of existing utility lines is shown along with the proposed extensions to the project. A typical note on a site drawing is the one regarding the removal of existing foundations. A complete site plan might also show such items as drives, parking lots, retaining walls, and other site improvements.

19.5.4 ISOMETRIC DIAGRAMS

Example D of Fig. 19.2 illustrates an isometric drawing of a piping diagram. Such diagrams are not normally drawn to scale. Rather, they are used to further explain layout and arrangement. The example chosen is a simple one showing the layout of a gas line. It enters the building through a sleeve in the wall, rises to the roof framing, and then connects to the furnace. A vertical drop is effected to make connection to the water heater. Size of pipe is given for each run.

19.5.5 DETAILS

Larger scale drawings showing a particular part of the project are called "details." Details are used to explain and describe a part of the project that may not be clear from the plans and elevations using a smaller scale. The scale used for details may vary all the way to full size. In this way more information is available to the construction forces. An anchor detail, shown as Example E in Fig. 19.2, serves as an illustration. This drawing is at one-half full size. It provides the builder with sufficient information to fully explain how this part of the work shall be performed. The location of the anchors is found on the plans.

19.6 DIMENSIONS

Dimensioning is the process of labeling the distance from one point or surface to another. This may be the distance of one side of a site, the depth of a beam, the width of a walk-way, or any other property that can be expressed in a lineal fashion. The dimensions for an architectural project are mostly expressed as feet, inches and fractions of an inch. For civil projects, the use of feet and decimals of a foot is common.

Most figures are placed above a dimension line, which in turn runs between two reference lines. The point of junction between the dimension line and the reference line is marked to indicate the extent of the dimension. Traditional practice is to use an arrowhead but some offices make use of a dot or a slash (see illustration).

FIG. 19.2 Drawing examples C, D, and E

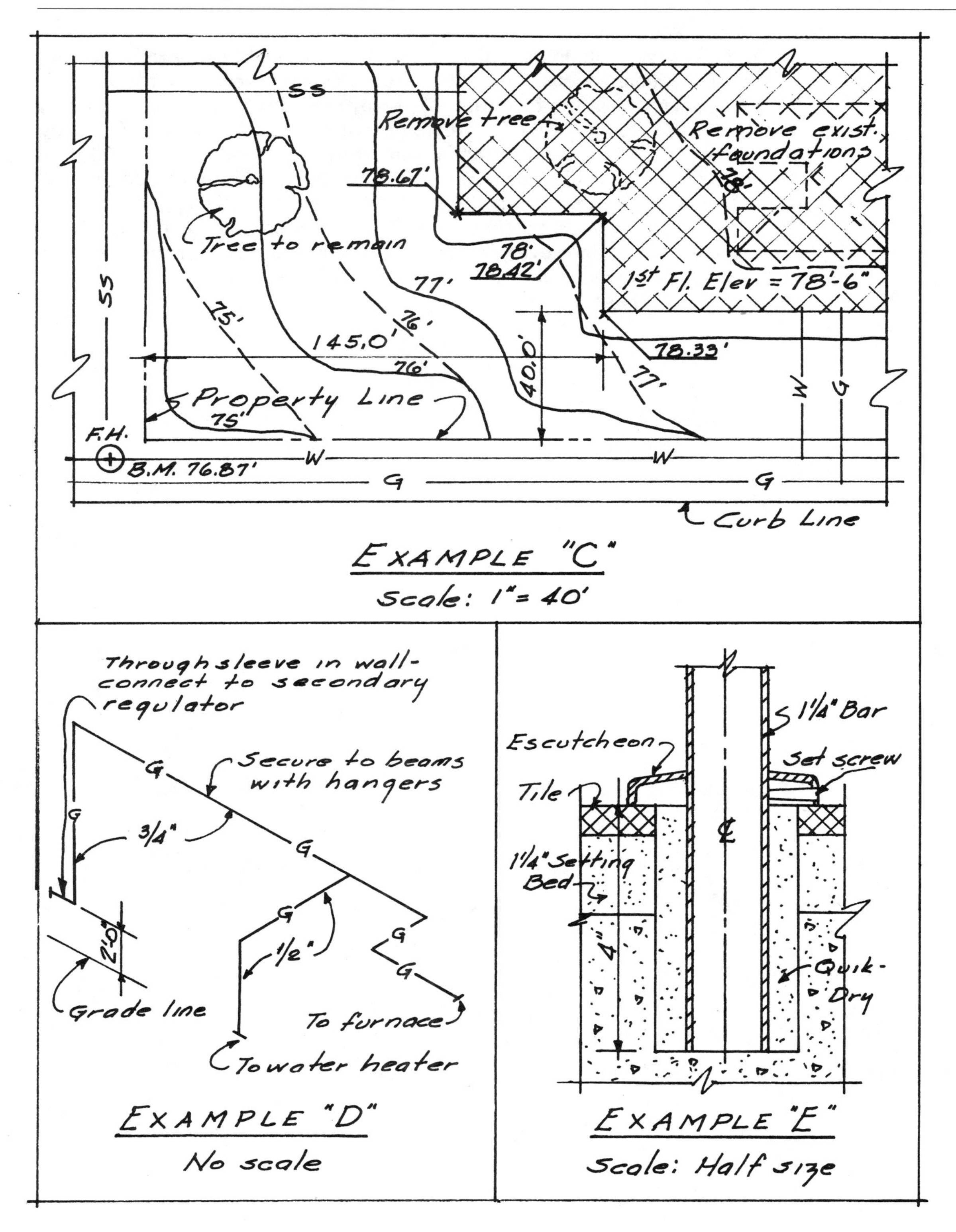

In most cases fractions smaller than 1/32 of an inch are never used, but usually no fractions smaller than 1/16 are used. This is in recognition of the limitations in accuracy of the field forces. Feet are indicated by a single prime (′) and inches by a double prime (″). Anything twelve inches or more is shown by both signs (7′–8½″) and dimensions under twelve inches are shown either as 0′–8½″ or as 8½″. If the dimension is on a large scale detail, it is common to drop the foot indication for anything less than one foot. Some architectural firms prefer to use modular dimensioning. This system eliminates the use of fractions of an inch since everything is based upon multiples of a module. Four inches is a desirable module but requires the cooperation of the architect, the contractor, and, most importantly, the manufacturer.

INDEX